BEARING THE HEAVENS

T0188093

This is a ground-breaking study of the astronomical culture of sixteenth-century Europe. It examines, in particular, the ways in which members of the nascent international astronomical community shared information, attracted patronage and respect for their work, and conducted their disputes. Particular attention is paid to the Danish astronomer, Tycho Brahe (1546–1601), known for his observatory Uraniborg on the island of Hven, his operation of a printing press, and his development of a third world-system to rival those of Ptolemy and Copernicus. Adam Mosley examines the ways in which Tycho interacted with a Europe-wide network of scholars, looking not only at how he constructed his reputation through print, but also at his use of correspondence and the role that instruments played as vehicles for data and theories. The book will be of interest to historians of science, historians of the book, and historians of early modern culture in general.

ADAM MOSLEY is Lecturer in History at Swansea University.

BEARING THE HEAVENS

*Tycho Brahe and the Astronomical Community of the
Late Sixteenth Century*

ADAM MOSLEY

CAMBRIDGE
UNIVERSITY PRESS

CAMBRIDGE UNIVERSITY PRESS
Cambridge, New York, Melbourne, Madrid, Cape Town,
Singapore, São Paulo, Delhi, Tokyo, Mexico City

Cambridge University Press
The Edinburgh Building, Cambridge CB2 8RU, UK

Published in the United States of America by Cambridge University Press, New York

www.cambridge.org
Information on this title: www.cambridge.org/9781107403659

First published 2007
First paperback edition 2011

A catalogue record for this publication is available from the British Library

ISBN 978-0-521-83866-5 Hardback
ISBN 978-1-107-40365-9 Paperback

For my parents

Contents

vii

Illustrations

Acknowledgements

It is only appropriate, in a book that is concerned with scholarly communication and collaboration, to acknowledge the debts that I have incurred in researching and writing it. Some of those debts have been financial. Accordingly, I gratefully acknowledge funding received from the British Academy; the Master and Fellows of Trinity College, Cambridge; the Isaac Newton Trust; Princeton University; and the Worshipful Company of Instrument Makers. A greater proportion of support received has been intellectual. A number of scholars have generously shared with me their time, expertise, encouragement, and – in several instances – their published and unpublished work. I would like to thank, in particular, my doctoral supervisors Nick Jardine and Liba Taub; Anthony Grafton, who stood, academically, *in loco parentis* during my time as a Jane Eliza Procter Fellow at Princeton University; and Sachiko Kusukawa.

Other scholars to whom I am individually indebted include Silke Ackermann, John Christianson, Mordechai Feingold, Owen Gingerich, Miguel A. Granada, Jürgen Hamel, Richard Kremer, Dieter Launert, Michel-Pierre Lerner, Bruce Moran, John North, Günther Oestmann, Alain Segonds, Gerard L'Estrange Turner, Karin Tybjerg, Steven Vanden Broecke, and Peter Zeeberg. It has been my great privilege to study and work at a number of institutions where the boundaries between professional collegiality, intellectual exchange, and friendly conviviality have been happily blurred. In addition to those already mentioned, I would like to thank the staff and students of the Cambridge University Department of History and Philosophy of Science, especially the members of the Cambridge Latin Therapy Group and *EPACTS*; the staff and students of the Program in History of Science of Princeton University 1999/2000, and the graduate intake to the History Department in that year; the Fellows of Trinity College, Cambridge; and the staff of the History Department of the University of Wales, Swansea. I am particularly grateful to the following individuals: Eric Ash, Alex Bueno-Edwards, Brooke Blower, Stuart Clark,

Jennifer Downes, Catherine Eagleton, Marina Frasca-Spada, Dan Healey, Tamara Hug, Jill Lewis, Volker Menze, Jo Miles, Clara Oberle, Richard Serjeantson, Kemal de Soysa, Andrew Taylor, and Adelheid Voskuhl. I have also benefited enormously from the collections and the expertise and helpfulness of the staff at the following institutions: the British Library; Cambridge University Library; Det Kongelige Bibliothek, Copenhagen; and the Whipple Museum, Cambridge.

I would also like to thank a number of other individuals who have, over the years, contributed to the completion of this book through the provision of welcome and necessary distractions: Stephen Balchin, David Chart, James Goodman, Ian Halverson, Jennifer Jellicorse, Marisa Lohr, Robin Oakley, Mike Pitt, Geoff Pradella, and Helen Steele.

Finally, I must thank my editors at the Press, and all those involved in the production of this book, including the anonymous readers.

It takes a great many people to support the production of what, somewhat unfairly, is known as a monograph – a fact that may usefully be borne in mind when reading this book. Nevertheless, just one person is responsible for any errors and omissions in the finished work, and that is the author. I look forward to the correction, criticism, and development of this study that publication provokes.

Abbreviations

ADB   Königliche Akademie der Wissenschaften, Historische
        Commission. *Allgemeine Deutsche Biographie*, Leipzig and
        Munich: Verlag von Duncter and Humblot. 56 vols.

CCC   Baldini, U. and P. Napolitani, eds., 1992. *Christoph Clavius:
        corrispondenza*, Pisa: Universita di Pisa. 7 vols.

CWE   Corrigan, B. and R. Shoeck *et al.*, eds., 1974– . *Collected Works
        of Erasmus*, Toronto and London: University of Toronto Press.

DBE   Killy, W., ed., 1995–2000. *Deutsche Biographische
        Enzyklopädie*, Munich: Saur. 11 vols.

DSB   Gillispie, C., ed., 1970–1990. *Dictionary of Scientific
        Biography*, New York: Charles Scribner's Sons. 18 vols.

KGW   Caspar, M. and W. von Dyck *et al.*, eds., 1938– . *Johannes
        Kepler, Gesammelte Werke*, Munich: Bayerischen Akademien
        der Wissenschaften.

NDB   Bayerische Akademie der Wissenschaften, Historische
        Kommission. 1953– . *Neue Deutsche Biographie*, Berlin:
        Dunker and Humblot.

OdG   Favaro, A. *et al.*, eds., 1890–1907. *Le Opere di Galileo Galilei*,
        Edizione Nazionale, Florence: G. Barbèra. 21 vols.

ODNB   Matthew, H. C. G. and B. Harrison, 2004. *Oxford Dictionary
        of National Biography*, Oxford: Oxford University Press. 60
        vols.

TBOO   Dreyer, J. *et al.*, eds., 1913–1929. *Tychonis Brahe Dani Opera
        Omnia*, Copenhagen: Nielsen and Lydiche. 15 vols.

CHAPTER I

Bearing the heavens

For he did not practice this art in peasants' huts, or in books, or in sweating rooms (as REGIOMONTANUS lamented of common Astronomers), but did not refuse to zealously pay attention to it in the sky itself, frequently with his very own eyes, using appropriate and well-constructed instruments; and he greatly advanced it by supporting skilled practitioners. In which heroic and truly Atlas-like course he steadfastly persisted, as long as he could regard the stars and the Sun (for the sake of which eyes were allotted to men, as a certain ancient philosopher appositely declared) – to the extent that he did not cease to contemplate this visible and temporal Theatre of Heaven until he crossed from the horizon of time into eternity and, with the aid of GOD, exchanged that eternal and invisible heaven with this other one. Wherefore who will rightly deny that it is entirely appropriate for the astronomical letters produced by so great an Atlas, a prince not only by virtue of his illustrious line, but also in this art, to claim for themselves the principal parts in this book?

Tycho Brahe on Landgrave Wilhelm IV of Hesse-Kassel,
Epistolarum astronomicarum liber primus (1596),
Dedication to Landgrave Moritz, (:)4r.[1]

In 1596, the Danish astronomer Tycho Brahe published a selection of his own correspondence. Although the title under which this book of letters appeared, *Epistolarum astronomicarum liber primus*, signified that the volume was to be the first in a series, it was the only volume of Tycho's correspondence to appear in his lifetime. For the most part, Tycho's sole book of *Epistolae astronomicae* consisted of letters that he had exchanged with an astronomically inclined prince, Landgrave Wilhelm IV of Hesse-Kassel, and with Wilhelm's court *mathematicus*, Christoph Rothmann. Wilhelm had died several years before the letters were published. Not inappropriately, therefore, the volume was cast as a memorial to the Landgrave, and

---

[1] *TBOO* VI, 13.4–18. The 'ancient philosopher' referred to by Tycho is probably not Ovid, as suggested by Dreyer in *TBOO* VI, 347. Cicero and Aristotle are both possibilities, but the most likely candidate is Plato. See Rantzau 1580, 9; Patrides 1982, 85.

dedicated to his son and heir Moritz. In the letter of dedication, and again in a poem of his own composition placed at the end of the work, Tycho praised his deceased correspondent by comparing him with a certain mythical figure: Wilhelm's death was the withdrawal from the Earth of a second Atlas; he was a man who not only ruled over his country, but who was also capable of holding up the heavens.[2] The association of the Titan with Mount Atlas, and hence with a king who had ruled over the inhabitants of north Africa overlooked by that mountain, was an ancient rationalisation of the myth of the giant who supported the heavens on his shoulders. The image of Atlas as bearer of the heavens was therefore an elegant way of referring to an astronomer who was also a prince.

This book takes its title from the idea of Atlas 'bearing the heavens'. It does so partly because its principal theme is the communication of astronomy in the early modern period. The phrase works just as well as a metaphor for the conveyance of astronomical theories, data, and techniques as it does as a description of the legendary task of the Titan. In the chapters which follow, four main modes of astronomical communication are considered: the exchange of letters, the production and use of books, the manufacture and transfer of ownership of instruments, and the movement from one site to another of individual practitioners. Investigation of these various forms of communication can shed considerable light on the study of the heavens in the early modern period. In particular, it reveals a great deal about one of the most striking features of the astronomical culture of the era, the emergence and development of an international astronomical community. The fundamental nature of the connection between modes of communication and the existence of a community is not difficult to grasp. Clearly, no such community could have existed in the absence of contact between individuals studying the heavens at different locations. But recognising this fact is only the first step in developing an understanding of how the astronomical community actually operated.

In this book, I have chosen to focus quite narrowly on the ways in which one of the best-known astronomers of the late sixteenth century engaged with the international astronomical community and other contemporary audiences. The centrality of Tycho Brahe to my study is another reason for using the phrase 'bearing the heavens' as its title. For as we shall see, Landgrave Wilhelm IV was not the only individual whom Tycho sought to represent as a latter-day Atlas – the image was one with which he himself was also keen to be associated. The reasons why Tycho considered himself

---

[2] *TBOO* VI, 340.10–11.

worthy of being viewed as an Atlas-like figure are closely related to the reasons why he makes a good focus for this examination of the late sixteenth-century astronomical community. His standing and reputation as an astronomer, both in the period and today, derive partly from the programme of high-quality astronomical work which he devised and pursued at Uraniborg, his observatory on the island of Hven, and partly from the nature of his engagement with other astronomers and their writings. Tycho, it could be argued, was as assiduous and meticulous with respect to the communication of astronomical material as he was about the labour of observing. This is one of the claims that this book will explore and develop.

A third reason for calling this work 'bearing the heavens' is that the motif that the phrase relates to, the task and role of the mythical King Atlas, provides some interesting insights into the culture of astronomy in the early modern period. This was a culture which valued more than just technical expertise in the study of the heavens, crucial as that was. Acquaintance with classical literature and imagery was also of importance to the scholarly elite, since it aided in the representation of astronomical endeavours and hence in the promotion of the art. A mythico-historical figure such as Atlas could be of considerable use when it came to making claims about the intrinsic nobility and importance of studying the heavens. The variety of ways in which this classical imagery was deployed is instructive. It appeared in the liminal verses and the prose sections of printed books, in letters and other manuscript texts, and in orations delivered on formal occasions. It also found expression in visual form, being employed in the context of astronomical diagrams and instruments whose practical roles were complemented by, or subordinate to, some symbolic or decorative function. The propagation of the 'heaven-bearing' motif therefore illustrates the full range of ways in which astronomical ideas were shared and conveyed. Admittedly, the communication of allegorical representations of astronomy via elegiac verse or highly ornamented, essentially decorative, scientific instruments might seem of little relevance to the evolution of technical astronomy. However, as recent scholarship has made clear, the principal site at which both literary virtuosity and artisanal splendour were appreciated, the early modern court, was also one of the key places where technical astronomy was sustained and developed.[3] Study of the period use of the image of Atlas 'bearing the heavens' reveals something of the importance to sixteenth-century astronomers of princely patronage and courtly aspirations. It can also illustrate the richness of the connections

---

[3] Biagioli 1993; Jardine 1998.

that existed between individual members of the international scholarly community. As well as reading one another's works, and sharing in a common intellectual heritage, sixteenth-century astronomers and writers on the heavens knew each other personally. They met, often as a result of travels undertaken for the sake of education or in search of employment, and subsequently corresponded; in a few cases, they were joined by a familial relationship. And even when they did not know one another directly, they were often separated from one another by just one or two intermediaries. Frequently, in fact, such mediated relationships were possible through more than one mutual acquaintance or chain of shared contacts. For this reason amongst others, therefore, consideration of the motif of Atlas as bearer of the heavens provides a point of entry into the subject-matter of this study.

I *COELIFER*: ATLAS AS BEARER OF THE HEAVENS

Long-haired Iopas, once taught by mighty Atlas, makes the hall ring with his golden lyre. He sings of the wandering moon and the sun's toils; whence sprang human kind and the brutes, whence rain and fire; of Arcturus, the rainy Hyades and the twin Bears; why wintry suns make such haste to dip themselves in Ocean, or what delay stays the slowly passing nights. (Virgil, *Aeneid*, I.740–746)[4]

The *Epistolae astronomicae* was not the first published work to present Landgrave Wilhelm IV as an astronomer of note. Inspired to study the heavens by Peter Apian's exquisite *Astronomicum Caesareum* (1540), Wilhelm had been tutored in mathematics by one of the sons of the instrument-maker and cartographer Gerard Mercator (1512–1594). He started making his own astronomical observations in the late 1550s and, *c.*1560, he established what has often been considered the first true observatory in western Europe at Kassel.[5] These astronomical activities were rapidly publicised: Wilhelm, his instrument-maker Eberhard Baldewein (*c.*1525–1593), and one of the instruments at the Kassel observatory, a large astronomical quadrant, were all mentioned by the mathematician Andreas Schöner (1528–1590) in his book *Gnomonice*, published in 1562.[6] Andreas, who was the son of the Nuremberg professor of mathematics Johannes Schöner (1477–1547), had acquired personal experience of the astronomical activity at Kassel during a brief period of employment there between 1559 and 1560, and was able to refer to specific observations carried out at the Landgrave's observatory.[7]

---

[4] As translated by Fairclough 1967–1969.
[5] Leopold 1986, 15–16; Hamel 1998, 9. See also *ADB* XLIII, 32–39; *DSB* XIV, 358–359.
[6] Schöner 1562, 89v, 90v, 93r. For Schöner, see Zinner 1979, 527–528.
[7] Leopold 1986, 16. For Andreas' father, see Coote 1888; Schottenloher 1907; *DSB* XII, 199–200.

Subsequently the educational reformer Petrus Ramus (1515–1572) praised the Landgrave's astronomical efforts, remarking in his *Prooemium mathematicum* of 1567 that it was as if Wilhelm had transported Alexandria to Kassel; as if the ancient astronomer Claudius Ptolemy, with his observational armillaries and rulers, had arrived in Germany.[8] In 1580, Heinrich Rantzau (1526–1599), Governor of the Danish Duchy of Holstein, lauded both Wilhelm and Tycho in the *Catalogus* of astrologers he compiled and published, also placing particular emphasis on astronomical instruments: Wilhelm, along with Tycho's royal patron King Frederick II, was commended for commissioning 'globes, clocks, and mathematical machines, from which can be seen the risings and settings of the signs and the heavenly houses, as well as the conjunctions of the Sun and the Moon and the rest of the planets, and the increases and decreases of the days'.[9]

All three of these authors were concerned with promoting the study of the mathematical arts: Schöner, presumably, because it was the basis of his livelihood, Ramus as part of his campaign for the reform of education in his native France, and Rantzau with particular reference to the pursuit of astrology.[10] In each case, mention of princely interest in mathematics may have been helpful to their cause. The idea that it would be was, at any rate, a major premiss of Rantzau's work; as its full title indicated, it largely consisted of a list of the emperors, kings and princes who had esteemed, honoured, and practised the astrological art. Tycho may or may not have been acquainted with Ramus' text. Having briefly met the Frenchman in 1570, he later professed to be unimpressed by his philosophy, and in particular by the nature of his calls for the reform of astronomy.[11] He may not have felt inclined, therefore, to seek out his publications. At some point in his life, however, he seems to have read Schöner's *Gnomonice*, and noted his mention of the Landgrave.[12] And he certainly knew Rantzau's work, most likely in the form of the augmented edition that was issued at Leipzig in 1584. The year after this version appeared, he addressed a poem

---

[8] Ramus 1567, 267: 'Guilielmus Landgravius Hessiae videtur Cassellas Alexandriam transtulisse: Sic Casellis artifices organorum observandis sideribus necessariorum instruxit, sic quotidianis per instructa organa observationibus oblectatur, ut Ptolemaeus ex aegypto in germaniam cum armillis & regulis venisse videatur.' On Ramus see *DSB* XI, 286–290; Margolin 1976; Sharratt 1976; Grafton and Jardine 1986, 161–209.

[9] Rantzau 1580, 29–30: 'Fridericus II, Rex Daniae, et Guilhelmus Landtgravius Hassiae hoc nostro tempore summo studio et cura, globos, horologia, et machinas mathematicas fieri curant, ex quibus ortus et occasus signorum ac domos caelestes, nec non coniunctiones Solis ac Lunae, caeterorumque planetarum incrementa ac decrementa dierum ac noctium, conspicere possunt . . .' On Rantzau, see *ADB* XXVII, 278–279; Steinmetz 1991; Lohmeier 2000; Oestmann 2004; Zeeberg 2004.

[10] Keller 1985, 351; Allen 1966, 84–85; Evans 1984, 263.

[11] *TBOO* VI, 88.25–89.12; Thoren 1990, 33–34.   [12] *TBOO* II, 40.1–8.

to Rantzau that registered his dissatisfaction with the somewhat slighting
description it gave of Uraniborg, his domicile and observatory on the island
of Hven, as a *specula*, that is a 'watchtower' or 'lookout post'.[13] Tycho had
good reason to feel aggrieved: by 1584 the patronage of King Frederick II, in
conjunction with his own not inconsiderable resources as a Danish noble,
had transformed Hven into a major centre for astronomy, and Uraniborg
was much more than a tower occasionally used for observational work.[14] If
it was the enlarged *Catalogus* that Tycho saw, then he may well have noticed
that it listed among the historic practitioners of astrology one, 'Atlas the
Moor, who is also called Iapetus, son of Libya; who ruled in Africa, taught
the theory of the sphere, and on a certain occasion predicted the future from
the stars'.[15] Nevertheless, the immediate model for Tycho's description of
Wilhelm in the *Epistolae astronomicae* as 'so great an Atlas' was no more
Rantzau's book than it was Ramus' or Schöner's. Inspiration came instead
from a more specifically eulogistic text, the funeral oration composed for
the Landgrave by Hieronymus Treutler (1566–1607), professor of rhetoric
at the Hessian University of Marburg.[16]

Tycho quoted the relevant section of the oration in his dedicatory letter
to Wilhelm's son Moritz: 'while he lived, our Prince, no differently than
Atlas, that most ancient king of Mauretania, supported the sky, as if on
his shoulders, and never laid down that very heavy burden of astronomical
matters'.[17] The analogy that Treutler was drawing was slightly richer than a
simple comparison of the physical effort of the Titan, bearer of the heavens,
with the intellectual effort of Wilhelm, their careful investigator. Notions
of kingship and the intimation of a shared interest in actual astronomical
labour were also involved. But even a reader who was insensitive to these
subtleties would have been able to see that Tycho employed the image
repeatedly in his edition of the letters. Besides the identification of Wilhelm
with Atlas in the dedication and in the memorial poem at the close of
the work, Tycho made reference in the preface of the book to Nicolaus
Copernicus (1473–1543) and 'that more-than Atlas-like work of his, *On*

---

[13] Rantzau 1580, 30; Rantzau 1584, 75–76; *TBOO* IX, 187–190, esp. 189.3–8; Nørlind 1970, 359–360.
[14] Thoren 1990, 144–219; Christianson 2000, 58–124.
[15] Rantzau 1584, 15: 'Atlas Maurus, qui & Iapetus dictus est, Libyae filius, qui in Africa imperavit, Sphaerae rationem docuit, & ex astris futura certo eventu praedixit.'
[16] For Treutler, see *ADB* XXXVIII, 585–587; Gundlach 1927, 318.
[17] Treutlerus 1592, 82: 'noster non aliter ac Atlas vetustissimus ille Mauritaniae Rex, coelum humeris quasi suis, dum vixit, sustinuit: nunquam gravissimam illam rerum astronomicarum sarcinam deposuit'; *TBOO* VI, 13.36–40.

the Celestial Spheres.[18] Somewhat more significantly, he also contrived to have himself compared to the Titan in five of the six poems with which the letter-book opened – poems that were written by individuals such as Albertus Voitus, professor of poetry at the University of Wittenberg, and Nathan Chytraeus (1543–1598), rector of the Bremen gymnasium and a highly esteemed writer of verse.[19] In the texts with which Tycho chose to frame his edition of his correspondence with the Kassel astronomers, therefore, reference to Atlas played an important role in linking Wilhelm, Tycho and, to some extent, Copernicus. Since, at the time of the work's publication, two of these three individuals were deceased, it would not have taken too much imagination on the part of a reader to recognise here the suggestion of a prestigious, successively inherited office, now occupied by the lord of Uraniborg.

While the prominence given to the Atlas motif in the *Epistolae astronomicae* should probably be attributed to its appearance in Treutler's speech, Tycho's awareness and use of this imagery predated his reading of the published oration. In a letter to Christoph Rothmann of 1589, for example, Tycho had exhorted the *mathematicus* to see to it that, by his 'Atlas-like labours', others might come to enjoy the fruits of the Landgrave's patronage of astronomy.[20] The Titan also appeared in letters sent to Tycho by others. The Scottish historian and poet George Buchanan (1506–1582), who was interested enough in the heavens to labour over a verse *Sphaera*, a work of didactic astronomical poetry, for the best part of three decades, mentioned him in 1575; quoting from Ovid's *Metamorphoses*, he declared of Tycho's *De nova stella* (1573) that the astronomer had set an example which encouraged others to 'aspire to ride the clouds, and to take position on stout Atlas' shoulder'.[21] And in 1587, the Rostock professor of medicine and astronomy, Heinrich Brucaeus (1531–1598), sent Tycho a letter in which he commended a young man who had sought to travel to Denmark particularly in order 'to greet you, who is called another Atlas, bearing the heavens on his shoulder'.[22] Long before the *Epistolae astronomicae* was published, moreover, Tycho had incorporated the figure of Atlas into one of his

---

[18] *TBOO* VI, 23.42–24.2.
[19] *TBOO* VI, 5–8. On Voitus, whose dates are not known, see Jöcher 1750–1751, IV, 1698. For Chytraeus, see *ADB* IV, 256; *DBE* II, 326; Lohr 1975, 712; Elsmann, Lietz and Pettke, 1991.
[20] *TBOO* VI, 200.10–12.
[21] *TBOO* VII, 21.34–22.1; Miller 1977–1985, II, 375. For Buchanan and his *Sphaera*, see Naiden 1952; McFarlane 1981, esp. 355–378.
[22] *TBOO* VII, 114.11–14. Brucaeus also employed this imagery in a later letter to Tycho; see *TBOO* VII, 142.17–20. For Brucaeus himself, see *ADB* III, 374–375.

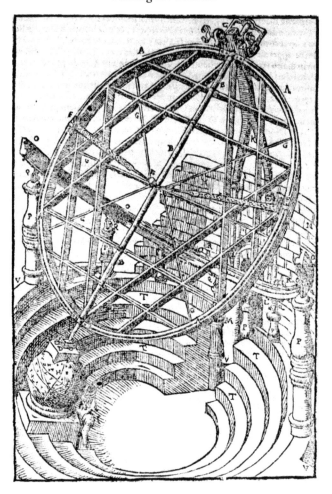

1.1 (a) Tycho's 'great equatorial armillary of one-and-a-half circles', as illustrated in his *Astronomiae instauratae mechanica* (Wandsbek, 1598), but reproduced here from the later 1602 'edition' of the work sold under the Nuremberg imprint of Levinus Hulsius.

observing instruments, the one he referred to as the great equatorial armillary of one-and-a-half circles (Fig. 1.1).[23] This instrument, and the significance of the Atlas mount with which it was equipped, were described in a 1591 account of the instruments on Hven which the Danish astronomer sent to the Landgrave.[24] It is even possible, therefore, that Treutler picked out

---

[23] *TBOO* V, 64–67. On the date of the instrument's fabrication see Thoren 1990, 173–174.
[24] *TBOO* VI, 276.36–278.41.

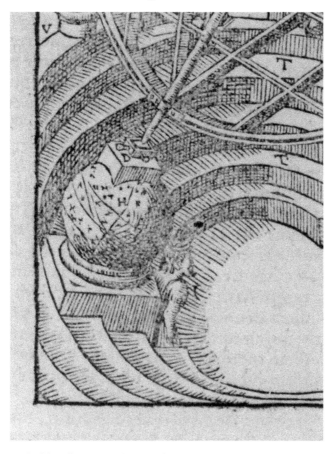

1.1 (b) Detail of this illustration, showing the instrument's mount: the crowned figure of Atlas, bearing the heavens in the form of a celestial globe. Courtesy of the Whipple Library, University of Cambridge.

the Atlas motif from documents which originated at Uraniborg. Although he had no particular reason to be aware of the Hessen astronomical communications during the lifetime of the Landgrave, it seems likely that, placed in the situation of having to praise his deceased prince for possessing such unusual enthusiasms, he had inspected this material in search of inspiration. Indeed, one reason for Tycho's citation of Treutler in the dedication to Moritz was that he had remarked in his oration that publication of the Landgrave's correspondence with the Danish astronomer would make evident to all how assiduously Wilhelm had practised astronomy, and how devoted he was to the elimination of the errors to be found in astronomical tables.[25] This statement, which Tycho used in justifying his decision to produce the *Epistolae astronomicae*, shows that Treutler had at least some knowledge of the contents of the letters.

It seems most likely, however, that Treutler arrived at his use of the Atlas motif without assistance from Tycho. The association of the Titan with the study of astronomy was certainly widespread enough by the time of Wilhelm's death. The primary vehicles for establishing and propagating that association were the classical texts, and the manuals derived from them, that encapsulated and interpreted ancient mythology. In particular, Virgil's *Aeneid* and Ovid's *Metamorphoses*, whether studied at first or second hand, rendered a range of descriptions of the ancient deities accessible to literate men, and to those who employed them. There were also multiple visual antecedents – increasingly so as classical mythology was mined by early moderns for use in emblems, medals and impresa, royal entry festivals and other courtly spectacles. The image shown in Fig. 1.2, for example, comes from *The Cosmographicall Glasse* (1559), a work by the English physician and astrologer William Cuningham (1531–1586).[26] The picture shows Atlas resting on one knee as he bears the heavy burden of the ten heavenly spheres, with a banner giving both his name and the epithet *Coelifer*, 'bearer of the heavens'. The plaque at the bottom paraphrases Virgil: 'He sings of the wandering moon and the sun's toils; of Arcturus, and the rainy Hyades and the twin Bears'. This text is slightly misleading if taken as a caption; in the *Aeneid* it was not Atlas, but his student Iopas, who sang of these things. Nevertheless, it serves to anchor the image to a classical source which connected Atlas and astronomy in a manner highly appropriate to the nature of the book. To judge from the catechismal form that Cuningham employed, the work was intended to serve as an introductory textbook, albeit for gentlemen rather than schoolboys. Its purpose was didactic, therefore, and its subject-matter was cosmography, the mathematical study of both the heavens and the Earth. In both respects it might be thought similar to the song supposedly taught to Iopas by Atlas.

The representation of the cosmos resting on the shoulders of the Atlas combines two forms of celestial schemata. One is that provided by the concentric circles representing the ten spheres of heaven and the elemental divisions of the sublunary realm: the *primum mobile*, crystalline heaven, firmament of the fixed stars, Saturn, Jupiter, Mars, Sun, Venus, Mercury, Moon, Fire, Air, Water, Earth. Such a conception of the cosmos can be found, so drawn, in numerous medieval and early modern manuscripts

[25] Treutlerus 1592, 83; *TBOO* VI, 14.2–8.   [26] Cuningham 1559, 50; Taylor 1954, 26–27, 172, 318.

1.2 Atlas as *coelifer*, bearing the heavens in the form of an armillary sphere. From W. Cuningham, *The Cosmographicall Glasse* (London, 1559), 50. Courtesy of the Whipple Library, University of Cambridge.

and books.[27] But the second form of heavenly scheme is the attempt to show, with an illusion of three-dimensionality, the conceptual divisions of the celestial sphere: the Tropics of Capricorn and Cancer, the equator, the polar circles, the line of the ecliptic (the Sun's apparent annual path about

---

[27] For some examples, see Whitfield 1995, 44, 50; Walker 1996, pl. XIII; Roob 1997, 43, 46.

the Earth), and the zodiac surrounding it. This component of the Titan's burden essentially constitutes a depiction of an astronomical instrument, a demonstrational armillary sphere.[28] The diagram includes not only a vertical degree scale, but also a wide horizontal band whose purpose, as shown, is rather unclear. In the case of a real armillary, supposing that one were the model for the picture, this band could have functioned as part of the cradle supporting the instrument, but in the absence of such a stand it seems entirely redundant. That two distinct representations of the heavens have been combined into one is strongly suggested by the fact that the putative celestial pole of the armillary fails to coincide with the axis of the Earth, even though an attempt has been made to show the celestial divisions at or near their true positions with respect to the degree scale.

Both forms of image used in the Cuningham picture had previously been combined with the *coelifer* motif. In the case of the simple concentric spheres, a putative precursor occurs in a work prepared by the Carthusian monk Gregor Reisch (*d.* 1525), the *Margarita philosophica*.[29] This book, first published at Freiburg in 1503, and repeatedly reissued, was an encyclopaedic introduction to philosophy and the seven liberal arts. It is as an illustration to the section on astrology that we see the woodcut of Atlas, standing within and supporting the nested spheres of the heavens (Fig. 1.3).[30] Similar images, with man positioned within the spheres of the universe, were also employed to represent the macrocosm–microcosm relationship, the notion that man and the cosmos were intimately connected.[31] A picture of an armillary-bearing *coelifer*, whose resemblance to the image in *The Cosmographicall Glasse* is slightly more obvious, can be seen (Fig. 1.4) on the title-page of *Quaestiones in libellum de Sacro Busto* (1552), an elementary astronomical text written by the Wittenberg-trained theologian and mathematician, Hartmann Beyer (1516–1577).[32] A representation so like the woodcut in Cuningham's work as to almost certainly be one of its exemplars, however, is that which appeared in Johannes Schöner's *Opusculum geographicum* of 1533 (Fig. 1.5). It was subsequently reproduced in Schöner's collected *Opera Mathematica*, a work owned by Landgrave Wilhelm IV and most likely known to Tycho as well; Wilhelm's copy of the second edition of 1561 was sent to him by its editor Andreas Schöner, the author's son and Wilhelm's one-time assistant.[33] In Schöner's version of the image,

---

[28] Bud and Warner 1998, 28–31.   [29] On Reisch, see *ADB* XXVIII, 117; Lohr 1980, 685–686.

[30] Reisch 1503, m2v.

[31] For examples drawn from Robert Fludd's *Utriusque Cosmi* (1617–1621), see Roob 1997, 542–543.

[32] See Bennett and Bertoloni Meli 1994, 113–114; *NDB* II, 203–204.

[33] Leopold 1986, 214.

1.3 Atlas standing within the heavenly spheres, overlooked by *Astronomia*. From G. Reisch, *Margarita Philosophica* (Freiburg, 1503), m2v. By permission of the Syndics of Cambridge University Library.

QVAESTI

ONES IN LIBELLVM

DE SPHAERA IOANNIS DE
Sacro Busto, in gratiam studiosæ iu-
uentutis collectæ ab HART-
MANNO BEYER, & nunc
denuo recognitæ,

1552.

FRANCOFORTI, ex officina Petri Brubacchij.

1.4 Atlas bearing the heavens in the form of an armillary sphere, as depicted on the
title-page of Hartmann Beyer's *Quaestiones in libellum de sphaera Ioannis de Sacro Busto*
(Frankfurt, 1551). Courtesy of the Whipple Library, University of Cambridge.

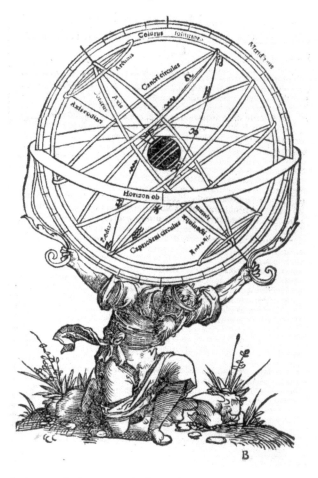

1.5 Atlas bearing the heavens in the form of an armillary sphere. From J. Schöner, *Opusculum geographicum* (Nuremberg, 1533), B1r. By permission of the British Library [BL 713.f.31].

the redundant band of Cuningham's sphere appears, and is labelled, as the instrument's horizon ring. The illustration still leaves some room for doubt concerning the fidelity with which the armillary is depicted; it would make sense for such a sphere to be movable within the cradle formed by the horizon and stand, but so far as can be seen from the image, this sphere is firmly fixed to the horizon. The three-dimensionality of the armillary circles is also somewhat poorly portrayed. Nevertheless, it seems quite likely that an actual instrument was the model for this image.

If the Schöner woodcut was based on a three-dimensional model, it is not impossible that this too was borne by the figure of Atlas. The elaborate and undoubtedly expensive instrument shown in Fig. 1.6 was constructed by the Augsburg artisan Christoph Schissler (*c*.1531–1608) in 1569.[34] Tycho, already greatly interested in the construction and use of astronomical instruments, was himself in Augsburg for much of that year; it was one of the cities at which he stayed for an extended period during the educational travels of his youth, and the site at which he erected the earliest of the great observing instruments of his career, the wooden *Quadrans Maximus*.[35] It is possible, therefore, that he saw this particular armillary sphere in the Schissler workshop.[36] The existence of Atlas-born armillaries dating from the seventeenth and eighteenth centuries attests to the longevity of the motif as an instrument-maker's conceit, and it seems likely that a systematic search of museum collections would turn up earlier examples.[37] Other types of instrument, particularly globes, were also provided with the same form of stand.[38] The most famous example, the so-called Atlante Farnese, is a marble celestial globe over 60 cm across that dates from antiquity.[39] Its existence indicates that early modern artisans were not the first to express the *coelifer* motif in three-dimensions. Moreover, since it was rediscovered in the mid-sixteenth century, when it entered the collections of Cardinal Alessandro Farnese, it is not implausible that it provided some inspiration for later artists and craftsmen; claims about its likely notoriety at the time of Tycho and Schissler, however, are not well supported by the historical evidence.[40] It seems much more likely that the depiction of the Atlas-borne

---

[34] Bayerisches Nationalmuseum, Munich, Inv. Nr. Phys. 27. See Zinner 1979, 503–520, esp. 511.

[35] Dreyer 1890, 29–34; Raeder, Strömgren and Strömgren 1946, 88–91; Thoren 1990, 30–36; Keil 1992.

[36] Thoren 1990, 30–31 n. 54, 36, asserts that Schissler was the Augsburg maker from whom Tycho commissioned the shell of his *Globus Magnus Orichalcicus*, but the evidence on which he bases this claim is not conclusive. Indeed, the letter of Johannes Maior to Tycho of 1 August 1576 strongly suggests that Schissler was not the maker of Tycho's instrument; see *TBOO* VII, 36.5–26.

[37] For example, Wh 0911 of the Whipple Museum of the History of Science, Cambridge, made by Fenig, was fashioned in the mid-eighteenth century.

[38] Turner 1987, pl. 8; Bott 1992, 548–549, nos. 1.47 and 1.48; Kejlbo 1995, 96–110.

[39] The Atlante Farnese is frequently described as dating from the second century BC; see, for example, Field 1996, 110. For the claim that it should properly be assigned to the second century AD, however, see Valerio 1987.

[40] For some suggested examples, see Bedini 1985, 23–36, figs. 2 and 5. Bedini does not, however, seem to distinguish between Atlas-figures and figures of Hercules. The claim that a description of the Atlante Farnese was published in Ulisse Aldrovandi's *Delle Statue Antichi* (1550) has been rejected by Valerio; he argues that the artefact was first described in print in the *Hercules Prodicius* (1587) of Stefano Winandus Pighius, who saw it in 1575. See Valerio 1987, 106–108, n. 10 and n. 22.

1.6 An elaborate armillary sphere made by Christoph Schissler of Augsburg in 1569. In a variation of the *coelifer* motif, the sphere is borne by Hercules, not Atlas. By permission of the Bayerisches National Museum [Inv. Nr. Phys. 27].

armillary in Schöner's works was the principal precursor for their use of this design.

Variations on the *coelifer* theme were also expressed in two and three dimensions in the early modern period. Hercules, for example, was a particularly fashionable iconographic figure of the European courts; his feats and labours made for flattering comparison with the achievements of monarchs and were well suited to forming the theme of elaborate festival pageants.[41] The hero's mythological association with the Titan, his temporary assumption of Atlas' burden, may well, therefore, have encouraged artists and artisans in their use of the 'heaven-bearing' idea. The Schissler sphere, in fact, with its serpentine ornaments and lionskin, is evidently a depiction of Hercules 'bearing the heavens' rather than Atlas.[42] An equally striking sixteenth-century representation of the celestial sphere being borne by Hercules occurs in an entirely non-astronomical context, on a ceremonial suit of armour and barding fashioned for King Erik XIV of Sweden between 1562 and 1564. The armour now resides in the Rüstkammer of the Dresden Staatliche Kunstsammlungen.[43] Also in Dresden, in the Grünes Gewolbe, is a globe pair in the form of decorative goblets, fashioned by the Augsburg goldsmith Elias Lenker in the late 1620s; in this instance Hercules carries the terrestrial globe and St Christopher the celestial one, on which perches the Christ-child.[44]

Like the Atlas-motif, the image of Hercules 'bearing the heavens' was enduringly popular: it was still employed, for example, in the decoration of a silver table commissioned for Prince Maximilian Wilhelm von Braunschweig-Lüneburg in the 1720s.[45] It seems unlikely, moreover, that a systematic search would fail to find other examples of St Christopher performing a similar function. More important than tracing all the instances of the *coelifer* motif, however, is understanding why these globe-bearing figures were so frequently crafted and drawn. To demonstrate that the idea of 'bearing the heavens' was not simply a decorative motif, it is necessary to return to the texts, both classical and early modern, that gave these images their meaning. As we shall see, the figure of Atlas could be used to represent astronomy and kingship, both separately and together; the Titan

---

[41] Seznec 1953, 25–26; Vivanti 1967; McDonald 1976; Strong 1984, 24, 54, 71, 83–84, 146; Tanner 1993, 236.

[42] Graves 1955, II, 91, 96.   [43] Schöbel 1975, 31, figs. 13 and 15.

[44] These objects, with inventory numbers IV 294 and IV 290, are illustrated in Seelig 1995, pl. 7.

[45] This table, made by Johann Ludwig II Biller of Augsburg, *c*.1725–1726, was paired with one showing the Fall of Phaethon. See Seelig 1995, 28–29, figs. 13 and 14. For some further examples, see Maurice and Mayr 1980, 301; von Mackensen 1991, 25, fig. 10.

was especially suitable, therefore, as a representation of a princely patron of astronomy or princely practitioner.

II ASTRONOMY AND KINGSHIP: DISCIPLINARY HISTORY AND PRINCELY PRACTICE

He [Perseus] held out from his left hand the ghastly Medusa-head. Straightaway Atlas became a mountain huge as the giant had been; his beard and hair were changed to trees, his shoulders and arms to spreading ridges; what had been his head was now the mountain's top, and his bones were changed to stones. Then he grew to monstrous size in all his parts – for so, O gods, ye had willed it – and the whole heaven with all its stars rested upon his head. (Ovid, *Metamorphoses* IV.655–662)[46]

Regarding the identification of Atlas with the study of astronomy, it can be difficult to determine which elements of the myth were considered most important. Clearly, the manner of the Titan's demise was a striking part of his story: Albertus Voitus, the author of the first poem in the *Epistolae astronomicae*, wrote explicitly of 'Atlas, whom credulous antiquity thinks to have metamorphosed, having been changed into a mountain'.[47] But did artists and artisans seize upon the same passage of the *Metamorphoses* for that detail, the beard, with which he was almost always represented? It is difficult to believe that this, any rate, was of great symbolic importance. What was significant, and why was it thought to be?

In the sixteenth and seventeenth centuries, comparisons between Atlas and contemporary astronomers could be made simply on the basis of the description of the Titan, by Vergil and others, as a teacher and practitioner of astronomy. Thus, on the title-page of Johann Stöffler's *Elucidatio fabricae ususque astrolabii* (1524), the author was likened in verse to Atlas, Orpheus, and other sages of antiquity.[48] Similarly, in the preface to his *Tabulae Prutenicae*, published in 1551, Erasmus Reinhold described Copernicus as a 'second Atlas, or a second Ptolemy'.[49] As we have already seen, however, an additional part of the attraction of the comparison lay in the fact that an astronomer's labours could be identified with the Titan's arduous task of holding up the heavens. This was the use to which Atlas was put in the lengthy dedicatory verse to Schöner's *Opera Mathematica*,

---

[46] As translated by Miller 1977–1985, I, 225.   [47] *TBOO* VI, 5.5–6.

[48] 'Quicquid Atlas Afris, quicquid Thracensibus Orpheus: / Quicquid apud Thebas creditur esse Linus: / Aegypto quicquid Vulcanus: quicquid ubique / Aut Chaldaeorum sunt monumenta virum: / Quicquid apud Gallos Druides: quicquid Zoroastes / Quicquid Persarum Gymnosophista fuit: / Omnia (crede) Stofler Germanus, origine Suevus, / Hic habet: exacto quae premit aere Cobel.'

[49] Gingerich 1978, 408.

written by his successor as Nuremberg professor of mathematics, Joachim Heller (*c*.1518–1590): 'Atlas, who has suffered so much under the starry mass, will delight at having been released from some part of his great burden by Schöner taking on a great weight'.[50] It was also the use which Galileo Galilei (1564–1642) made of the Titan when he asserted, in a letter of 1611, that determination of the periods of the newly discovered satellites of Jupiter would be 'a truly Atlantic', that is, an Atlas-like, 'labour'.[51]

On other occasions, however, particular emphasis was given to Atlas' status as a regal individual. Cuningham, in *The Cosmographicall Glasse*, wondered:

how many sundry Artes, secrete Sciences, and wonderful Ingens, through well spending of tyme, did the auncient Philosophers in their dayes invente? Archimedes devisyd glasses, with whiche the Siracusians might burne their enemies farre distant, on the seas from them. Ptolemaeus, Atlas and Alphonsus (being kinges) founde out the marvellous course and sondry motions, of the super celestial bodies: writyng sundry volumes of them, to the great comfort of such, as ar lyvng at this presente. Apollo first founde Physicke the repayrer of health. And in lyke manner, some one thing, and some another; of whose Godly travelles so many precious monumentes yet remayne: yea and the Authors themselves being dead so many hundred years sence are as freshe in the minds of man, as it were but yesterday, such is the reward of vertuouse travell.[52]

'Ptolemaeus, Atlas and Alphonsus (being kinges)' is a line that propagates multiple confusions about the history of astronomy. The Alexandrian mathematician Claudius Ptolemy (*fl.* AD 150), author of the *Almagest*, was arguably the greatest astronomer of antiquity, but he was not one of that dynasty of Ptolemies who ruled over Egypt. King Alfonso X of Leon and Castile (1221–1284), though certainly a patron of astronomy, and the commissioner of a body of astronomical and astrological texts, was not really the author of 'sundry volumes' on heavenly topics: those astronomical works which he had played a role in having written and compiled were never widely known, and what passed as the Alfonsine Tables for calculating celestial motions, and hence secured him his reputation for astronomy, were actually the products of later Parisian scholarship.[53] Atlas, we would now say, was no more King of Mauretania, or an astronomer of note, than he was a heaven-bearing giant. Yet as we previously saw with respect

---

[50] Schöner 1551, a5r–a5v: 'Ergo Atlas tot siderea sub mole labores / Qui tulit, ipse aliqua gaudebit parte levatus, / Tanti oneris magnum pondus subeunte Schonero.' On Heller, see Doppelmayr 1730, 54–55.

[51] *OdG* XI, 80.25–35, Galileo to Belisario Vinta, Secretary of the State of Tuscany, 1 April 1611: 'fatica, veramente atlantica'. See also Shea 1990, 60.

[52] Cuningham 1559, 2.   [53] Procter 1945; Poulle 1988; Gingerich 1990.

to Rantzau's *Catalogus*, early modern promoters of the mathematical arts found royal associations particularly attractive. What better way could there be to emphasise the intrinsic nobility of astronomy than to construct a lineage of astronomer-kings, of individuals who were both practitioners and princes?[54] Disciplinary histories like Rantzau's, in which the pedigree and utility of a form of study were displayed simultaneously, were popular with scholars of various kinds seeking to win greater recognition for their subject, and were commonly featured in the dedications and prefaces of textbooks. In many fields, indeed, the early modern period witnessed the gradual transition from historical accounts presenting the emulation of antiquity as the highest form of achievement, to those which acknowledged recent advances in knowledge, and the concept of progress.[55] Yet study of the past taught not only the absolute value of particular achievements, but also the role that they could play in establishing a lasting reputation. By writing that, 'such is the reward of vertuose travell', Cuningham made clear how the period concern with posterity informed the didactic role of his truncated history of astronomy.

It is striking to the modern reader that Cuningham, like Rantzau, presents Atlas as if he were an actual historical figure. In fact, there was no difficulty about rejecting the notion of an Atlas who held up the firmament, since Plato and Aristotle had firmly dismissed this idea as mere superstition.[56] But the place actually accorded to the Titan in early modern histories of astronomy depended on whether his tale was rejected as pure myth, or whether it was demythologised and given a more rational slant. In the *Margarita philosophica*, for example, the *magister* of Reisch's dialogue, when asked by his pupil about the origins of astronomy, replied that certain of the Greeks advanced the view that Atlas was the first astronomer. These unnamed authors favoured the poetic view that he was a giant who 'holds up the sky with his head at one Pole, his feet at the other, his right hand to the east, his left hand to the west'.[57] This was the image that was presented in the illustrative woodcut. But, the *magister* went on to explain, Atlas was actually a real man, skilled in the art of astrology, who gave his name to a mountain. Because the ignorant thought that the mountain was so tall as to touch the vault of heaven, they said that the heavens were supported by Atlas.[58] In Reisch's text, therefore, application of a relatively

[54] Cf. the term prince-practitioner, coined by Moran 1978.
[55] Burke 1969; Kelley 1970; Black 1982; Pumfrey 1991; daCosta Kaufmann 1993, 164–167.
[56] Plato, *Phaedo*, 99c; Aristotle, *De Caelo*, II.1.
[57] Reisch 1503, m1r: 'Athlantem gigantem esse maximum qui capite in uno polo & pedibus in reliquo manu dextra in oriente ac sinistra in occidente coelum sustineat.'
[58] Reisch 1503, m1r, quoting *De civitate Dei*, XVIII.8; see Green 1960–1972, V, 386–387.

mild scepticism to the ancient sources allowed Atlas to retain his place as a father of the discipline. Other early moderns, including Regiomontanus (1436–1476), and Tycho himself, took a more critical position. However, the treatment of Atlas by authors as diverse as the Italian diplomat Polydore Vergil (c.1470–1555), the physician and astrologer Girolamo Cardano (1501–1576), and the Jesuit mathematician Christoph Clavius (1538–1612), suggests that there was no simple chronological progression in the way that the myth was interpreted.[59]

Reisch's acknowledged authority for his euhemeristic explanation of the *coelifer* myth was St Augustine of Hippo. But he could also have cited the first-century BC author Diodorus Siculus, who accounted for the myth of the 'heaven-bearing' Atlas in similar terms and extended his explanation to cover Hercules as well. Atlas had perfected the science of astrology and was the first to reveal the doctrine of the sphere; he taught this lore to Hercules as a reward for his rescue of Atlas' daughters from pirates; consequently when Hercules brought the doctrine to the Greeks, 'he gained great fame, as if he had taken over the burden of the firmament which Atlas had borne, since men intimated in this enigmatic way what had actually taken place'.[60] It was this fuller interpretation of the tale which Regiomontanus addressed in the history of astronomy that he delivered as the introduction to a course of lectures he gave at Padua in 1464: 'They say that Hercules bore the heavens on his shoulder for Atlas, either because he studied astronomy under Atlas, or because he was regent for a while in Atlas' absence. But Hipparchus of Rhodes was the first father of this art.'[61] Thus, even as he dismissed these explanations as myth, Regiomontanus mentioned both astronomical learning and the burdens of kingship as reasons for the connection that was made between Atlas and Hercules. The two were already associated through the *coelifer* motif.

Regiomontanus' oration was published by Johannes Schöner in 1537, together with al-Farghani's *Rudimenta astronomica* and al-Battani's *De motu stellarum*.[62] It could well, therefore, have been one of the models Tycho employed for his own disciplinary history of astronomy, also delivered as the introduction to a course of university lectures. Tycho gave these lectures

---

[59] Vergil *On Discovery*, I.XV, in Copenhaver 2002; Grafton 1999, 128–130; Clavius 1585, 3–4.
[60] *The Library of History*, IV.27, as translated in Oldfather 1933–1967.
[61] Regiomontanus 1537, bv: 'Herculem pro Atlante coelum humeris suis sustinuisse aiunt, sive quod sub Atlante astronomiam didicerit, sive quod in regno eius absentis praefectus aliquamdiu fuerit. Hipparchus tamen Rhodius huius disciplinae primus parens...' The oration is discussed in Swerdlow 1993a.
[62] See Bennett and Bertoloni Meli 1994, 26–27.

at Copenhagen in 1574.[63] Like Regiomontanus, Tycho excluded Atlas from the list of the discipline's forefathers, recounting instead a transmission of astronomical expertise from Abraham to Copernicus via Pythagoras, Timocharis, Hipparchus, Ptolemy, and al-Battani.[64] As we have already noted, however, rejection of the Titan as a real historical figure who practised astronomy did not translate into a reluctance to make use of him symbolically. The ways in which Tycho did so at Uraniborg are made clear in the *Synopsis of the Astronomical Instruments which the Dane Tycho Brahe has had placed here and there on the Island of Hven*, a document that was sent to Landgrave Wilhelm in 1591 and reproduced in the *Epistolae astronomicae*.

Describing the instrument furnished with the *coelifer* mount, the unnamed assistant of Tycho who prepared the *Synopsis* wrote that:

The lower end of the axis of the round armilla is revolved on a steel plate which is skilfully provided with screws, so that it can be positioned exactly and correctly according to the axis of the world, and this on top of a large rectangular Gothic stone which, placed about two ells below the earth on a firm stone foundation, projects more than two ells from the ground; on the lower part of this a skilfully moulded image of Atlas, King of Mauretania, can be seen, as if he is occupied with bearing the heavens on his shoulder. For indeed, above his crown on this same stone, there is the hemisphere of a globe marked with golden stars; which, touching a certain part of his crown, he seems to carry. By which it is signified hieroglyphically that kings and princes justly ought to support and generously sustain this sublime and truly royal art of astronomy, as that Atlas and many others once did; and in a former age ALPHONSUS, King of Spain, supported this as much it was in him to do so, with a famous generosity that hardly ever ought to be forgotten.[65]

In this fashion, one of the observing instruments used on Hven was equipped to act as a visual allegory, reminding those who visited the island both of astronomy's need for the patronage of princes, and of the long mytho-historical tradition of aristocratic involvement in the study of the stars. In fact, the great equatorial armillary of one-and-a-half circles was far from being the only instrument on Hven which combined a symbolic significance with a practical function. The very first object described in Tycho's later and more famous account of his observing equipment, the *Astronomiae instauratae mechanica* (1598), was decorated with a *memento mori* featuring a skeleton, and another was decorated with figures of *Urania*, *Geometria*, and *Arithmetica*.[66] For reasons which have a great deal

---

[63] Dreyer 1890, 73–78; Thoren 1990, 78–92.
[64] *TBOO* I, 145–173. The oration is discussed in Jardine 1988, 262–264.
[65] *TBOO* VI, 277.39–278.10.
[66] Raeder, Strömgren and Strömgren 1946, 12–15, 40–42; Segonds 1994.

to do with Tycho's conception of himself as a noble astronomer, his observatory and instruments were apparently designed with almost as much care
paid to the messages they conveyed as to their suitability for astronomical
work.

With respect to the functionality of instruments supported by a *coelifer*
mount, however, Tycho's great equatorial armillary of one-and-a-half circles
is somewhat unusual. For the majority of similarly borne objects, the instrumental function was very much subordinated to the role played by the object
in the courtly culture of display. The Schissler armillary sphere pictured
above, for example, is so elaborate as to be almost useless as a demonstrational device. But the Atlas motif may itself have contributed to the appetite
for these objects, for a reason suggested by Regiomontanus' *Oration*, and
supported by sources such as a late seventeenth-century manuscript in
the Österreichische Nationalbibliothek in Vienna. Taking the form of a
Historische Beschreibung of the life of Emperor Charles V, and dedicated to
his great-great-grandson, Emperor Leopold I, this manuscript includes a
visual portrayal of Charles as a heaven-bearer himself.[67] The *coelifer* motif
could be used, this picture indicates, to represent the burdens and triumphs
of kingship in a context otherwise free from astronomical significance.[68] It
was not the first time the conceit was employed in this way: when Charles V
abdicated his position as King of Spain and Holy Roman Emperor, a medal
was struck which showed the global burden of rule passing from Charles,
as Atlas, to the Hercules-figure of Philip II. Since the globes and armillaries produced for popes and princes were, quite independently, symbols of
worldly dominion, their support by the mythical Titan, or his proxy, can
only have given them added significance in the allegory-conscious courts of
early modern Europe.[69] Clearly, both Treutler and Tycho were consciously
exploiting the double-metaphor of the heaven-bearer when they praised
Wilhelm for his princely support of astronomy.

A further use of Atlas was also made on Hven, one that combined all of
these themes of succession, kingship, astronomical competence, posterity,
and disciplinary history. This instance was also published in the *Epistolae
astronomicae* as a result of having been included in the manuscript *Synopsis*.
The unnamed author of this account explained that in the centre of
Stjerneborg, Tycho's underground observatory adjacent to Uraniborg, there

---

[67] von Philippovich 1969–1971. See also Tanner 1993, 138–139.

[68] It ought to be noted, however, that Charles V was frequently among those cited as prominent patrons
of astronomy; see, for example, Rantzau 1580, 29; Clavius 1607, 10.

[69] For a contemporary account that is explicit on the allegorical significance of globes, see the description
by Paulus Fabricius of one of the triumphal arches prepared for Rudolf II's entry into Vienna in
1577, as transcribed and discussed by daCosta Kaufmann 1993, 141–142 and 209–210.

was a room decorated with eight portraits, each one furnished with an identifying couplet. Six of these portraits represented astronomers singled out by Tycho as particularly praiseworthy: Timocharis, Hipparchus, Ptolemy, al-Battani, King Alfonso X of Castile, and Copernicus. The seventh was Tycho himself, whose verse ran: 'How many stars I have subjugated to a rule, from the ancient's observations and my own, it will be up to posterity to judge'.[70] Nearby was depicted a large diagram of the Tychonic world-system, the arrangement of the Sun, Moon and planets advanced by Tycho as an alternative to the systems of Ptolemy and Copernicus. Pointing at this diagram with one hand, the Tycho portrayed in this gallery held between his fingers a sheet of paper on which was written *Quid si sic?*, 'What if it were thus?'. It was, the author of the *Synopsis* explained, 'As if he were saying to those older astronomers depicted around him "What does this invention seem like to you?"'[71] To make this conversation with his predecessors even more poignant, the eighth portrait represented the *Tychonides*, Tycho's male offspring. The lines under this final panel read: 'May GOD grant you, venerable Father, and your lofty inventions, a succeeding generation to continue in the role of Atlas'.[72] Not blushing to count himself among the best-known princes and practitioners of astronomy of nearly two millennia, Tycho cherished the hope that his astronomical fiefdom and prowess might be something which his children would inherit. Indeed, since he had married a commoner, an astronomical patrimony represented one of the best chances his children possessed for retaining something of the status that Tycho had enjoyed as his birth-right. This was one of several reasons why Tycho might have wanted to emphasise the intrinsic nobility of studying the heavens, and the tradition of princely support for astronomers, through the *coelifer* motif.

III TYCHO BRAHE, PRINCE OF ASTRONOMERS

To whom, I ask, should this little commentary on the nature of the heavens be dedicated rather than to you, indefatigable examiner of the Heavens and Atlas? To whom is it not sufficiently evident that you have inherited, to the eternal glory of the name, the nobility of descent from the very ancient and distinguished family of the Brahes; but by your heroic and Herculean labours in the science of the stars, in which you excel, you wanted to place the very beautiful crown of immortal fame on your family. (Cunradus Aslachus to Tycho Brahe, *De natura caeli* (1597))[73]

---

[70] *TBOO* VI, 275.7–8.   [71] *TBOO* VI, 276.19–25.

[72] *TBOO* VI, 275.9–11. On Tycho's hopes that his astronomical enterprise would be continued by his heirs, and his attempts to ensure that Uraniborg would remain in the control of his family, see Christianson 2000, 125–130, 140–142, 193–194.

[73] Aslachus 1597, A3r: 'Et cui quaeso potius, haec de Caeli natura commentiacula, quam inscribi deberet? Cui non satis visum est ad perennem nominis gloriam ex perantiqua & splendida Brahorum

Tycho Brahe's use of the *coelifer* imagery, and his high opinion of his own astronomical significance, did not go unnoticed. Several of the many assistants who helped him carry out his programme of astronomical research subsequently made reference to the way that he represented himself at his observatory and in the *Epistolae astronomicae*. Some of them, no doubt, were strongly encouraged to do so; the laudatory poem by Tycho's future son-in-law, Franz Tengnagel, which appeared in Tycho's *Mechanica*, and invited comparison of him with Timocharis, Ptolemy, Copernicus, Atlas and Hercules, can probably be attributed directly to the influence of the master on the pupil.[74] Others, however, employed the same technique at a considerable distance, in miles or years, from Tycho and his household. The use by Cunradus Aslachus (1564–1624) of references to Atlas and Hercules in the dedication of his *De natura caeli* to Tycho, and his connection of these figures with the royal status conferred by immortal fame, was written several years after this one-time assistant at Uraniborg had left to make his own way in the world.[75] Johannes Pontanus (1571–1639), another Uraniborg worker, wrote in 1617, long after his former master's death, of 'the magnificent and Illustrious Tycho Brahe who is now deservedly celebrated with the title of a second Atlas'.[76] And Christian Longomontanus (1562–1647), Tycho's second-most successful astronomical pupil, described Tycho in his *Astronomica Danica* (1622) as a 'distinguished man, the greatest astronomer of all time, and like another Atlas'.[77] A preliminary verse contributed by Caspar Bartholin (1585–1629), Longmontanus' brother-in-law and like him a professor at the University of Copenhagen, similarly praised Tycho as a second version of the Titan.[78]

Tycho was also lauded by some of his contemporaries for being an astronomical prince. In July 1591, the professor of mathematics at the University of Bologna, Giovanni Antonio Magini (1555–1617), closed a letter to Tycho by referring to him as *Astronomiae Princeps*, the prince of astronomy.[79] In later letters, both Johannes Kepler (1571–1630), the most famous of all of Tycho's assistants, and the eccentric Chancellor of the state of Bavaria,

familia generis nobilitatem accepisse; sed, pulcherrimam nominis immortalis coronam familiae tuae imponere voluisti.' See also the reference, on p. 43 of Aslachus' work, to Tycho as 'that other Atlas of our age' ('altero illo nostræ tempestatis Atlante Tychone Brahe').

[74] See *TBOO* V, 93–95, particularly 94.10 and 24.

[75] See Christianson 2000, 252–253, under 'Aslakssøn, Cort'.   [76] Hues 1639,) (r.

[77] Longomontanus 1622, *4r: 'Dominum Tychonem Brahe . . . hunc eximium virum, summum omnium temporum Astronomum, & velut alterum Atlantem . . .'

[78] Longomontanus 1622, **3v: 'Astris illuxit primus Tycho Danicus ille / Sol Patriae & coeli: nobilitatis apex. / Sol novus illuxit stellis Atlasque secundus: / Sic Soles binos Danica terra dedit.' For Longomontanus and his relationship to Bartholin, see Christianson 2000, 313–319.

[79] *TBOO* VII, 304.17–19.

Georg Johann Herwart von Hohenburg (1553–1622), flattered Tycho by calling him *princeps mathematicorum*, the prince or chief of mathematicians.[80] To call someone the *princeps* of their particular vocation was by no means unusual: Tycho's correspondent Buchanan was similarly called the prince of poets, and Aristotle was sometimes called the prince of philosophers.[81] But given Wilhelm's actual status as Landgrave of Hesse, Tycho's praise of him as a prince also of the art of astronomy, and the suggestion in the *Epistolae astronomicae* that Tycho was in some sense his successor in astronomical endeavour, the title 'prince of astronomers' possessed an added quality in the context of the letter-book. This did not entirely work to Tycho's benefit. Two of the poets who compared Tycho to Atlas at the opening of the work also addressed him as the prince of his discipline.[82] Despite the fact that Tycho did not actually claim the title himself, this piece of bespoke self-promotion made it possible for Nicolai Reymers Baer (1551–1600), called Ursus, to mock him for boasting that he was the Prince of Astronomers.[83] This remark was just one of the ways in which Ursus, whom Tycho had accused of plagiarising his new system of the world, sought to undermine the Dane's claim to astronomical and mathematical excellence.[84]

Tycho's high opinion of himself was not, however, merely the product of hubris and flattery. Had it been, then Tycho's astronomical reputation would hardly have survived the judgement he called upon himself, that of posterity. Yet in the modern edition of his works, the Danish Society of Language and Letters approvingly referred to the description of Tycho, by the German mathematician and astronomer Friedrich Wilhelm Bessel (1784–1846), as a 'king of astronomers' and remarked that he may 'rightly be considered a prince'.[85] The several reasons why Tycho is remembered by astronomers and historians have been mentioned already. First and foremost there is the observational work that he and his assistants carried out on Hven over many years. This work provided the data which Kepler used not only to calculate the *Tabulae Rudolphinae* (1627), the culmination of the Tychonic project to achieve an empirical reform of astronomy, but also in discovering his three so-called laws of planetary motion. It was the accuracy of Tycho's observations which made them so useful to Kepler, and that accuracy was enabled by the instruments which Tycho constructed and employed at Uraniborg and Stjerneborg: by the number of them, by

---

[80] *TBOO* VIII, 14.24–26, 157.3–7.   [81] *DNB* III, 188; Chamber 1601, 15.
[82] *TBOO* VI, 6.19–21, 8.1–5.
[83] See Ursus 1597, A3r, B2r, B4r, F1v, F2v. One such instance is discussed by Jardine 1988a, 35.
[84] See Jardine 1988a, 29–57.   [85] *TBOO* I, I.

their size and precision, and by the care and persistence with which they were used. For his data, his instruments and his observatory, Tycho has often, indeed, been considered the greatest of pre-telescopic observational astronomers.

To historians, moreover, if not to astronomers themselves, Tycho's work in cosmology has long been thought of as significant for subsequent developments. By retaining an immobile Earth, about which the Sun and Moon were supposed to rotate, at the centre of the universe, but having the remaining heavenly bodies orbit the Sun, Tycho proposed a system of the universe acceptable to those who thought that the Ptolemaic geocentric system was no longer tenable yet considered the Copernican hypothesis either physically absurd or theologically objectionable (Fig. 1.7). This system, and variants of it, became increasingly prominent following the heavenly discoveries made with the aid of the telescope and after the Catholic condemnation of heliocentric astronomy in 1616.[86] And it was in connection with his geoheliocentric cosmology that Tycho's infamous argument with Ursus began – a dispute strongly implicated in explanations for the collaboration of Tycho and Kepler in Prague, and one that led directly to Kepler's composition of a major manifesto of the so-called 'new astronomy' of the seventeenth century. Since antiquity, practitioners of astronomy had laboured under constraints imposed on their subject by natural philosophers and metaphysicians: they were charged with the task of mapping and modelling the cosmos mathematically without presuming to make their own inferences about the physics of celestial motion or the underlying nature of the heavens.[87] It was this traditional distinction between mathematical astronomy and cosmology, the natural philosophical study of the heavens, that Kepler dispensed with in works such as his *Astronomia nova* (1609) and attacked in his unpublished *Contra Ursum*, or *Defence of Tycho against Ursus*. In doing so, however, Kepler was escaping from bonds already loosened by his immediate predecessors, Tycho included. One of the contributions to this process for which Tycho is often remembered is his demonstration, by mathematical analysis of the comet of 1577, that there were no real orbs in the heavens as claimed by scholastic natural philosophers.[88] Both the nature of that demonstration and the process by which Tycho acquired the credit for making it are less straightforward than has sometimes been recognised. Nevertheless, it remains the case that he played an important part in increasing the natural philosophical authority of mathematical astronomy.

---

[86] Schofield 1981.   [87] Jardine 1988a, 225–227.   [88] Rosen 1984, 1985a; Barker and Goldstein 1995.

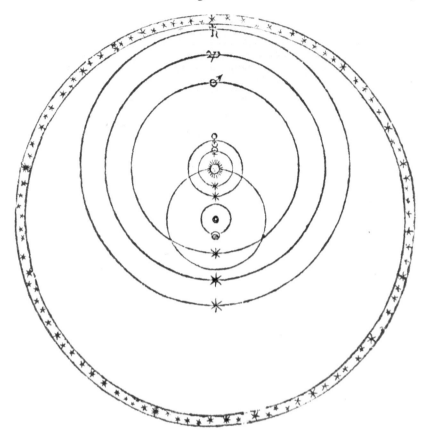

1.7 Tycho's geoheliocentric world-system, as illustrated in his *De mundi aetherei recentioribus phaenomenis* (Uraniborg, 1588), 189. Courtesy of the Whipple Library, University of Cambridge.

Tycho's acknowledged role in the transformation of early modern astronomy might seem reason enough to focus on him in any account of the study of the heavens concentrating on the late sixteenth century. But this is not a book about a solitary practitioner, and neither the technical details of Tycho's astronomy, nor the story of his life, are its principal subject. It is significant, therefore, that Tycho has also been singled out by historians for reasons which relate more strongly to this book's themes of collaboration and community. The use he made of printing technology, and his abilities as organiser and manager of the researchers at Uraniborg, have both been highlighted in recent scholarly literature. Ownership of a printing press enabled Tycho to issue texts, such as the *Epistolae astronomicae* and the *Mechanica*, without requiring that he submit to the advice or the interests of a commercial book-producer.[89] Possession of the fiefdom of Hven, granted to him by King Frederick II, allowed him to develop what some have called the first modern research institute, the first centre for postgraduate training in the natural sciences.[90] The emphasis placed on these aspects of Tycho's working life complements the older scholarship by focusing attention less on his well-known contributions to astronomical and cosmological theory and more on the ways in which astronomy was collectively studied and practised. Such an approach is better suited to accounting for, rather than demonstrating, Tycho's contemporary and historical significance. It locates him within the overlapping technical, courtly, and literary cultures of the period and illuminates the extent to which, if at all, the directions he took in his astronomical and cosmological work were determined by his context.

In this book I have tried to extend that approach by using Tycho Brahe as a window onto the nature and activity of the international astronomical community. His suitability for the role derives partly from his important place in the history of astronomy and the recognition he received from his contemporaries, partly from his management of his own printing press and position as the 'dean' of an early modern 'research institution', and partly from some of the other aspects of his working life we have glimpsed in this chapter. Besides his activities as an author and publisher, these include the correspondence he conducted, with individuals ranging from princes and poets to professors of mathematics, and the use he made of astronomical instruments, not only as technical tools, but also as embodiments of ideas and vehicles for self-promotion. Each of these means of communication will be studied more closely in the following chapters.

[89] Christianson 2000, 85.   [90] Thoren 1985.

CHAPTER 2

Tycho Brahe's astronomical letters

> If there is something that can be said to be characteristic of this genre,
> I think that I cannot define it more concisely than by saying that the
> wording of a letter should resemble a conversation between friends.
> For a letter, as the comic poet Turpilius skilfully put it, is a mutual
> conversation between absent friends, which should be neither unpol-
> ished, rough, or artificial, nor confined to a single topic, nor tediously
> long.
>
> Desiderius Erasmus, *De conscribendis epistolis* (1522)[1]

To judge from the number of his letters which survive, Tycho Brahe was not,
by the standards of his age, an especially prolific correspondent. Whereas the
letters of an Erasmus or a Melanchthon are numbered in their thousands,
fewer than 500 letters sent or received by Tycho during his lifetime are
presently known to historians. Even so, these documents have been deemed
valuable enough that a number of individuals have thought it worth the
effort to make them more widely available. Tycho himself, as we have
seen, published one volume of correspondence and planned to produce
more. Some of his letters were later printed by Jacob Langebek in 1747,
by F. R. Friis in various collections issued between 1875 and 1909, and by
Fr Burckhardt in 1887. In 1926, Wilhelm Nørlind published a number in
Swedish translation. By that date, the production of Tycho's *Opera Omnia*,
edited and annotated by John Dreyer and others under the auspices of the
Danish Society of Language and Letters, was nearing its end, and the three
volumes entirely devoted to Tycho's correspondence had already appeared.
With the issue in 1928 of volume XIV of the *Opera*, containing a miscellany

---

[1] As translated in *CWE* XXV, 20. That a letter was a conversation between absent friends was a
commonplace of the period: see Brandolinus 1549, 9 and 364 (the latter is part of an edition of
Erasmus' *Brevissima maximeque compendiaria conficiendarum Epistolarum formula*, first published in
1521); Hegendorff 1545, 2. The idea is, however, earlier in origin. Constable 1976, 13, notes its use by
Ambrose and medieval epistolographers, and an even earlier expression of the notion that a letter
should be like a conversation is to be found in letter 75 of Seneca's *Epistulae morales*. See Gummere
1917–1925, II, 136–137.

of documents pertaining to Tycho's life and career, all of the Tychonic letters
then known became available in print for the very first time.

Despite all this editorial labour, however, Tycho's letters have attracted
little in the way of scholarly analysis *as* correspondence. Tycho's biographers have made good use of the letters in their reconstructions of his life,
and there have also been a handful of articles and monographs that draw
on portions of the correspondence in order to establish Tycho's relationship with certain of his rivals or associates, or to investigate his intellectual
development.[2] But scant time has been expended considering his letters as
a form of scholarly production in themselves, or in exploring the role that
epistolary communication might have played in his astronomical enterprise. This problem is not one of Tycho scholarship alone, for similarly
restricted use has been made of the correspondences of his younger contemporaries, Johannes Kepler and Galileo Galilei, both of which have also
long been available to historians in the standard editions of their works.[3]
The value of the letters of early modern astronomers as historical sources
continues to be recognised in the effort devoted to making them accessible;
the correspondences of the Jesuit mathematician, Christoph Clavius, and
John Flamsteed (1646–1719), England's first Astronomer Royal, have both
recently been published.[4] It is somewhat surprising, therefore, that the
period significance of astronomical correspondence has not attracted more
in the way of critical commentary.[5] Clearly, these letters were not written
solely for the benefit of modern historians.

In this chapter, I propose to show that correspondence was an important component of the culture of early modern astronomy. The letters
exchanged by astronomers played a crucial role in constituting and maintaining an international community of scholars interested in the study of
the heavens. These letters often shaped the theories and observing strategies employed at different locations, and sometimes performed the role
of *calibrating* the astronomical work carried out by different practitioners.
Letters on astronomical subjects also served other functions, of considerable significance to their authors, that reflect features of epistolary practice
common to courtly and scholarly letter-writing of the period more generally. Astronomers faced similar difficulties to other writers in managing the
delivery of letters, and they were schooled in the same conventions of epistolography and etiquette. These conventions, although they constrained

---

[2] Moran 1982; Rosen 1986; Jardine 1988a; Barker and Goldstein 1995; Helfricht 1999.
[3] *KGW* XIII–XVIII; *OdG* X–XVII; but see Baumgartner 1988.
[4] *CCC*; Forbes, Murdin and Willmoth 1995–2001.
[5] See, however, for the eighteenth century, Widmalm 1992.

the structure and language of the letter, also granted it great flexibility as a means of self-presentation. By exploring the interplay of these various elements with respect to a specific body of correspondence, I aim to show how, as a form of communication capable of 'bearing the heavens' between sites and individuals, the letters of early modern astronomers should be read and understood.

I TYCHO BRAHE AND THE REPUBLIC OF LETTERS

The fact that historians have paid relatively little attention to the letters of astronomers might not seem to require much by way of explanation. Many deserving scholarly projects have failed to be carried out for no reason other than the want of someone with time and energy to pursue them. There are, however, several reasons that help to account for the fact that historians have not considered the study of sixteenth-century astronomical correspondence a matter of urgency. One of the most significant, I think, is that intellectual historians in general have been somewhat slow to appreciate the character and function of the early modern letter. A number of preconceptions about letters have tended to cloud historians' perception of their value and significance, particularly in comparison with other types of writing.[6] These include the belief that letters are casually composed and ephemeral documents, of a more personal nature than an author's published works, and the supposition that, as essentially private forms of communication, they were typically written for a readership of one. Sometimes, it is true, intellectual historians consult letters and diaries precisely because of the suspicion that they will prove a better guide to writers' thoughts and beliefs than the possibly self-censored material issued for wider consumption. More frequently, however, they delegate the task of reading correspondence to those undertaking to produce a biographical study. In this way, they suppose, the circumstantial details of an individual's domestic situation and emotional life, both thought of as useful aids to an analysis of an individual's 'real' scholarly output, will eventually emerge, along with basic chronological facts, in a more convenient form.

This view of the character and significance of correspondence does not conform well with the facts. Most importantly, of course, it does not fit with what is known about early modern epistolary culture and practice. But it does not even match most historians' own day-to-day experience of the scholarly and professional uses of letters and e-mails. The false consciousness

---

[6] Mosley, Jardine and Tybjerg 2003, 422–425; Henderson 1993, 2002.

of historians with respect to correspondence may derive from the powerful effect on the imagination of eighteenth- and nineteenth-century models of letter-writing, as propagated through epistolary novels and similar literary sources. But whatever its origin, it, and the prejudices that accompany it, can be difficult to dispel. John Dreyer himself, although he oversaw the publication of Tycho's scholarly correspondence, seems to have persisted in the view that letters were important primarily insofar as they shed light on an author's publications.[7] While it is true that they can be used for this purpose, this view of early modern letters hardly reflects their full historical value, and it fails to acknowledge their true period significance.

A second reason for the comparative neglect of the letter-writing activities of a sixteenth-century figure like Tycho may well be the emphasis given, in standard accounts of the early modern development of the scientific enterprise, of later and larger correspondence networks. These highly centralised networks, associated with figures such as the Minim friar, Marin Mersenne (1588–1648), and Henry Oldenburg (1619–1677), the Bremen-born emigré to England, have rightly been associated with the rise of scientific academies, societies, and journals.[8] But the importance assigned to the coalescence of scientific communities about such institutional bodies and communication forums may have had the effect of eclipsing earlier stages in the process of these communities' formation. In comparison with Oldenburg, for example, prior attempts at communication by letter have sometimes been characterised as of little significance: they were, we are told, 'more or less informal, and restricted to small circles of personal contacts'.[9] As a commonplace of the history of science, the story of the rise of correspondence as a tool of seventeenth-century scientific 'intelligencers' has perhaps dampened any enthusiasm for pursuing the kind of study undertaken here.

Tycho Brahe was, it is perfectly true, not a correspondent like Oldenburg or Mersenne. He did not consciously make himself into the central figure of an epistolary network, nor was he the secretary of any scientific society. He was not, therefore, an acknowledged focus for the accumulation and dissemination of natural philosophical and mathematical material by epistolary means. Even so, communication by letter worked well enough for Tycho to be an important tool in the pursuit of his astronomical endeavours. That his circle of contacts was relatively small was partly compensated for by the fact that most of those with whom he exchanged letters also corresponded with others, both within and outwith his own epistolary network.

---

[7] Dreyer 1924, 303. [8] Ornstein 1913, 166–169; Boas Hall 1965. See also Lux and Cook 1998.
[9] Eamon 1985, 343. See also Vickery 2000, 67–87.

As well as contributing to the cohesion of the community, this allowed communication with fellow scholars to take place at one or two removes. The apostate Hungarian bishop Andreas Dudith (1533–1589), for example, who had settled in Wroclaw after converting from Catholicism, was not one of Tycho's correspondents; but contact of a sort between the two men was made possible via their mutual acquaintances. Thus in 1588, Tycho sent Dudith a copy of his *De recentioribus phaenomenis* and solicited his opinion of it through Bartholomaeus Scultetus (1540–1614), a resident of Görlitz and Tycho's one-time tutor in mathematics.[10] When, as happened in 1589, the Prague-based Imperial Physician Thaddaeus Hagecius (1525–1600), a correspondent of both men, informed Tycho of Dudith's demise, the Danish astronomer replied that he was sorry to hear it. He had, he wrote, 'decided to contract a friendship with him through letters; and, at the request of Praetorius, to send him a certain student of mine with a skilfully wrought instrument for astronomical observations. For he, when he was alive, requested this of Praetorius.'[11] The Altdorf mathematician Johannes Praetorius (1537–1616) was another of Dudith's correspondents; and on this evidence, although no extant letters between them are known, he may have been one of Tycho's as well.[12] Praetorius also corresponded on astronomical topics with the Bavarian chancellor Johann Georg Herwart von Hohenburg (1553–1622), who himself exchanged letters with Tycho and with two other of his correspondents, Johannes Kepler and the Leiden classical scholar Joseph Scaliger (1540–1609).[13] The density of interconnections is such that the extended intellectual community formed by the combination of personal acquaintance and epistolary contact is easier to represent graphically than to put into words. Even this, however, is only selectively possible; if the attempt were made to add more of the known contacts of individuals already shown on Fig. 2.1 to the diagram, the task of representing the connections between them would quickly become prohibitively complex.

Tycho's scholarly correspondents occupied a range of positions in early modern society. Several were professional mathematicians, in the sense of holding a teaching post in mathematics at a school or university; besides

---

[10] *TBOO* VII, 123.11–16. On Dudith, see Costil 1935; for Scultetus, Helfricht 1999. That Tycho received instruction from Scultetus is noted by Thoren 1990, 17.

[11] *TBOO* VII, 182.35–37, 214.27–33. For Hagecius, see Hellman 1944, 184–193.

[12] Szczucki and Szepessy 1992– , VI, 189, 390. It should be noted, however, that Tycho and Dudith shared another mutual correspondent in the form of Joachim Camerarius, who also knew Praetorius. See Szczucki and Szepessy 1992– , VI, 84, 895.7–10; *TBOO* VII, 42. For Praetorius, see Westman 1975b, 289–292; Müller 1993; Folkerts 1996.

[13] See *KGW* XIII, 131–140, 147–150, 177, 205–206; Günther 1889; *NDB* VIII, 722–723.

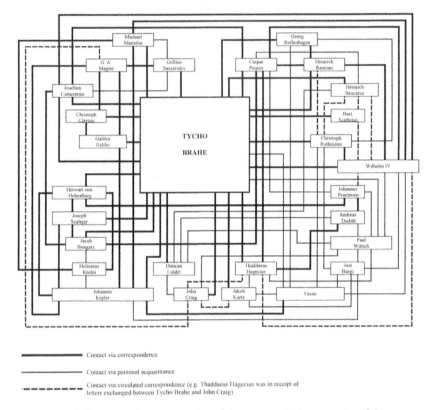

Contact via correspondence

Contact via personal acquaintance

Contact via circulated correspondence (e.g. Thaddaeus Hagecius was in receipt of letters exchanged between Tycho Brahe and John Craig)

2.1 A diagrammatic representation of the astronomical community of the late sixteenth century.

Scultetus and Praetorius, this group included Giovanni Antonio Magini (1555–1617), professor of mathematics at Bologna,[14] and Heinrich Brucaeus, whose chair at Rostock was also in medicine. Astrology was an important component of contemporary medical practice and theory, and provided one of the most powerful motivations for studying the heavens in the period, so it is hardly surprising that practising physicians were also among those with whom Tycho corresponded.[15] In addition to Hagecius, this category included Joachim Camerarius the younger (1534–1598), who practised in Nuremberg, and the Scotsman John Craig (*d. c.*1620), with whom Tycho became involved in a protracted epistolary dispute over the

[14] Favaro 1886; *DSB* IX, 12–13; Clarke 1985.
[15] Chapman 1979; Tester 1987, 222–224; Kusukawa 1993.

nature of comets.[16] Other individuals, like Herwart von Hohenburg, possessed some interest in astronomy, astrology, or some other branch of mathematical or medical scholarship, but served their princely masters in other capacities: Heinrich Rantzau was the governor for the Danish crown in Schleswig-Holstein, Jacques Bongars (1554–1612) was a diplomat in the service of France's Henri IV, and Jakob Kurtz (*d.* 1594) was Prochancellor at the Imperial Court.[17] There were also those whose interests or expertise extended as far as matters of immediate relevance to Tycho's work, but whose careers and principal contributions to scholarship lay in rather different fields: such were Scaliger, editor of the verse *Astronomica* of Marcus Manilius, and an expert on chronology, and the Augsburg humanist Hieronymus Wolf (1516–1590), who besides being a Greek scholar, Byzantinist, and the librarian of the Fugger banking dynasty, was an enthusiastic early correspondent of Tycho's on the subject of comets and eclipses.[18] Two other librarians with whom the Danish astronomer exchanged letters were Hugo Blotius (1533–1608), the Dutchman who headed the Imperial Library in Vienna, and Paul Melissus (1539–1602), who made his name as a poet but was later appointed as the librarian of the Elector Palatine in Heidelberg.[19] Tycho's interest in Melissus derived partly from the books in his care, partly from his talent as a writer of verse: astronomical texts and laudatory poems were two of the things he most frequently sought from those with whom he corresponded.

Of course, describing Tycho's correspondents in this way can be slightly misleading. Not all of these individuals corresponded with Tycho over the same period of time, or with the same intensity of interest. Some of the exchanges were more fleeting than others, even perfunctory, and some of them were not greatly concerned with the communication of astronomical material. It would be equally unhelpful, however, to place too much emphasis on the vocations and competencies of the individuals with whom Tycho was in contact. Early modern astronomers certainly possessed the notion of an expertise shared by some and not all; that, after all, was the import of Copernicus' famous statement in *De revolutionibus* that mathematical material was written for mathematicians.[20] But Tycho and other mathematical experts did not only consider themselves members of an

---

[16] On Camerarius, see *DBE* II, 269. For Craig, see Mosley 2002b; *ODNB* XIII, 950–951.

[17] For Bongars, see Hagen 1874; on Kurtz, von Gschliesser 1942, 140–141.

[18] For Scaliger, see Grafton 1983–1993; for Wolf, *ADB* XLIII, 755–757.

[19] On Blotius, see Louthan 1997, 53–84; Staikos 2000, 436–442. For Melissus, see *ADB* XXI, 293–298; *NDB* VII, 53.

[20] Copernicus 1543, 4v: 'Mathemata mathematicis scribuntur . . .'

elite group defined by their superior knowledge of astronomy; they were also citizens of the *respublica litterarum*, that larger imagined community which united all scholars, even those separated by both territorial boundaries and confessional differences. To communicate with others in the Republic of Letters was considered partly an obligation to be fulfilled, partly a means of establishing and securing one's own reputation.[21]

In order to show just how important letter-writing was to Tycho's astronomical practice, it has seemed sensible to concentrate in this chapter on a subset of his total correspondence. The choice of letters on which I have decided to focus respects the selection which Tycho himself made when he singled out his exchanges with the Kassel astronomers as the first volume of *Epistolae astronomicae* to be collated and published. There are a certain number of disadvantages to this choice. Partly because of interest in Landgrave Wilhelm IV of Hesse-Kassel as a princely practitioner of the sciences, and partly because of their status as a body of material which Tycho published himself, the Hven–Kassel exchanges have attracted more attention to date than any other portion of his extant correspondence.[22] It might seem, therefore, that, by choosing to focus on them here, I am compounding the neglect of those letters which remained in manuscript until significantly later. There is also the problem that, since not all of the letters Tycho published have survived independently, the use of them as sources prior to a thorough analysis of the process of transforming them from manuscript to print requires a great deal concerning their reliability as historical documents to be taken on trust. Finally, there is the fact that Wilhelm IV, as an enthusiastic prince with considerable resources to devote to astronomy, and Christoph Rothmann, as a mathematician employed by him to pursue a particular programme of observational work, cannot be considered representative of Tycho's correspondents in general.

Against these various difficulties, however, can be set the desirability of reasserting the importance of the Hven–Kassel exchanges as a set of manuscript exchanges, given that this has been somewhat occluded by the attention given to the letters in their later printed form. The volume of *Epistolae astronomicae* has often been considered a contribution to Tycho's dispute with the Imperial Mathematician Nicolaus Reymers Baer, but it

---

[21] For a reference to the *Respublica litterarum* within Tycho's correspondence, see for example *TBOO* VII, 330.8–11. On the Republic of Letters, its status as an imagined community, and the notion of mutual obligation between its 'citizens', see Goldgar 1995, esp. 1–53; Bots and Waquet 1997, esp. 11–27, 117–119; Findlen 1999; Burke 2000, 19–20. Feingold 2002, esp. 1–45, situates Jesuit scholars within this wider Republic.

[22] Moran 1978, 1980, 1982; Westman 1980a, 98–99; Westman 1980b, 477; Rosen 1984, 1985a; Gingerich and Westman 1988; Thoren 1990; Barker and Goldstein 1995; Gingerich and Voelkel 1998.

needs to be studied as the product of two separable stages if the reasons for its appearance are to be fully appreciated. It seems better, therefore, to first consider the exchange of letters over the six or so years of the active correspondence, and only then, informed by a knowledge of their contents, turn to the subsequent labour of remaking them into a single printed volume. This second task will be deferred to Chapter 3, when we come to consider the role of books in astronomical communication and community-formation in the later sixteenth century. At that point it will also be possible to consider some of the ways in which Tycho's other correspondents assisted his labours.

II EPISTOLARY ORIGINS: THE BEGINNING OF THE HVEN–KASSEL EXCHANGES

Moreover, I would like even the kindly reader to be warned and asked not to search carefully and rigorously for charming expressions and a fine elegance of Latin prose in those letters which we ourselves wrote, or revised for others. For besides the fact that we, from an early age, have paid more attention to the learning and examining of things, than of words, these letters were for the most part spontaneous, and set down by me, or dictated to my amanuenses, at various times, during other more serious occupations; so I scarcely had the leisure to read them over, let alone adorn them with elegant and nice prose, even if I had very much wanted to produce such, or indeed had taken it upon myself to do so. (Tycho Brahe, *Epistolae astronomicae* (1596))[23]

According to Tycho's own account, two events triggered the correspondence between Kassel and Hven.[24] The first, a celestial phenomenon, was the comet that was seen in the skies above northern Europe during the months of October and November of 1585. The second occurred in January or February of the following year, when the young nobleman Gerard Rantzau (1558–1627) visited the palace of King Frederick II of Denmark, and, finding that Tycho was also staying there, delivered to him two letters with which he had been entrusted.[25] One of these letters had been addressed to Tycho by Gerard's father Heinrich, but the other was a copy of a letter of which Heinrich himself had been the designated recipient. Its author thanked Rantzau for a letter and book he had been sent at the end of September

---

[23] *TBOO* VI, 23.5–15, *Praefatio generalis.* [24] *TBOO* VI, 33.19–29, 34.7–19.

[25] As commander of the citadel of Kronenburg, Gerard Rantzau was well known to his near neighbour Tycho, and had previously visited him on Hven in December 1584 or January 1585. Tycho described this visit in the letter that he wrote to Gerard's father Heinrich shortly afterwards. See *TBOO* VII, 89.13–33.

1585, and mentioned that he had been busy with astronomical matters, in particular the observation of the comet.[26] 'It is small and hairy all over', he wrote, 'so that we consider it to be the type of comet that one calls Circean, and that are wont to appear at the time of grand conjunctions'.[27] Prompted by his curiosity to enquire whether the comet had been seen in Denmark, and in particular whether Tycho Brahe had observed the phenomenon and could be persuaded to send an account both of it and his observational methods, the writer of this letter, Landgrave Wilhelm IV, included a brief note on its position and movement at the time that it had first been perceived at his observatory in Kassel.[28] By passing this letter on to Tycho, Heinrich Rantzau and his son became the first, and most distinguished, of several intermediaries who helped to facilitate the transfer of data between these two noble astronomers.[29]

Tycho responded enthusiastically to the relayed request of the Landgrave. Replying directly to Wilhelm, he recounted how he had often thought of writing to him, as he still retained fond memories of the time he had spent at Kassel several years previously.[30] This was a reference to a visit made in 1575, during the last of his youthful travels around Europe.[31] At that time, Wilhelm's observatory had been in existence for one-and-a-half decades, and he had already collaborated with a number of talented individuals: in

---

[26] *TBOO* VI, 31.12–39.

[27] *TBOO* VI, 31.39–32.3. The term 'Circean' as a description of a comet is obscure, and not noted by either Hellman 1944 or Dall' Olmo 1980. Circe was described by Homer in *The Odyssey*, e.g. XI.8, as 'fair-tressed', and in Ovid's *Metamorphoses*, XIII.968, as the daughter of the Sun. These descriptions may have been sufficient to sustain a corruption or misreading of the word *hirceum*, goat-like; in Pliny's taxonomy in the *Naturalis Historia*, II.90, comets ringed with tufts of hair are named 'Goat comets', a name which derives from Aristotle's *Meteorologica*, I.341b. See Lee 1952, 30–33; Rackham 1938–1963, I, 232–235. Another possibility, suggested to me by Nicholas Jardine, is that the adjective, used of Circe's homeland Colchis, implies hairiness because this was also the territory of the Golden Fleece. As John North has pointed out to me, one such instance of the word occurs, suggestively enough, in Valerius Flaccus' *Argonautica* VI.426. But whatever the term's origin, the association Wilhelm makes between comets and planetary conjunctions was not so unusual. That comets were caused by conjunctions was one of the ideas discussed by Seneca, albeit critically, in his *Quaestiones naturales*, VII.12; see Corcoran 1971–1972, II, 251–255. 'Grand conjunction' (*maxima coniunctio*) is the term normally used of a conjunction that sees the planets in question at the same point on the ecliptic at which they were first in conjunction; this point would be the spring equinox according to some conceptions of the configuration of the heavens at Creation. In the case of Saturn and Jupiter, whose conjunctions are normally called 'great' and occur approximately every twenty years, the periodicity of 'grand' conjunctions is 960 years. See Pomian 1986, 36–38; Smoller 1994, 21.

[28] *TBOO* VI, 31.30–39, 32.3–9.

[29] A comprehensive account of Heinrich Rantzau's role as a communications facilitator and patronage-broker in a variety of fields is still awaited. See, however, Fuhrmann 1959; Evans 1984; Steinmetz 1991, esp. 117–141; Witt *et al.* 1999; Lohmeier 2000; Zeeberg 2004.

[30] *TBOO* VI, 34.9–16.

[31] See Dreyer 1890, 78–81; Thoren 1990, 92–96. Tycho's own account of his time in Kassel appears in the dedicatory preface of the *Epistolae astronomicae*, *TBOO* VI, 10.13–35.

particular, Victorin Schönfeld (1525–1591), whom he appointed as professor of mathematics at the University of Marburg, Andreas Schöner, and Eberhard Baldewein, his instrument-maker and astronomical assistant.[32] But after he became a ruling-prince in 1567, Wilhelm found he had less time for observational work; and it seems that, apart from a brief flurry of activity accompanying the appearance of the new star of 1572,[33] the observatory had remained underused for several years prior to the Danish astronomer's visit.[34] Accordingly, Tycho encouraged his host to acquire further skilled assistance to pursue the study of the heavens.

Although Tycho's sojourn in Kassel was very brief, as it was cut short by a tragic death in Wilhelm's immediate family,[35] he seems to have made enough of an impression for it to be commemorated in a portrait of the prince and his consort (Fig. 2.2).[36] Moreover, Tycho later wrote that it was an embassy of Wilhelm's to King Frederick, following this meeting, that persuaded the Danish monarch to support his astronomical ambitions.[37] But despite these ties, no correspondence had been initiated before 1586. In replying to the letter forwarded by Rantzau, Tycho said that he had intended to write, yet had thought that, with an 'interval of so many years', and the 'administering of so much state business', Wilhelm might have forgotten him.[38] However,

since your Highness did not scorn to greet me kindly, via the aforementioned letter sent to MASTER RANTZAU; and moreover, having seized the occasion of the recently seen comet, invited me to write to you, I resumed my intention more boldly, and having congratulated myself a great deal that this opportune moment to fulfil my desire of so many years had so suitably arisen, I decided not to put off sending some letters to your Highness any longer, relying on that kind invitation of your Highness to excuse my boldness.[39]

---

[32] For the employment at Kassel of Schönfeld, Schöner, and Baldewein, see Leopold 1986, 15.

[33] This 'new star' or *nova*, was actually a supernova, but I shall continue to refer to it in the same terms as it was described in the period. For an explanation of the phenomenon in modern terms, see Baade 1945.

[34] Schönfeld, though he continued to collaborate with Wilhelm, was based in Marburg, as was Baldewein, whereas Schöner was only in the Landgrave's service for a short period between 1558 and 1560. Another court *mathematicus*, Johannes Ottonis, is known to have served Wilhelm from 1572, and he and Schönfeld worked with Wilhelm on observing the 1572 nova. But although he was at Kassel until his death in October 1576, he cannot have satisfied Tycho that the observatory there was being used to its fullest. See Leopold 1986, 17–18.

[35] As Tycho explained in the *Epistolae astronomicae*, a daughter of Wilhelm's died after he had been there one week, and not wishing to intrude on the Landgrave's grief, he departed for Frankfurt. See *TBOO* VI, 10.30–35.

[36] On the portraits of Wilhelm and Sabine, see Kirchvogel 1967; von Mackensen 1988, 18–19.

[37] Tycho made this claim in the *Epistolae astronomicae*; see *TBOO* VI, 11.3–21. He did not, however repeat it in the autobiographical section of the *Mechanica*, *TBOO* V, 108.36–109.17.

[38] *TBOO* VI, 34.17–21.　　[39] *TBOO* VI, 34.22–29.

WILHELM VON GOTS GNADEN
LANDGRAF ZV HESSEN.
GRAF ZV CATZENELNBOGEN
DIETZ. ZIGENHAIN .
VND NIDDA.

ÆTATIS SVÆ.XXXXV.
ANNO.M.D.LXXVII.

2.2   The matched portraits of Wilhelm and his wife Sabine, painted in 1577, probably by
the artist Caspar van der Borcht. The figure at the bottom right of the portrait of Wilhelm
is thought to be Tycho; he is partially obscured by a sextant. To the left is visible the
calculating globe employed in the Kassel observatory. In the picture of Sabine, an
azimuthal quadrant and a torquetum are visible. By permission of the
Astronomisch-Physikalisches Kabinett, Staatliche Museen Kassel.

Tycho continued his letter with a brief description of the manner in which
he had himself come to notice the comet. Momentarily, he admits, he had
foolishly thought it a star. But from 18 October onwards, it had been 'very
carefully observed with instruments excellent, and subject to no error'.[40]
The most significant of the observations, Tycho stated, 'set down on their

---

[40]   *TBOO* VI, 34.29–35.

2.2 (*cont.*).

own piece of paper (so that this letter was not enlarged too much by their inclusion here) I am sending to be studied at your Highness' leisure'.[41] He requested that the Landgrave might in turn send some of the observations made at Kassel, and proposed further exchanges of material, 'so that from the mutual comparison performed, the certainty of the instruments and the observations might be determined that much more accurately'.[42] Indeed, in addition to the cometary observations, he sent several other documents

---

[41] *TBOO* VI, 34.41–35.4. The document mentioned was also reproduced in the *Epistolae astronomicae*. See *TBOO* VI, 41–47.
[42] *TBOO* VI, 35.27–35.

as well: 'a picture of the building in honour of *Astronomia* built by me in recent years on this island of Hven, at not inconsiderable expense'; an engraving of his observing sextant, including a description of its mount; another of an equatorial armillary;[43] and a solar ephemerides calculated for the longitude of Hesse.[44] For, as he put it, 'I would like these labours of mine to be subject to the judgement of your Highness in particular, since I know that your Highness, a distinguished master of this *scientia*, can judge these matters soundly'.[45]

Apparent in this opening to the correspondence are a number of features characteristic of the Hven–Kassel exchanges in general. Most notably, the transmission through and alongside the letters of various astronomical productions – including tabulated observations, prose accounts of celestial phenomena, and instrument designs – was to continue for much of its six-year duration. Indeed, Wilhelm's request for observations from Hven, and Tycho's hope that, by obliging the Landgrave, he would in turn acquire some of those made at Kassel, suggest that these men anticipated a real benefit from the exchange of data obtained at their different locations. Tycho's comment, that such exchange could provide a basis for the determination of the certainty of their instruments and observations, indicates why: comparison of observations made at the two sites would enable a form of calibration of their equipment and techniques. As we shall see, calibration by letter was not an entirely straightforward process, but it was a continuing objective of the Hven–Kassel correspondence.

The early correspondence also demonstrates aspects of sixteenth-century epistolary culture significant for understanding the exchanges between Hven and Kassel. As a taught branch of rhetoric, the art of letter-writing had been established in the eleventh century under the name of the *ars dictaminis*. Assimilating letter-writing to oratory, and rapidly adopted by the scribes who served the expanding chanceries of Europe, this *ars* fixed the formulae and structure of letters for approximately two centuries.[46] In the fourteenth century, rediscovery of the less formal epistolography of the ancients, including Cicero's letters and those of the younger Pliny, laid the foundation for the *ars epistolica*, a humanist reappraisal of the nature of epistolary communication. Among the many formularies, form-letter books, and manuals of letter-etiquette newly produced in the fifteenth and sixteenth centuries, was the *De conscribendis epistolis* of Desiderius Erasmus

---

[43] *TBOO* VI, 37.8–17.   [44] *TBOO* VI, 38.32–36.   [45] *TBOO* VI, 40.16–19.
[46] Gerlo 1971, 104; Witt 1982, 3–7; Henderson 1983a, 90; Camargo 1991.

(1466–1536).[47] This influential text, first published officially in 1522,[48] was known throughout Europe by the time of its author's demise, its popularity having been augmented by the availability of published exemplars of Erasmus' own correspondence with scholars such as Thomas More, and Guillaume Budé.[49] This popularity persisted: Erasmus' rhetorical works were part of the curriculum of the University of Copenhagen during Tycho's time there as a student.[50]

In discussing the letter, Erasmus considered not only the oratorical categories that had served medieval officials, but also a *genus familiare* – the familiar letter exchanged between friends.[51] Furthermore, he mentioned a dialectical form of epistle to be used by scholars discussing or debating learned information.[52] Unlike early exponents of the *ars epistolica*, Erasmus chose not to distinguish the letter from the oration on the basis of its *salutatio*, one of the five parts of the letter described by the medieval rhetoricians, but instead emphasised its characteristic of flexibility.[53] Choices about the form of a letter were to take into account both the demands of the subject-matter, and consideration of what would please the recipient; so the letter-writer need not be constrained by the structural divisions of the medieval *dictatores*, nor restricted to slavish imitation of classical authors and authorities. Issuing from the pens of Erasmus and his contemporaries, letters became an important instrument for displaying humanist erudition. And indeed, Erasmus, among others, not only recognised the role of correspondence in presenting a public and historically recoverable self, but also utilised this understanding in deliberately manufacturing, and manipulating, his fame and identity. For him, the letter was flexible enough to accommodate the subject-matter, the interests of the recipient, *and* the

---

[47] On Erasmus as humanist and letter-writer, there is a large literature. See, for example, Jardine 1993, and the works cited in note 49.

[48] For the publishing history of this text, see Jardine 1993, 160–164.

[49] Gerlo 1971; Bietenholtz 1977; Henderson 1983b; Jardine 1993, 14–20. Erasmus' first collection of letters, the *Epistolae aliquot illustrium virorum ad Erasmum et huius ad illos*, was published at Louvain in 1516.

[50] Christianson 1967, 200, does not specify the individual texts used, but both *De conscribendis epistolis* and *De duplici copia verborum ac rerum commentarii duo* (1521) are acknowledged as popular rhetorical works by Sonnino 1968, 234.

[51] *CWE* XXV, 71–73.

[52] 'There is a class of letters quite common among scholars, in which they carry on a reciprocal scholarly exchange among themselves, when they wish to learn about some topic, or reply to an enquirer, or dispute some point on which they fail to agree. Of this class, as it is so varied, no set account can be given. I shall merely point out some examples.' As translated in *CWE* XXV, 254. On the exact relationship between the medieval categories of deliberative, demonstrative, and judicial letter, and those added by Erasmus, see Henderson 1993, 150–151.

[53] Gerlo 1971, 104–105; Fumaroli 1978; Witt 1982, 8–9; Henderson 1983a, 91–94; Jardine 1993, 149–150.

requirements of authorial self-representation; so much so that modern readers of his published letters are warned against viewing them as neutral media
transmitting historical facts about his life, and his relations with others.[54]
But printing was not the only route to epistolary manipulation of one's
own reputation. As happened with the forwarding of Wilhelm's letter by
Rantzau, even manuscript letters could be copied and circulated.[55] Renaissance writers are likely to have kept this in mind when they set their pens
to their paper.[56]

Tycho was no Erasmus, stylistically. But his letters do show a concern
for certain proprieties, stemming partly from the humanist *ars epistolica*,
partly from the more conservative traditions of the early modern courtier.
His very first letter to the Landgrave was addressed, 'TO THE MOST
ILLUSTRIOUS PRINCE AND LORD, LORD WILHELM, LANDGRAVE
OF HESSE, COUNT IN CATZENELLEBOGEN, DIETZ, ZIEGENHAIN,
AND NIDDA: PRINCE AND LORD MOST MERCIFUL'.[57] In *De conscribendis epistolis*, Erasmus had argued that this part of the letter, the *salutatio*, had become too formulaic, over-burdened with meaningless extravagance. He had also expressed distaste for what he called the 'absurd practice'
of addressing individuals in the plural, and objected to the liberal sprin-
kling of titles, instead of first names, in the body of a letter.[58] But the
majority of his epistolary exemplars were addressed to fellow scholars, and
he admitted that some concession to the dignity of kings and popes was
perhaps only prudent. 'Young men should not so much be untaught this
vice as encouraged to scorn it altogether', he wrote of pluralisation; but
added, 'at least in dealing with those with whom no risk is involved'. And
in the same section: 'let us allow them what they want. Let it be a sign of
politeness to flatter one's betters with this form of respect, and let it be the
mark of an uncouth and impolite person not to pay this respect.'[59] Thus,
while many writers of epistolographic manuals followed Erasmus' lead in
suggesting that letter-writers stick to clear and simple formulae of greeting,
tables and lists specifying the form of address appropriate to individuals of
a particular rank or occupation continued to be provided. Kings, princes
and dukes were to be called most Christian, most potent, most illustrious,
celebrated; barons and city senators were noble, generous, or excellent; theologians, distinguished and most wise. Not omitting to identify suitable

[54] Jardine 1993, 147–174.
[55] For this phenomenon see, with respect to Erasmus, Bietenholtz 1977; Jardine 1993, 18, 117. With
respect to Galileo and his Church opponents, see Blackwell 1991, 69–72, 107, 196–197, 203, 206, 211.
[56] Bietenholtz 1977, 63.   [57] *TBOO* VI, 33.13–18.   [58] *CWE* XXV, 50–55, 45–50, 60.
[59] *CWE* XXV, 45, 47.

epithets for physicians, lawyers, and mathematicians, the manual-writers ultimately arrived at women, for whom they proposed such titles as most modest, most chaste, or most virtuous.[60]

The form of Tycho's *salutatio* to the Landgrave was thus highly conventional. It was also employed with great consistency throughout the epistolary exchanges between Hven and Kassel. In replying, Wilhelm, and later his son Moritz, made repeated use of the formula, 'To our especially beloved Tycho Brahe, Lord of Knudstrup' (later emended in Tycho's Latin rendition to include the epithet 'most noble'), to which was added the self-description, 'Wilhelm [or Moritz], by the grace of God Landgrave of Hesse, Count in Catzenellebogen, etc.'[61] The consistency of usage is probably due, in part, to the functional role the *salutatio* played in designating the recipient of the letter; a role that is obscured by all the published editions of the correspondence, including Tycho's own, but becomes apparent upon inspection of the manuscripts.[62] Thus the first letter sent from Hven to Kassel, by no means an unrepresentative example, was written on two folio sheets sewn together to form a booklet of eight pages, the last of them blank.[63] This pamphlet was folded into a small packet and sealed with wax, and the formula of greeting was then written on the exposed part of the final page. In physical location and practical significance, therefore, formulaic *salutationes* were the precise equivalent of the names and addresses applied to modern envelopes.

Nevertheless, the *salutationes* adopted by the Hven–Kassel correspondents for the exterior of their letters manifest to us, as they manifested to their contemporaries, a difference in status that had important consequences for what appeared within these epistolary parcels. The inferior of a

---

[60] These examples are extracted from Conrad Celtis' *Methodus Conficiendarum Epistolarum*, as printed in Brandolinus 1549, 396–397. Cf. Erasmus in *CWE* XXV, 57–59, and Hegendorff 1545, 30–32. For an epistolographer who rejected all such flourishes, one must turn to Juan Luis Vives (1492–1540). He, in his *De conscribendis epistolis*, spoke out against the teachings of Erasmus, Guillaume Budé, and Thomas More, arguing that: 'No one adds in books to M. Tullius, most eloquent orator: or to Aristotle, most intelligent philosopher. Therefore it is apparent that nothing is more serious or more becoming than to write to the Emperor or a King without an adjective; as is certainly indeed approved by men clever and talented.' See Brandolinus 1549, 316: 'neque enim libris M. Tulli adiicit quisquam, oratoris facundissimi: vel Aristotelis, philosophi acutissimi. Itaque deprehensum est nihil esse gravius, aut decentius, quam ad Caesarem sine adiecto scribere, aut ad Regem, quod prudentibus et ingeniosis viris valde probatur.' For a discussion of this text, see Henderson 1983a.

[61] *TBOO* VI, 48.11–17. This descriptive formula was standard for rulers throughout the German lands.

[62] Copies of Tycho's correspondence with Wilhelm and his son Moritz are preserved in the Marburg Staatsarchiv as collections 4a.31.14 and 4a.39.70.

[63] Although it should be noted that it is not readily apparent *when* the sheets were sewn together. The page-sizes of this quire, approximately 330 × 210 mm, are fairly representative of the rest of the correspondence.

ruling prince, Tycho behaved towards the Landgrave as a subject, or client, to his lord or his patron. At the same time, however, he was of higher standing than Rothmann, whom he greeted as the 'most erudite and distinguished Court *mathematicus* of the most illustrious Landgrave',[64] and who responded by saluting him as the 'most noble and most distinguished Lord Tycho Brahe, most pre-eminent *mathematicus* of our age'.[65] Contrary to the advice of Erasmus, Tycho was always addressed by Rothmann as 'Your Excellency' (*Tua Excellentia*) within the body of the letter, just as he typically called Wilhelm 'your Highness' (*Tua Celsitudinis*).

This tripartite hierarchy meant that Tycho's exchanges with Wilhelm differed from those with Rothmann for reasons other than the mere fact that, more removed from the day-to-day running of the Kassel observatory, the Landgrave was less capable of being specific about its operations than his assistant and *mathematicus*.[66] As the lord and patron of the institution, he rightfully took credit for its successes, yet avoided blame for its shortcomings. Indeed, the extent to which Tycho could, within the bounds of courtesy, challenge and contradict the assertions of Rothmann, depended very much upon the extent to which they could be identified as his views, rather than those of the Landgrave. Tycho was restricted to making requests of Wilhelm, but could and did command his *mathematicus*. On the other hand, Rothmann held a rather precarious licence to treat with Tycho, without deferring to him, by virtue of acting for Wilhelm. As we shall see, however, this did not entirely protect him, as an individual, from the Danish astronomer's propensity for intellectual bullying.

The status-differential between the Danish astronomer and Wilhelm may also lie behind the delicate negotiation of the transition between a mediated correspondence, in which the Landgrave and Tycho were linked by their mutual acquaintance with Rantzau, and the direct transfer of letters between the observatories at Hven and at Kassel. Tycho's reasons for not previously writing to the Landgrave sooner sound, to the modern ear, like excuses from a poor and lazy correspondent. But his plea that his boldness in writing now be excused points to the possibility that it would have been a breach of etiquette for him, the inferior in status, to initiate epistolary contact with Wilhelm, notwithstanding their cordial meeting almost eleven years before the start of these exchanges.

---

[64] *TBOO* VI, 85.1–7.   [65] *TBOO* VI, 54.9–14.

[66] For this reason it is misleading to write of the letters as offering 'an unusual view of Tycho's interactions with two people who, *between them*, were roughly his social and professional equals' (my italics); see Thoren 1990, 271.

In addition to his sensitivity to courtly manners, Tycho displayed a keen awareness of the requirements of the epistolography of the humanists. He referred explicitly to the conventions of the letter-genre, often apologetically. Thus, near the end of his first letter to Wilhelm, he wrote, 'But I see that I, while conferring more eagerly with your Highness about astronomical matters, have transgressed the laws of the letter with too much prolixity, and have kept your Highness occupied in reading this longer than is appropriate'.[67] Brevity had been, for Erasmus and others, one of the letter's distinguishing features,[68] but Tycho was frequently to admit to an inability to write tersely over the course of the six-year correspondence;[69] and when he came to print the letters in the *Epistolae astronomicae*, he made the same apology in the work's general preface. In a passage also interesting because it displays an attempt to associate the correspondence with both of the new epistolary categories, he wrote:

But if sometimes in these letters of ours, especially in those with which we make reply to the Landgrave's, and certain others, I have perhaps been more prolix than the laws of familiar letters allow, then I ask that a fair-minded examiner takes it in the better part; and at the same time, it should be taken into account that these letters do not concern commonplaces, as others written familiarly, but typically treat difficult matters, and the very innermost parts of the divine Astronomy, and are for the most part dogmatic; and therefore they discuss things concerning this matter with more prolixity, sometimes speaking for them, sometimes against.[70]

In fact, with the preface also intended to apply to future books of correspondence that Tycho hoped to publish, the astronomer was acknowledging a fault for which he frequently offered up *pro forma* apologies;[71] they read like requests for forgiveness from an individual who knows that his supposed transgressions have actually been acts of excessive generosity.

Tycho's comment on length was part of a larger *apologia* for the stylistic shortcomings of the letter-book's contents. He asked that the reader forgive him for having paid more attention to the subject-matter than to literary studies. 'I prefer to admire and praise philologists of this and earlier ages, than to imitate them',[72] he asserted, before going on to argue that 'Mathematical Truth delights in plain words; since even the style of Truth is plain in itself'.[73] But such disclaimers were themselves rhetorical devices of great antiquity,[74] and Tycho was drawing on classical sources in

[67] *TBOO* VI, 40.19–22.   [68] *CWE* XXV, 21. On brevity as a stylistic ideal, see Constable 1976, 19.
[69] See, for example, *TBOO* VI, 74.42–75.6, 225.2–5. Brevity was also a concern for Rothmann; see *TBOO* VI, 54.24–30, 154.33–35.
[70] *TBOO* VI, 24.3–10.   [71] E.g. *TBOO* VII, 63.20–22, 82.36–38, 109.24–26, 121.19–20, 141.7–9.
[72] *TBOO* VI, 23.15–20.   [73] *TBOO* VI, 23.32–37.   [74] See Janson 1964, esp. 51–52.

expressing them; by calling the language of truth plain he was following Seneca, who had himself been quoting Euripides.[75] Moreover, Tycho noted that good style was not incompatible with the treatment of astronomical material, when he conceded that, 'that extraordinary Franconian Johannes Regiomontanus, outstanding among the Germans of the last century, luminary of the University of Vienna, executed his writings in a very elegant and clear Latin style'; a style that, 'in that dark age, when humane letters had clearly not yet taken root among the Germans, was not the least admired'.[76] Regiomontanus was an astronomer of considerable reputation among the practitioners of northern Europe, and since Tycho wanted to be equally admired, his appearance in the preface at this juncture may represent something more than just passing acknowledgement of an illustrious predecessor.[77] But whether Tycho did, or did not, wish to invite some comparison, his employment of these rhetorical devices indicates that he was neither ignorant of the conventions of authorship and the letter-genre, nor unwilling to draw them to the attention of his readers. This suggests that, in analysing his correspondence with the Kassel practitioners, as we are about to do, we should not suppose that it lacks all claim to the sophistication and flexibility identified in the epistles of his more literary contemporaries.

III WEDDINGS, MERCHANTS, AND OTHER HINDRANCES TO GOODWILL

But since, up to now, we have obtained no reply from our gracious request for one, and this opportune embassy to you was then presented to us, we could not pass up the chance to graciously remind you of our previous letter; consequently our gracious request to you is that, since our observations that pertain to the recently appearing comet and the new star near Cassiopeia, and yours, are very nearly in agreement, would you not only, on the one hand, send to us your opinion of our observations of the fixed stars, but also [on the other] share your own observations of the fixed stars that you have obtained both before and since, so that we may compare them with ours. (Wilhelm to Tycho, 26 August 1586)[78]

Despite the delight expressed by Tycho when first writing to Wilhelm, the correspondence between Hven and Kassel did not proceed smoothly.

---

[75] Gummere 1917–1925, I, 328–329; Kovacs 2002, 258–259.   [76] *TBOO* VI, 23.32–42.

[77] On Regiomontanus, see Zinner 1968; *DSB* XI, 349–352; Swerdlow 1990, 1993a, 1996, esp. 188–195, and 1999; Brown 1990. For his reputation amongst astronomers of the succeeding generations, see Thorndike 1923–1958, V, 332–377; Pantin 1999.

[78] *TBOO* VI, 59.14–25.

Having sent some observations to Tycho in April of 1586,[79] Wilhelm and Rothmann wrote to him again in August of that year, reminding him of his obligation to respond to them in kind. Of the two letters, Rothmann's conveys more of the surprise felt by the Hessen observers at the lack of any response, perhaps because it would have been demeaning for Wilhelm to have expressed the true extent of his frustration. Rothmann wrote that he was certain the contents of his first letter, including his eager requests for a reply, would be fresh in Tycho's mind. The Landgrave, on the other hand, was beginning to suspect that their letters had never reached Uraniborg, and was growing impatient with the excuse, offered up daily by Rothmann on Tycho's behalf, that the distance between Hven and Kassel was likely to cause significant delay. Accordingly, Rothmann concluded, 'I ask your Excellency, still more, not to defer his reply concerning the welfare and advancement of astronomy any longer'.[80] Even this request was not sufficient to provoke a response from Tycho, however, and Wilhelm wrote yet another reminder that October.[81] But it was not until January 1587 that Tycho finally set pen to paper, sending to Kassel, along with some of his own observations, a lengthy explanation for the delay his correspondents had experienced:

> But the fact that I have not replied to this in so great an interval of time, and have neglected to satisfy the desires of your Highness for so long, although I was twice reminded strongly of this matter by other letters from your Highness in the meantime, is not for the reasons which your Highness might suppose, namely that it happened because of some negligence [on our part] or through the pains bestowed [on this reply], but rather it is the case that the excessive distance between us, and the few opportunities to send letters to you, have kept me from writing until now.[82]

Europe in the sixteenth century was just beginning to experience the emergence of an international postal service, but the carriage of private mail was still a spin-off, albeit a lucrative one, of the organisation of couriers and routes for conveying diplomatic despatches securely and rapidly, rather than a service in itself.[83] In the absence of proximity and access to a regular courier, would-be correspondents were forced to rely on other means to get their letters to recipients. Thus, Tycho lamented that his best opportunity to send material to Kassel came only twice a year, when Danish merchants travelled to the book-fair at Frankfurt. Since the Kassel letters had arrived shortly after the spring fair of 1586, he had been hoping to reply in the

---

[79] Wilhelm's reply to Tycho's first letter was written on 14 April 1586, as was the accompanying letter from Rothmann. See *TBOO* VI, 48–51 and 54–58.

[80] *TBOO* VI, 61.6–21.   [81] *TBOO* VI, 63.2–18.   [82] *TBOO* VI, 63.30–35.

[83] Housden 1903; Harlow 1928, 59–80; Allen 1972; Falk and Abler 1985; Siegert 1999, 4–19.

autumn, and had even gone so far as to contract with a bookseller to take the package to Wilhelm's secretary.[84] However, his composition of the letter had been forestalled by an unexpected visitation of royals to Hven, led by Queen Sophie, and the opportunity had escaped him.[85] The long-awaited reply was therefore carried by another bookseller, setting off to Frankfurt in the following spring.[86] It was the necessity of awaiting a suitable courier that had prevented Tycho from responding more promptly.

The very first letter from Hven to Kassel had been carried by a *domesticus* of Tycho; that is to say, by Petrus Jacobi (1544–1599), a trusted member of his household.[87] He had been able to bring back Wilhelm's response without delay or difficulty. But this was an arrangement that Tycho was unwilling or unable to continue, and, as the correspondence proceeded, a variety of different solutions to the problem of transporting letters between the two sites were adopted. Often, as in Tycho's second letter, the correspondents depended upon the movement of Danish merchants travelling to and from Frankfurt.[88] However, a system that depended upon accommodating the schedules of men travelling for trade and profit was often unsatisfactory, and frequently less than expeditious: whereas Tycho's own courier delivered the first letter only forty days after it was written at Uraniborg, the second took seventy to travel exactly the same distance.[89]

On 12 August 1587, Tycho wrote to Rothmann that, 'Contrary to my every expectation, the man through whom, at the spring fairs of Frankfurt, I last wrote copiously to your Most Illustrious Prince and yourself about astronomical matters, and so, I hope, satisfied your expectations to the utmost, carried back no letters at all when he returned to us'.[90] This was not in itself a cause for complaint, as Tycho's messenger had reported that Wilhelm had been occupied with 'a certain wedding' and Rothmann had also been busy with preparations for these same celebrations.[91] As was made clear in a later letter, Wilhelm had been about 'the giving away in marriage of

---

[84] *TBOO* VI, 63.36–64.7.

[85] The other visitors included the Queen's mother, Elizabeth of Denmark, and her father Duke Ulrich of Mecklenburg. See *TBOO* VI, 64.8–24; Dreyer 1890, 138–139; Christianson 2000, 112–113.

[86] *TBOO* VI, 64.32–37.

[87] *TBOO* VI, 40.27–30. On Jacobi (Peter Jacobsen of Flemløse), see Dreyer 1890, 117–119; Thoren 1990, 92 no. 35, 210–215; Christianson 2000, 277–280.

[88] It is apparent from explicit statements in the letters themselves that merchants were used by Tycho as couriers on several occasions. See *TBOO* VI, 105.20–26, 121.36–122.10, 148.29–35, 203.4–9, 204.14–18, 296.7–10, 307.27–32.

[89] This information is derived from manuscript annotations of the form 'Praesent. Cassel am 10 Aprilis Anno 86' on the letters preserved at Marburg, recorded in the *Opera Omnia* and checked by the present author. For the cases cited, see *TBOO* VI, 349 n. to 33.13, 352 n. to 63.19.

[90] *TBOO* VI, 105.8–13.   [91] *TBOO* VI, 105.13–20.

his illustrious daughter, and in the splendid preparation of the wedding';[92] a good reason for him not pursuing his astronomical interests, although Rothmann complained of time having to be spent 'quite vainly on the vainest of matters'.[93] But Wilhelm had also promised Tycho a reply soon, to be carried by his own courier.[94] Hence, he wrote to Rothmann:

> Up to now I have willingly been satisfied with this report, and from day to day I have eagerly awaited a reply from the Most Illustrious Prince and also from you; [but] since I see that this has been delayed beyond the time hoped for, I could not, with the opportunity to write provided by this bookseller HEINRICH (through whom, serving the same function, I sent the earlier one), not send you a letter and remind you of the reply you have hitherto omitted.[95]

The same bookseller was once more travelling to Frankfurt, and had agreed to deliver this reminder to Kassel. If the Landgrave's courier could not carry a reply to Tycho, then a message could also be carried back by this merchant.[96] However, 'he should find the letters prepared in readiness as soon as he returns to Kassel from the fairs of Frankfurt; for as it might cost him his companions and transport, to cause him some delay there will hardly be convenient for him'.[97] It was a point that Tycho made on more than one occasion.[98] Employing merchants as carriers constrained the frequency with which letters could be exchanged, extended the time it took a letter to reach its destination, and, if attempts were made to avoid these difficulties, threatened to inconvenience the correspondents by forcing them to adapt to the necessary haste of men of commerce.

Given these problems, it is hardly surprising that other messengers were used whenever it proved possible. In 1588, one of Tycho's assistants, Gellius Sascerides (1562–1612),[99] set out from Hven carrying not only letters for Rothmann, but also copies of the astronomer's *De recentioribus phaenomenis* for the Kassel *mathematicus*, for Michael Maestlin (1550–1631) in Tübingen, and for Giovanni Antonio Magini in Bologna.[100] Sascerides reached Kassel on 11 July, whereas the letter he bore was dated to the summer solstice; a

[92] *TBOO* VI, 162.7–19.
[93] *TBOO* VI, 110.8–14. The nature of Rothmann's involvement in the preparations is not clear, but it was not uncommon for court *mathematici* to act as advisors for the staging of masques and the other entertainments that attended a major wedding.
[94] *TBOO* VI, 105.13–20.   [95] *TBOO* VI, 105.20–26.   [96] *TBOO* VI, 105.29–35.
[97] *TBOO* VI, 105.29–35, *cont.*   [98] E.g. *TBOO* VI, 148.31–36.
[99] On Gellius, see Dreyer 1890, 121–122; Thoren 1990, 192, 212, 356–362; Christianson 2000, 351–353.
[100] *TBOO* VI, 133.23–29. For Maestlin, see *DSB* IX, 167–170; Westman 1972b; Methuen 1996a. For Magini, see the references in note 14. After meeting, Magini and Sascerides corresponded for several years; their letters were published in Favaro 1886.

journey time of only thirty days in total.[101] That was considerably better than the six months it took Tycho's letters of 21 February 1589 to reach the Hessen astronomers.[102] And as that delay left Wilhelm without news of Tycho's most recent activities, once he learned, hunting at Homburg in July of that same year, that Nicodemus Frischlin (1547–1590), poet of the Count Palatine, was intending to travel to Denmark, he induced him to go via Kassel with instructions for Rothmann to compose something to be delivered to Uraniborg.[103] Having done so, the *mathematicus* entrusted to Frischlin, in addition to his own letter, a copy of the one sent to him by Wilhelm, so that Tycho would understand the urgency of this request by the Landgrave.[104] Although Frischlin never made it to Uraniborg, these letters did, presumably after having been passed on to some other courier.[105] As such examples show, communication between Hven and Kassel was conducted on a largely *ad hoc* basis, making use of whatever opportunities arose on whatever occasions. Very similar accounts could be given of Tycho's exchanges with other correspondents.[106]

At times, the interruptions in the course of the correspondence, whether caused by the preoccupations of the participants, or the problems of conveying letters from one site to the other, led to displays of bad temper. In February 1590, for example, Tycho complained to Rothmann that he had, in his previous letter,

replied very well to your last one, and did so as soon as I received it; and asked you to write back to me quickly, so that I, in turn, could again prepare a response to it at the time of the Frankfurt Fairs. But I see that you are more sluggish in replying, and too slow, and always pleading other occupations as your excuse. But although I am probably detained by no fewer, I do not neglect to write.[107]

Such an outburst might prompt us to wonder why Tycho, Rothmann, and Wilhelm struggled at all to maintain their correspondence. What value did they perceive in this exchange of letters that motivated them to keep writing, and that provoked irritation and anger when replies were delayed, and their anticipation was frustrated? As we shall see, one plausible explanation for

---

[101] *TBOO* VI, 149.9–13. It is thirty days, because the date given is for the Julian calendar; by the Gregorian calendar, Sascerides arrived on 21 July.

[102] See *TBOO* VI, 360 n. to 162.1, 181.17–20. Here 'diebus' is an error for 'mensibus', as Dreyer observed in *TBOO* VI, 361 n. to 181.19.

[103] *TBOO* VI, 184.11–19. On Frischlin, see *DBE* III, 492; Midelfort 1992, 26–28.

[104] *TBOO* VI, 183.19–24. It is, of course, the forwarding of Wilhelm's letter to Rothmann that accounts for its presence in the *Epistolae astronomicae*.

[105] *TBOO* VI, 198.9–11.

[106] See, for example, *TBOO* VII, 72.30–35, 75.24–29, 93.9–15, 93.21–27, 98.34–99.1, 102.34–37.

[107] *TBOO* VI, 203.4–11.

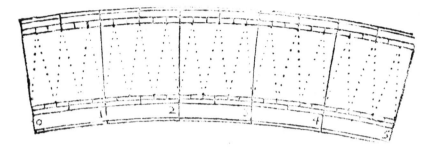

2.3 The method of transversal scale division, as illustrated in Tycho's *De mundi aethereis recentioribus phaenomenis* (Uraniborg, 1588), 461. Courtesy of the Whipple Library, University of Cambridge.

such behaviour from three committed astronomers was a perception on their part that correspondence was a form of communication that, as well as having the potential to alter one another's practice and theory, was of substantial benefit to the restoration of astronomy that they sought to accomplish.

IV EPISTOLARY CALIBRATION: INSTRUMENTS, REFRACTION AND COSMOLOGY AT URANIBORG AND KASSEL

When the Most Illustrious Prince received your writing on the comet, I was not allowed to examine it, his Most Illustrious Highness forbidding that I did so until I had also produced my own writing on this same comet. At which point, when we saw that observations obtained at places lying so far apart corresponded with each other to the very minute, it is scarcely possible to relate with how much delight we were flooded. (Christoph Rothmann to Tycho Brahe, 14 April 1586)[108]

In his letter to Rantzau of October 1585, a copy of which Tycho also received, Wilhelm had explained that he had been busy not only with observing the comet, but also because, 'at the instruction of Paul Wittich, we have greatly improved our mathematical instruments'.[109] Wittich, an itinerant mathematician, had spent some time at Hven, and this fact prompted Tycho to wonder whether the improved Kassel instruments were modelled on those at Uraniborg, 'particularly with respect to the subdivisions of degrees through transverse points, and the particular method of constructing the *pinnacidia* as parallel slits, especially suited to the night-time observations of the stars'.[110] By employing divided lines drawn diagonally between the upper and lower limits of adjacent arc-divisions (Fig. 2.3), Tycho had been

[108] *TBOO* VI, 54.24–30.   [109] *TBOO* VI, 31.17–20.   [110] *TBOO* VI, 35.38–36.2.

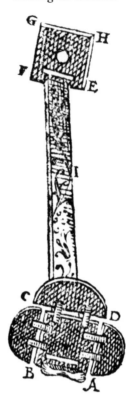

2.4 The instrument-beams with the *rimulae* or slit-sights, as illustrated in Tycho's *De mundi aethereis recentioribus phaenomenis* (Uraniborg, 1588), 462. The slits at the near end could be adjusted via a single screw to accommodate celestial objects of different sizes. By sighting the target object along BC–FG and AD–HE simultaneously, the parallax error that resulted from using 'pinhole' sights was avoided. The hole at the far end of the beam was for observing the Sun by projecting its image onto the base-plate. Image courtesy of the Whipple Library, University of Cambridge.

able to graduate circular scales beyond previous limits of accuracy; and by using *rimulae*, or slit-sight mechanisms, on his instruments (Fig. 2.4), he had eliminated the parallax errors associated with sighting upon a distant object through a so-called pinhole.[III] These were important instrumental innovations, essential to the production of the accurate astronomical data for which Tycho is so famous, and they were first employed on the *quadrans minor* (Fig. 2.5), an instrument completed in 1580, the very year of

[III] See Tycho's description in the *Mechanica*, *TBOO* V, 153–155; also Chapman 1989, 73–75; Thoren 1990, 152–157. Goldstein 1971 discusses precedents for the use of transversal divisions on instrument-scales.

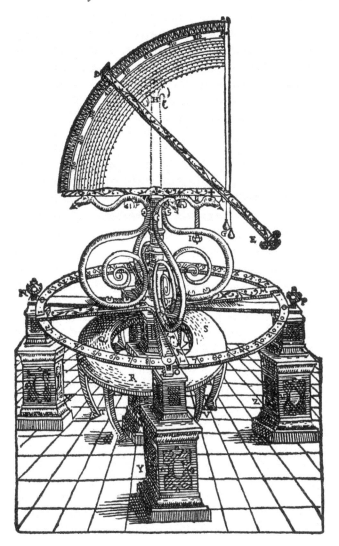

2.5 Tycho's *quadrans minor* (or *mediocris*), as illustrated in the *Astronomiae instauratae mechanica* (Wandsbek, 1598). Courtesy of the Whipple Library, University of Cambridge.

Wittich's visit.[112] The mathematician had left Hven after a stay of just three months; and although he had departed promising to return very shortly,

---

[112] See Thoren 1990, 152–153. On the discrepancy between Tycho's name for this instrument in his observing logs, 'Q. min', and its appearance in the *Mechanica* as the 'Q. mediocris', see Thoren 1973, 43 n. 19.

Wilhelm's letter was almost the first thing the Danish astronomer had heard of him subsequently.[113] Tycho thus had good reason to suspect that the new designs being used at Kassel were the ones employed at Uraniborg and, since Wilhelm had not indicated otherwise, that Wittich had dishonestly claimed credit for them as if they were his own.

It may well have been this suspicion that motivated Tycho to use his own servant as the courier for the first letter sent to Wilhelm.[114] Jacobi was capable of recognising Tychonic instrument designs, and carrying news of them back to Uraniborg. If this was his task, however, then, he carried it out with Wilhelm's consent, since the Landgrave wrote in his letter of 14 April 1586 that, 'as far as our instruments are concerned, we let your attendant see them all, [and] he can give you a report about them'.[115] This inspection did not forestall an epistolary conversation about the instruments at the two sites, their design, and their origin, but the discussion quickly moved away from a dispute about issues of property and priority, alchemising into a much richer debate about the methodology and theory proper to the construction of a renewed observational astronomy. The differences and similarities in the practices and philosophies of the two observatories became a topic of great significance to all three of the participants in the Hven–Kassel exchanges.

Wilhelm's very first letter to Tycho expressed his pleasure at the agreement of the cometary observations that had been made at their respective centres of astronomy. Rothmann had compared his own results with those conveyed by Jacobi, and discovered that they 'scarcely differed from each other by a minute'.[116] This, Wilhelm wrote, was 'truly a great thing, and a sign that both our instruments and yours are right and good, and that the observers were diligent and had a good view'.[117] Neither Rothmann nor Tycho disagreed with Wilhelm about drawing support for their individual observations from this mutual agreement, so it is apparent that a notion of calibration, or of corroboration through independent confirmation, was familiar to all of them. That this was not just an intuitive understanding is suggested by Rothmann's report that Wilhelm had not allowed him to study Tycho's treatise on the comet until he had written his own. The principle

---

[113] Tycho stated that this was the first he had heard of Wittich after his departure in his letter to Rothmann of 20 January 1587. In fact, Tycho learnt from Hagecius in 1582 that Wittich had delivered in Prague certain letters that had been entrusted to him to convey. See *TBOO* VI, 89.29–34; *TBOO* VII, 68.21–25.

[114] Although in a letter to Rantzau of 23 March 1586, Tycho stated that a matter had arisen at Frankfurt that required the attention of one of his trusted servants; hence Jacobi had been sent, and would deliver the letters to Kassel on his way there; see *TBOO* VII, 101.37–102.7.

[115] *TBOO* VI, 49.34–3.   [116] *TBOO* VI, 48.24–33.   [117] *TBOO* VI, 48.24–33, *cont.*

was clearly being used to evaluate, and discriminate between, more than raw data alone.

As a methodology, the comparative principle and its application may have owed something to Renaissance humanism. Earlier proponents of astronomical reform, including Regiomontanus and Copernicus, had *both* participated in the humanist project of recovering and restoring texts through critical comparison of manuscript sources *and* called for discrepancies between the mathematical models of astronomy and the apparent phenomena to be identified and removed.[118] The approaches and objectives of these two enterprises intersected at precisely the point when the agreement of more than one set of predictions was assessed observationally: when text, compared with text, was also compared with what was actually visible. Comparisons of this sort were certainly an important impetus to reform: Tycho himself later claimed that it was the manifest discrepancies between different sets of planetary tables that had attracted his attention as a youth, and set him on the path of seeking the restoration of astronomy.[119] They were also a crucial evaluative technique. In his letter to Rothmann of 17 August 1588, for example, Tycho wrote of comparing the lunar eclipse which occurred in March of that year with predictions from both the Alfonsine and Prutenic Tables, the *Tabulae eclipsium* of Georg Peurbach (1423–1461), and the *Ephemerides* of Maestlin, and of finding a discrepancy of at least half an hour in each.[120] Epistemologically, however, such comparisons depended on a more fundamental notion, one that underpinned astronomical practice itself: the belief that the stability and regularity of the heavenly bodies and motions, and their amenability to accurate observation and mathematical analysis, were such as to make a universally valid astronomy achievable in principle. Some sixteenth-century scholars interested in the heavens may have rejected this proposition,[121] but the Hven–Kassel correspondents were clearly not among them. Had they not considered that it was possible to perfect astronomy, or at least significantly improve upon its current unsatisfactory state, they would not have devoted the effort they did to pursuing its reform.

This should explain, therefore, why a technique of investigation more readily associated with the later experimental philosophy of the

---

[118] On Regiomontanus as a humanist, see Zinner 1968, esp. 51–52, 154–155; Rose 1975, 90–117; Pedersen 1978b, 172–185; Swerdlow 1993a. For Copernicus, see Rose 1975, 118–142; Czartoryski 1978, 364–365; Westman 1990, esp. 175–178; Barker 1999. The example of the Byzantine neo-Platonist George Gemistus Plethon (*c.* 1360–1452), suggests that a humanist reform of astronomy wholly dependent on existing textual authorities might also be pursued. See Mercier 1998; Tihon 1998.

[119] *TBOO* V, 107.7–16.   [120] *TBOO* VI, 142.22–28.

[121] E.g. Nicodemus Frischlin, as described by Jardine 1987, 90–91.

seventeenth century was familiar at both Uraniborg and Kassel, and indeed
was employed at these sites not only in assessing the current state of mathematical astronomy, but also in developing the tools required to improve
it.[122] By 1586, when his correspondence with the Landgrave commenced,
Tycho had installed over a dozen observing instruments at Uraniborg, and
had assessed their performance relative to one another, when possible, as
well as by the consistency of their individual results.[123] The observing instruments at Kassel are less well documented; but it seems clear that they were
also several in number, and were likewise evaluated by mutual comparison
of the data they generated.[124] To move from this calibration of instruments
at each site independently, to the calibration of the equipment at two separate locations, required only the modest assumption that local conditions
would not introduce sufficient variation, given the precision to which they
were working, to make exact agreement impossible. The exchange of data
would, at any rate, test that possibility, and hence establish if there were
practical limits to the accuracy of a universal astronomy. It would also provide an opportunity to uncover systematic errors at one site or the other that
might otherwise have remained undetected. The Hven–Kassel correspondents had good reason, therefore, to welcome the exchange of astronomical
material.

One small point deserves to be clarified immediately. It has been asserted
that discovery of the existence at Kassel of instruments of comparable accuracy to the ones at Uraniborg 'seems almost to have made Tycho sick'; that
it was a cause of great concern that the Landgrave had the 'inclination,
ability, and the means to do the same thing Tycho wanted to do'.[125] But in
his second and third letters to Kassel (written to Wilhelm and Rothmann
respectively) Tycho wrote of these instruments very graciously indeed. By
then, the letters of the Landgrave and his *mathematicus*, as well as the
reports of Jacobi, had encouraged Tycho to believe that Wittich, although
he had indeed conveyed certain principles of design to Kassel, along with
the name of the sextant, did 'not arrange everything suitably enough for
the Most Illustrious Prince, just as he had seen it here'.[126] Indeed, 'how

---

[122] On calibration and replication in experimental philosophy, and its difficulties, see Shapin and
Schaffer 1989, esp. 225–226. With respect to the Kassel observatory, it is worth noting the comment
of Moran 1978, 62–63, regarding 'the development of Baconian observational values before Bacon'.

[123] Thoren 1973.

[124] The Kassel observers are known to have employed quadrants, sextants, clocks, and torqueta. See
Leopold 1986, 16, 6; von Mackensen 1988, esp. 22, 128, 135–137, 156–157; Granada, Hamel and von
Mackensen 2003, 76–84. The care with which the accuracy of these instruments was assessed is
suggested by scattered comments of Rothmann, as well as analyses of the resultant data. See Leopold
1986, 26–29; Hamel 1998, 46–64.

[125] Thoren 1990, 267.   [126] *TBOO* VI, 89.42–90.1.

brilliantly he accomplished' the improvement of the instruments, Tycho remarked ironically in his letter to Rothmann, 'you learnt by experience, but not without the loss of many observations'.[127] Apparently, after Wittich's departure from Kassel, the *mathematicus* and the Landgrave's clockmaker, the talented Jost Bürgi (1552–1632), had cause to revise the instruments once more, removing the errors introduced by Wittich's imperfect relation of the designs used at Uraniborg.[128] Credit for the perfection of the Landgrave's instruments could therefore be shared between Kassel and Hven.

In commenting on what contribution Wittich *had* made to the Kassel developments, moreover, Tycho noted that, 'certainly in this way he accomplished something not displeasing to me; since I greatly desire many men to stand forth, in as many places as possible, to investigate with me the appearances of the stars with instruments that are choice, and not at all fallacious'.[129] While it could be argued that Tycho, nausea in check, was simply making the best of a *fait accompli*; there are good reasons to suppose, even if the issue of calibration is discounted, that he was quite sincere in hoping that other observers with whom he was in contact would come to be equipped with more accurate instruments. Both meteorological conditions that prevented observation at one site on a particular occasion, and the simple fact that it was impossible to observe everything from a single location, frequently rendered an astronomer dependent upon the testimony of others for information about celestial events and phenomena.[130] The validity of such recourse to others' observations was well entrenched within the astronomical tradition; analysis of long-term phenomena, like the precession of the equinoxes, often required the use of data supplied by quite distant predecessors. But a practitioner like Tycho, who was concerned with achieving the highest accuracy possible, faced the prospect of having to rely on the observations of astronomers, both ancient and modern, with less exacting standards and less precise apparatus. Something of the potential for frustration can perhaps be gathered from a letter that Tycho wrote to Scultetus in December 1590:

---

[127] *TBOO* VI, 90.13–15.

[128] *TBOO* VI, 55.35–56.24. Bürgi is discussed further in subsequent chapters, but for biographical details see Leopold 1986, 19–34; von Mackensen 1988, 9, 21–37.

[129] *TBOO* VI, 89.34–37.

[130] Tycho remarked on the problem of poor weather obstructing observations in the preface to the *Epistolae astronomicae*; see *TBOO* VI, 19.20–26. For an example within the Hven–Kassel exchanges of weather-related difficulties rendering reliance on others' observations helpful or necessary, see Tycho's discussion of the lunar eclipses of September 1587 and March 1588, *TBOO* VI, 142.6–28. On the evaluation of testimony in the early modern era, see Hacking 1975, 20–52, some of whose remarks pertain specifically to astronomy.

I have recalled that you once wrote to me that you keep a large hodgepodge of celestial observations put together at some point in Leipzig. I would like, therefore, that anything of this sort which you have in your archives be fetched out, transcribed and sent to me. About which matter I think I have written to you previously. For I am not unaware that those observations are not accurate, having been taken with either no instruments or ones that are erroneous. Nevertheless, since frequently the position would have been determined by sight alone, through straight lines or other dispositions with respect to the fixed stars, I, who have accurately corrected places of the fixed stars, will derive some planetary positions this way, and determine and set right many other things in these observations, with careful circumspection employed in each case. From which they shall emerge, if not exactly, not too inaccurately to be used wherever it is necessary.[131]

The problem of acquiring data from such sources occasionally led Tycho to take steps to assist others in acquiring better instruments and learning how to use them.[132] It therefore seems likely that any discomfort Tycho experienced in 1586 was less with the *fact* of the transfer of Uraniborg design principles, which eventually furnished the Kassel observatory with instruments of greater accuracy, than with the way it was accomplished.

In his letter to Wilhelm of 18 January 1587, Tycho explored in some detail the agreement between the two observatories' results. The coincidence of the cometary observations was all the more pleasing, he wrote, because no similar agreement had occurred in other treatises, such as those that concerned the comet of 1577. This could be attributed, he stated, to the observational and calculating errors made by their authors.[133] There were small differences between Rothmann's observations and his own; but these were most likely due, he thought, to a slight disagreement about the positions of the fixed stars that had been used as points of reference.[134] The key point about the comet, its complete or almost complete lack of detectable parallax, was confirmed by both.[135] 'And plainly, therefore, in examining

---

[131] *TBOO* VII, 302.4–16.
[132] The individuals whom he assisted include the Imperial physician Thaddaeus Hagecius and the Imperial Prochancellor Jakob Kurtz, as well as Gellius Sascerides and, through Sascerides, Magini. For the aid granted to Hagecius, Kurtz, and Magini, see Chapter 4. For the case of Sascerides, see Tycho's letter to Joachim Camerarius of 21 October 1590, in which he reveals that he was meeting the costs of his former student's construction and use of an astronomical instrument, precisely because certain observations were easier to make in Italy than at Uraniborg; see *TBOO* VII, 277.31–37 and the discussion in Chapter 4.
[133] *TBOO* VI, 65.7–10. On other works on the comet of 1577, see Hellman 1944, 118–306.
[134] *TBOO* VI, 65.10–16.
[135] *TBOO* VI, 65.16–19. As Tycho stated in his accompanying letter to Rothmann, he had determined a parallax of one to one-half minute of arc for the comet, which he thought a difference of almost no significance. See *TBOO* VI, 86.4–7.

the truth of the matter we agree on this thing, that this comet was not born in the elemental orb, but in the uppermost ether, contrary to what up to now the Peripatetics and almost all of the philosophers, and even *mathematici* themselves, had declared was settled and certain'.[136] In this respect, the 1585 comet was like the nova of 1572, which had also been observed by both parties and had likewise been identified as a celestial, rather than a meteorological, phenomenon.[137] Indeed, observation and analysis of other recent comets had led Tycho to conclude that *all* comets were ethereal objects and not elemental.[138] That this claim conflicted with the cometary observations of Regiomontanus and Johannes Vögelin (*fl.*1532), and the opinions of others regarding the more recent appearances, did not trouble him at all.[139] The determination of such scant parallaxes as comets displayed required the use of far greater skill than anyone else had been able to apply.[140]

Examination of the Kassel observations of the fixed stars for consistency with one other, and then for agreement with his own values, led Tycho to assert that the sextants at the two observatories were attuned to within ten seconds of arc.[141] But since these instruments were used to establish unknown stellar positions by triangulation on stars with predetermined coordinates, agreement on the relative distances between the stars on the celestial sphere did not preclude disagreement about their absolute positions.[142] Tycho noted that the two sets of data differed slightly, with a discrepancy in stellar declination of almost two arc-minutes resulting when Rothmann's value for the inclination of the equator to the Kassel horizon was applied to his observed meridian altitudes.[143] The accuracy with which he made his own observations, and the diligence evidently employed by the observers at Kassel, left Tycho momentarily nonplussed by this result. What, he claims to have wondered, could be the cause of this two-minute discrepancy?[144] However, one obvious place to look for the source of the disagreement was in the values assumed for the observatories' latitudes. Knowing that Rothmann's determination of Kassel's latitude had been based on observations of the meridian altitudes of circumpolar stars made with a quadrant, Tycho suggested that either the sights on this instrument were

---

[136] *TBOO* VI, 65.20–23. On Wilhelm's observation of the nova, and discussion of the phenomenon with other astronomers of the Empire, see Methuen 1997.

[137] *TBOO* VI, 65.23–26.   [138] *TBOO* VI, 65.26–31.

[139] See the lengthier discussion of Regiomontanus' and Vögelin's observations in Chapter 3.

[140] *TBOO* VI, 65.31–39.   [141] *TBOO* VI, 66.1–15.

[142] See Tycho's description of his sextants in his *Mechanica*, in *TBOO* V, 68–79; also Dreyer 1890, 349–354; Thoren 1990, 288–296.

[143] *TBOO* VI, 66.21–28.   [144] *TBOO* VI, 66.28–36.

misplaced by two minutes of arc, or that it had been mounted incorrectly, so as to not be quite vertical. Indeed, an error of only one minute of arc would account for the two-minute discrepancy, if the same instrument had also been used to determine the altitudes of the stars, from which the declinations were calculated.[145]

In fact, Rothmann *was* to decide that the Kassel quadrant was displaced by one minute from the vertical. As an instance of instrumental error diagnosed at a distance, Tycho's analysis would be rather striking, were it not for his own admission that he already knew of the problem with the quadrant at the time that he offered it. Tycho was writing in January 1587, and Rothmann had informed him of the discovery in his letter of the previous August.[146] Although Tycho claimed that he had deduced the probable source of the error before receiving that letter, we have no way of determining the truth of his statement from the extant correspondence. But two things suggest that we might give him the benefit of the doubt. The first is Tycho's stated intention to send an earlier letter to the Landgrave, the plan that had been thwarted by the unexpected visit of Queen Sophie and her retinue. Had that letter been written, it should have discussed these observations. Second there is the fact that the problem with the quadrant should not have been very difficult for Tycho to uncover, given knowledge of the observing technique employed and the error's apparent ubiquity throughout the Kassel results.

Indeed, whether we grant credence to Tycho's chronology or not, the most remarkable element of the whole discussion is that he should have bothered to write about such a small difference at all. In a tabulated comparison of fifty-two observed star positions with the values in published tables, a comparison that he appended to his letter of 18 January 1587, the mean discrepancies were much higher: fifty-one and twenty-one minutes for the Alfonsine and Copernican stellar longitudes, and twenty-three minutes for both sets of latitudes.[147] In this context, the few minutes at stake between the Hven and Kassel determinations would seem to be negligible. In a noteworthy statement of his own astronomical standards, however, Tycho declared that, 'I consider all heavenly observations should be carried out precisely, to the extent that they may be perceived by the human senses;

---

[145] *TBOO* VI, 66.36–67.1.

[146] Rothmann wrote that the quadrant had been sited incorrectly by Wittich and Bürgi, and that it had taken a considerable effort on his part to persuade Bürgi and Wilhelm of the error; see *TBOO* VI, 61.26–33.

[147] *TBOO* VI, 82–83.

so much so, that if anything in them is deficient even in the slightest, I hold them to be worthless'.[148] And in the letter he wrote to Rothmann two days later, he stated that, 'I desire that those things which arise from me on astronomical matters are either nothing at all, or so very precise that the human senses do not achieve greater exactness'.[149] The frequent use of the qualifiers detectably (*sensibiliter*) and undetectably (*insensibiliter*), throughout the correspondence, attests to the fact that, for Tycho, the limit of what could be visibly distinguished was the target accuracy of astronomy.[150] In Tycho's lifetime, prior to the invention of the telescope and micrometer, and thanks in part to his innovations in instrument sights and scales, that limit stood at about one minute of arc under the most favourable conditions.[151]

Tycho performed a comparison of the Kassel latitudinal and longitudinal stellar positions, first with one another, and then with his own. He pointed out that small discrepancies had resulted from the determination of these values through measurements of altitude, the interval between stars, and the use of a celestial globe to avoid the difficult transformation calculations.[152] He also possessed a brass-covered globe, six-foot in diameter and constructed for this purpose and for similar tasks of data recording and retrieval, he noted; but it was scarcely capable of preserving an accuracy of one-and-a-half minutes. For this reason, he wrote, it was better to determine stellar positions with 'the scrupulosity of Geometry and Arithmetic'.[153] Although never explicitly instructing Wilhelm to change the methods employed at Kassel, Tycho seems to have been suggesting that the practices of the Hessen astronomers be altered to conform with those employed at Uraniborg.

A difference of five or six arc-minutes in the Hven and Kassel stellar longitudes, calculated without the use of a globe, was to become a more serious issue than the matter of the globe errors. But in the letter of January 1587, Tycho initially restricted himself to noting that:

in the latitudes is found a difference almost no greater than that which could be caused either from the longitudes having been taken a little differently, or from the different definition of the obliquity of the ecliptic, but principally from the declinations, disagreeing with mine because of the two-minute error of your

---

[148] *TBOO* VI, 67.25–27.   [149] *TBOO* VI, 102.10–13.
[150] *TBOO* VI, 69.5–7, 87.19–21, 90.39–91.7, 97.17–22, 97.34–38, 99.2–6, 124.15–19, 128.10–15, 136.26–34, 138.6–21, 140.39–141.2, 141.4–7, 141.18–30, 146.36–147.1, 172.2–5, 174.7–13, 174.24–28, 204.7–8, 205.22–28, 323.5–13, 326.21–24.
[151] Chapman 1983b, 134–136.   [152] *TBOO* VI, 68.8–13.   [153] *TBOO* VI, 68.18–26.

perpendicular. Certainly even the difference of longitudes in this, to say nothing of the latitudes, is not very important, if we compare the observations of the ancients and certain of the moderns.[154]

Here we see Tycho himself acknowledging that, compared to the discrepancies between other observed and calculated values, the longitude difference between the Hven and Kassel results was virtually negligible. Tycho professed to find this most surprising; he was able to deduce that at Kassel the right ascensions had been determined by timing the meridian transits of base stars with a clock, a method that he thought was far from reliable.[155] Although he also possessed, 'four clocks of that sort, of different sizes and precisely constructed, which show not only each minute, but also each second', he considered it, 'useless to trust to them in so delicate a matter; as just 4 seconds, which are easily lost in many hours, produce an error in longitude of one whole minute'.[156] He listed a number of mechanical problems that could easily lead to such a slight error in timing.[157]

Tycho went on to describe the method for obtaining stellar longitudes that he considered preferable. This was a variation of the technique of Hipparchus, Ptolemy and Copernicus, in which the Moon, as a body visible during both night and day, was used to relate observations of the Sun to the stars.[158] There were, for Tycho, significant problems with that method: to give good results, it required a sound understanding of lunar parallax and, because it depended upon observations of the Sun near the horizon, of atmospheric refraction as well.[159] For this reason, Tycho had opted to determine stellar longitudes by making daytime observations of Venus. With the position of the planet determined from the known location of the Sun, it could then be used to give the stellar coordinates; corrections could be made for parallax and, Tycho claimed, at the altitudes at which all these celestial bodies had been observed, the refraction was negligible.[160] But if there were any small distortions due to refraction, these could be detected through other observations, and the truth of the stellar positions confirmed through repeated use of this 'infallible method', cross-checked

[154] *TBOO* VI, 68.30–36.
[155] *TBOO* VI, 68.36–42. On Tycho's attempts to employ clocks, see Dreyer 1890, 324–325; Thoren 1990, 157–159.
[156] *TBOO* VI, 68.42–69.5.
[157] Tycho mentioned as problematic distortions in the clock-parts, and the variation in weight of the pendulum cord. See *TBOO VI*, 69.5–10.
[158] *TBOO* VI, 69.17–21. See Ptolemy's *Syntaxis* IV.1, VII.4. As Dreyer observed, *TBOO* VI, 352 n. to 69.19, Tycho is wrong to attribute the use of this method to Copernicus.
[159] *TBOO* VI, 69.21–34. On Tycho and atmospheric refraction see, in addition to the discussion below, Moesgaard 1988; Gingerich and Voelkel 1998, 21–23.
[160] *TBOO* VI, 70.8–27.

with the intervals between stars arrived at using the sextant.[161] Although not explicitly stated, the provision of this lengthy description of his own observing method and procedure of checking again suggests that Tycho was seeking to convince the Kassel astronomers to change to the preferred Uraniborg practice.

Mention of the correction for parallax prompted Tycho to explain that he had investigated the parallax of both Venus and Mars in the early 1580s. These observations were cosmologically significant. Whereas determination of the parallax of Venus had the potential to decide the ancient question of whether the planet orbited 'around or below the Sun', study of Mars provided a way of discriminating between the two world-systems currently available.[162] According to the Copernican hypotheses, Mars approached almost a third part nearer to the Earth at opposition than the Sun. Necessarily, this would result in a greater parallax for the planet; and if such were found, it would be incompatible with the Ptolemaic world-system, and indicate that, 'either the Earth is turned about by an annual motion, by which all the epicycles of the planets are dispensed with; or another arrangement of the heavenly revolutions has to be looked for than has hitherto been constructed'.[163] In other words, if the Ptolemaic system were disproved observationally, then either the system of Copernicus, or some new world-system, would be the best candidate for an accurate representation of the heavens and their motions. Tycho's remarks here are important for a number of reasons. According to Tycho's understanding of the matter, the difference in Martian parallax predicted by the two systems would be slight: around two minutes of arc. It is striking, therefore, that he supposed he could detect it. But rather than specify whether any difference in parallax had been observed, Tycho committed himself only to discussing these matters 'more clearly and fully' on another occasion.[164] This evasiveness is all the more notable because, in a letter written to Brucaeus in 1584, Tycho had claimed that he had in fact demonstrated in 1582 that Mars did *not* come closer to the Earth than the Sun, and therefore the *Copernican* system had been disproved observationally.[165] Evidently aware, therefore, that such parallax determinations were difficult and depended on the values one assumed for atmospheric refraction, Tycho had either already reinterpreted his observations of 1582 or stood ready to do so in the light of a new attempt to observe Martian parallax to be undertaken later in 1587.[166] But his comments to Wilhelm strongly suggest that he was already persuaded of

[161] *TBOO* VI, 70.22–25.   [162] *TBOO* VI, 70.27–34.   [163] *TBOO* VI, 70.38–41.
[164] *TBOO* VI, 70.41–42.   [165] *TBOO* VII, 80.2–16.
[166] Thoren 1979, 61–62; Gingerich and Voelkel 1998, 16–17.

the merits of his own planetary scheme, the geoheliocentric world-system, which would give Mars, Venus, and all the other planets, circumsolar paths, and would also render Mars, at a certain point in its orbit, closer than the Sun to the Earth.

One result of the work on the stellar positions was the realisation that the stars had apparently altered not only their longitudes over time, as could be explained by the precession of the equinoxes, but also their latitudes. Wilhelm had made this point in his letter of 14 April 1586, charging Tycho, 'as the most distinguished *mathematicus* of this age', with determining whether the cause was an erroneous printing of the values determined in antiquity, or some proper motion of the stars as individual celestial bodies.[167] Tycho asserted emphatically that the stars were not liable to any such movement.[168] This was evident from observations of those stars that had been noted by Ptolemy to lie on a straight line, and that in some cases had previously been observed so by Hipparchus. If these stars possessed some individual motion, then they would not be found, almost 1,400 years later, to retain the same configuration.[169] Tycho therefore stated that the coordinate discrepancies should be attributed either to corruption of the text through careless copying, or to the poorer accuracy of the ancient observations, although a small alteration in the obliquity of the ecliptic also contributed, as Rothmann had previously asserted.[170] Actually, Tycho does not seem to have conceived of, or investigated, a change in the ecliptic before he received Rothmann's letter, a fact that he omitted to mention at the time, and one that he certainly left out of his published discussions of the topic. By the time he printed the *Mechanica* in 1598, it had become his discovery.[171]

So that Wilhelm and Rothmann would be able to more easily compare their star catalogue with his, Tycho attached the list of fifty-two star positions, adjusted to the beginning of the year in which he was writing.[172] In making this adjustment, Tycho used a value for the rate of precession of the equinoxes not taken, he said, from either the Alfonsine Tables or the determination of Copernicus, but derived instead from a comparison of his own observation of the star of Spica with that made by the Polish astronomer

---

[167] *TBOO* VI, 49.29–34.   [168] *TBOO* VI, 71.24–26.
[169] *TBOO* VI, 71.26–38. As Tycho stated, Ptolemy's discussion is to be found in the *Syntaxis* VII.1.
[170] *TBOO* VI, 72.5–11. In his letter to Rothmann of 20 January 1587, Tycho argued in much greater detail that the obliquity of the ecliptic had altered since the time of the ancients. See *TBOO* VI, 92.23–100.3, and the discussion in Chapter 3.
[171] As pointed out by Thoren 1990, 291–293.
[172] *TBOO* VI, 72.22–27. This document was reproduced in the *Epistolae astronomicae*; see *TBOO* VI, 82–83.

in 1525. This observation, however, he had first corrected according to the determination by his student Elias Olsen Morsing of the true elevation of the pole at Copernicus' Frauenburg.[173] The result was a rate of precession of one degree in seventy-three years.[174] It was necessary that the Kassel observers be told this figure, since otherwise computation of their own stellar positions for the same temporal base might introduce phantom discrepancies between the two observatories. But, Tycho confessed:

I judge that precise knowledge of the apparent motion of the eighth sphere is scarcely fully examinable by any of the mortals. For as yet it has not completed one whole revolution, and the fragmented little bits of its course from Timocharis, Hipparchus, Ptolemy, al-Battani, and Copernicus through to us, do not suffice; and even if they did, they are not established sufficiently precisely.[175]

There was therefore, in Tycho's opinion, an important restriction limiting the capacity of any one observer to accomplish a reform of astronomy – a restriction that could only be overcome by persistent study of the heavens, to a consistently high standard, through many generations. Consciousness of this limitation perhaps help to explain why Tycho, Rothmann and Wilhelm seem collectively to have been particularly attracted to the humanist practice of addressing posterity. It might also provide a connection Tycho's intellectual vision and his desire to bequeath his astronomical role to his children.

Near the end of his reply to Wilhelm, Tycho apologised for his prolixity, and pledged that 'the remaining things, if any, that are missing from here, I shall reveal in the letter to the *mathematicus* of Your Highness'.[176] The accompanying letter did indeed cover much of the same topics in somewhat greater depth. At the same time, however, Tycho was more dogmatic than he had been when addressing the Landgrave. Thus, when discussing the Kassel method for observing lunar eclipses, he wrote:

But in measuring these times very accurately, you should not rely on a clock too much, unless it keeps a revolution very precisely, and in the intervening hours is moved equally throughout, and shows separately each single minute, and even the seconds also.[177]

And:

But it ought to be noted that in a total eclipse, it is better to observe the moment of the entry of the whole body of the Moon into the terrestrial shadow, and of the

---

[173] *TBOO* VI, 72.37–73.7. On Olsen, and his journey to Frauenburg, see Dreyer 1890, 122–125; Moesgaard 1972a, 34–35; Thoren 1990, 194–196; Christianson 2000, 323–324.
[174] *TBOO* VI, 73.11–14.   [175] *TBOO* VI, 73.18–23.
[176] *TBOO* VI, 75.3–6.   [177] *TBOO* VI, 101.24–27.

first exit of the same, than the other two moments at the very beginning and the absolute end.[178]

Even more clearly than his letter to Wilhelm, Tycho's letter to Rothmann makes evident his concern to address, and eliminate, differences in the material technology, observing techniques, and methods of data reduction employed at Uraniborg and Kassel. The subsequent exchanges reveal the extent to which Wilhelm and Rothmann were prepared to accede to this agenda, and the point at which they began to resist it. But in addition to their concern with practical matters, the letters also illustrate how the project of restoring astronomy through observation increasingly legitimated the advancement of cosmological arguments. They expose how Tycho's attempts to calibrate the instruments and mathematical techniques of the Kassel observatory shaded into his efforts to dictate to its practitioners which was the true arrangement of the planets in the cosmos.

As we have seen, a shift from observational fact to cosmological claim was already manifest in the exchanges of 1587. Lack of parallax indicated that the 1585 comet was a celestial and not an atmospheric phenomenon. Further cosmological consequences were drawn out in Tycho's analysis of Rothmann's treatise on this comet. He praised the work for considering it 'mathematically, according to choice and infallible observations, with equal care and exactness', unlike similar texts by other German-speaking writers.[179] But Rothmann had not restricted himself to drawing geometrical conclusions from his data; and after asking that he be forwarded a copy of the completed manuscript, along with observations of solar elevations and meridian altitudes of the fixed stars, Tycho stated that:

what you asserted in the portion of it sent to me, that the whole airy heaven consists not of solid matter, indeed is nothing other than air itself, I would readily grant you, provided you understand that the air which is above the Moon is far more subtle than this element, to the extent that it deserves the name of most fluid and subtle ether rather than that of elemental air. For I have not, for many years, been of the opinion that some orbs really exist in heaven, and that the matter of the heavens is hard and impenetrable, and the stars are rolled around merely according to the motion of the orbs. Since many absurd things follow from this; and, if nothing else, the many comets observed precisely by us in recent years to pursue a course in the ether, that were not following the lead of any orb, refute this sufficiently.[180]

[178] *TBOO* VI, 102.2–5.   [179] *TBOO* VI, 85.34–38.

[180] *TBOO* VI, 88.4–15. Rothmann's treatise was unfinished when Tycho first saw it, lacking three chapters from the end; on 21 September 1587, Rothmann sent him these chapters, pointing out that one on the astrological significance of comets still remained to be written. See *TBOO* VI, 54.30–35, III.12–22. The manuscript was eventually published in Snellius 1619, still lacking this final chapter.

This passage is one that has received much attention from historians of astronomy anxious to defend, or deny, Tycho's right to be considered the astronomer who 'dissolved' the impenetrable celestial orbs of the pre-modern cosmos.[181] In his *De recentioribus phaenomenis*, Tycho repeatedly asserted that the supralunary course of comets indicated the non-existence of hard spheres of this kind and derived essential support from this fact for his new system of the world.[182] If celestial spheres were solid and impenetrable, and constituted the mechanism by which the planets were moved in their orbits, then the intersection of the paths of Sun and Mars seen in the Tychonic world-system would clearly be impossible. Yet the *De recentioribus phaenomenis* was issued with an imprint of 1588, postdating Tycho's receipt and perusal of Rothmann's manuscript treatise, and there are good reasons to believe that it was put together rather swiftly. Certainly, the decision to include the chapter that announced Tycho's new world-system seems to have occurred later than March 1586.[183] No earlier statement has been found to confirm that Tycho had, as he insisted, long been persuaded of the fluid nature of the heavens. Indeed, on his own later testimony, he believed in the reality of impenetrable spheres for 'some time' after the observations of Martian parallax in early 1582, and only abandoned the notion after a study of the courses of 'certain comets' – meaning, it would seem, those of 1577, 1580, 1582, *and* 1585.[184] It has therefore seemed plausible to suppose that Tycho owed more to Rothmann in respect of the 'dissolution' of the spheres, and hence in the formulation of his new system of the world, than he was prepared to admit.[185]

The issue of Tycho's debt to Rothmann is a complicated one, and exploration of it necessitates a substantial digression from the correspondence proper. In his German treatise on the comet of 1577, thought to have been produced in 1578, Tycho had been extremely modest with respect to the cosmological claims that he advanced on the basis of his mathematical analysis.[186] Determination of the fact that it was a celestial rather than a sublunary object dispensed, in Tycho's view, with the Aristotelian doctrines that comets are meteorological phenomena and that nothing new ever arises

---

[181] The literature on this topic is now substantial: Donahue 1975, 1981; Rosen 1976, 1984, 322–324, 1985a, 26–31; Swerdlow 1976, esp. 130; Aiton 1981, 96–97; Jardine 1982, esp. 170–173; Grant 1987, 1994, esp. 395–397; Lerner 1989, esp. 273–279, 1996–1997; Barker 1991; Barker and Goldstein 1995, esp. 395–397; Barker and Goldstein 1988; Granada 1996, 45–53; 1999, 2002, 2004, 2006; Randles 1999.

[182] *TBOO* IV, 3–378, esp. 155–159.   [183] Thoren 1979, esp. 55–56, 63–64.

[184] *TBOO* VII, 129.12–130.25, 408 n. to 130.23.

[185] Rosen 1985a, 26–31; Barker and Goldstein 1995; See also Lerner 1996–1997, II, 58–66.

[186] *TBOO* IV, 379–396; as discussed and translated by Christianson 1979.

in heaven. It also raised questions about the true nature of comets and substance of the heavens, but without appearing to settle them. Thus Tycho mentioned, without endorsing, the idea that the heavens were fiery rather than ethereal – a doctrine that he associated with Paracelsus and his followers.[187] He noted that this suggested the possibility of corruption and generation in the heavens, but not that a Paracelsian cosmology might also require or imply that the heavens were fluid.[188] Indeed, he showed a marked preference for retaining a more orthodox cosmological doctrine, suggesting that, 'it should all the more be regarded as a miraculous portent that such a new birth comes forth in the heavens, which are composed of the most subtle, most translucent, and most imperishable of all materials', when he could instead have rejected imperishability as having been called into question.[189] And he followed this remark with scepticism about man's capacity to know more, affirming that, 'We have no real knowledge of the matter or nature of the whole heavens, sun and moon, nor what causes their wonderfully adroit motion, though they have stood and been visible since the beginning of the world'.[190] With respect to the power of mathematics to inform cosmology, Tycho therefore remained rather conservative at this stage of his analysis.

By January 1587, Tycho was prepared to go further. In his letter to Rothmann, and subsequently in *De recentioribus phaenomenis*, he claimed that comets demonstrated the fluidity of the heavens and the non-reality of the spheres. Contrary to what has sometimes been suggested, however, these claims were not directly derived from his principal subject of analysis.[191] Tycho did not demonstrate through his observations that the comet of 1577 would have passed through successive planetary spheres, thereby proving that these spheres were not real, and indeed he could not have done so given the vanishingly small parallaxes such a method would have required him to show. While he did demonstrate that the comet exhibited much less parallax than it ought to have done had it been located at lunar distance or below, he was able to quantify the comet's parallax at only one point in its course.[192] From that determination, he placed the comet no closer than 300 terrestrial radii away from the Earth: not far, he asserted, from the orbit of

---

[187] *TBOO* IV, 381.31–383.11; Christianson 1979, 132–133.

[188] *TBOO* IV, 382.40–383.11; Christianson 1979, 133. On the relationship between Paracelsian cosmology and Tycho's understanding of it as expressed in this treatise, see Shackelford 2002, esp. 55 n. 31, where statements by Paracelsus indicative of a commitment to a fluid heaven are recorded.

[189] *TBOO* IV, 383.18–40; Christianson 1979, 133.

[190] *TBOO* IV, 383.18–40 *cont.*; Christianson 1979, 133.

[191] Swerdlow 1976, 130; Thoren 1979, 61.   [192] *TBOO* IV, 94–107, esp. 104.31–105.41.

Venus, but also somewhere between the Sun and the Moon.[193] At this stage Tycho was already thinking in geoheliocentric terms, and he introduced his new planetary system in order to construct, on the basis of his one parallax value and certain other assumptions, a cometary hypothesis from which he was able to calculate daily distances from the Earth.[194] This hypothesis, which depended on the assumption of geoheliocentricity, placed the comet on a circumsolar path slightly larger than the similar orbit of Venus. Rather than demonstrating the incompatibility of the comet's course with real planetary orbs, therefore, Tycho's book demonstrated its compatibility with a new world-system, in which, of necessity, real planetary spheres had already been dispensed with. The comet's successive positions were not related to the putative spheres in either a Copernican or a Ptolemaic system and, in conformity with his earlier epistemological modesty, Tycho made no positive claims about its material substance, even if he now affirmed that the ethereal heavens were definitely 'most fluid and simple'.[195]

Tycho did not himself claim to have offered a definitive disproof of celestial spheres in the *De recentioribus phaenomenis*, and seems instead to have deferred such a demonstration to a later publication, treating other comets, that never actually appeared.[196] The closest that he came to revealing what this demonstration might have looked like occurred in the tenth chapter of the work, in the context of his discussion of Michael Maestlin's *Observatio & demonstratio cometae aetherei*.[197] Like Tycho, Maestlin had attempted to construct a cometary hypothesis conformable with the actual appearance of the 1577 phenomenon as observed from the Earth. Unlike his Danish contemporary, however, he did so by utilising Copernican hypotheses, and by associating the comet with one of the planetary spheres. Tycho criticised Maestlin's ingenious addition of a cometary epicycle to the deferent of Venus not only on the grounds that it failed to fully save the appearances, but also on the basis that such an attempt, motivated as it was by the desire not to introduce new spheres into the cosmos, was predicated on the existence of objects that comets themselves had shown not to be real.[198] Tycho's reasoning, insofar as it can reconstructed from his rather

[193] *TBOO* IV, 105.22–34.
[194] *TBOO* IV, 159.25–170.42, 177–179. The chief constraint on Tycho's construction of his cometary hypothesis was probably its greatest elongation from the Sun of approximately 60°. See *TBOO* IV, 159.25–32; Westman 1972b, esp. 19–20. I am grateful to Steven Vanden Broecke for illuminating discussions of the relationship between Tycho's cometary analysis and his world-system. See, on this, Vanden Broecke 2006.
[195] *TBOO* IV, 159.1–10.   [196] *TBOO* IV, 159.1–10, 223.38–224.4.
[197] Maestlin 1578; see also Hellman 1944, 146–159; Westman 1972b; Thoren 1979, 61–64.
[198] *TBOO* IV, 221.42–222.39.

brief analysis, seems to have been that cometary courses were sufficiently similar to planetary motions to be mathematically modelled by combinations of circles, but dissimilar enough that they could neither be added to, nor modelled by, the same combinations as were used to construct planetary hypotheses. In particular, Tycho claimed that the prolonged retrograde motion of the comet of 1580, and its variation in latitude, were such as could not be modelled by any planetary or planetary-like sphere, whether Ptolemaic or Copernican.[199] This is what Tycho appears to have meant when he wrote in his letter to Rothmann that comets demonstrated the non-reality of the spheres because they failed 'to follow the lead of any orb'.

While such an argument might have been conceived by Tycho independently of Rothmann, it could also have been inspired by one of the points found in Rothmann's cometary treatise. Rothmann argued that comets, as objects in the heavens, could not be contained within planetary spheres if both spheres and comets were solid bodies; at the same time, he claimed that comets did not move in ways consistent with being moved by the planetary spheres that, if these were real, they would have inhabited.[200] Tycho's reasoning, as displayed in *De recentioribus phaenomenis*, bears a strong resemblance to the second of these arguments. For Rothmann, however, the evidence supplied by the motion of supralunary comets was only part of the case for the non-reality of the spheres and the fluidity of the heavens, and was discussed very briefly. Rothmann also claimed that the heavens were composed entirely of air, rather than ether, and hence held that there was no material basis for an elaborate celestial machinery. Rothmann's argument for this claim was an optical one: the existence of a qualitative boundary between the terrestrial and celestial regions of the world would lead one to expect a refraction of the rays of the planets and the stars at high altitudes that, unlike the refraction of celestial bodies viewed near the horizon, was not actually observed.[201] Refraction had to be the product,

---

[199] *TBOO* IV, 222.39–223.28. C.f. Thoren 1979, 61–64.

[200] Snellius 1619, 117–118: 'ipse tamen cometarum motus firmissimo est argumento, sphaeras Planetarum corpora solida esse non posse. Fieri enim potest, ut corpus solidum admittat dimensionum penetrationem. Sic tu in corpore tuo per parietem transire nequis. Non enim duo corpora simul in eodem loco Physico esse possunt. At toties iam a tot artificibus observatum est, firmissimeque ex Geometria demonstratum, cometas non tantum in regione Elementarum, verumetiam supra Lunam in sphaeris Planetarum subsistere longeque alium motum habere, quam habent Planetae ipsi, in quorum sphaeris subsistunt. Cum igitur nec penetratio dimensionum esse possit, nec unius sphaerae partes dispari ac dissimili motu moveri: manifestum est Planetas in nullo alio corpore quam aere pendere . . .'

[201] Snellius 1619, 104–105: 'si refractio ista esset ab orbibus coelestibus, non tantum usque ad 15 aut 20 ab horizonte gradus, verum (quemadmodum Alhazen & Vitellio in dictis locis demonstrare conantur) usque ad verticem duraret, adeoque omnium observationum certitudo turbaretur necesse esset.'

Rothmann claimed, of dense vapours rising up from the Earth.[202] Similarly, comets were to be seen as the product of terrestrial vapours elevated to a great height, providentially compressed into bodies, and displayed to mankind by the illumination of the Sun.[203] The prominence given in Rothmann's treatise to the optical evidence was such that his argument proceeded rather from the non-existence of the celestial orbs to the validity of his determination of comets' supralunary location and account of their origin, than from observations of the 1585 comet to the dissolution of the spheres. Thus, if Tycho was converted to the doctrine of fluidity by his reading of Rothmann, that conversion involved more than a straightforward acceptance and appropriation of Rothmann's key argument. Tycho would have needed to read the work closely enough to note the claims regarding cometary motion, disentangle them from the optical arguments, and elaborate them into the case he hinted at in his letter of 1587 and his later published work.

Rothmann's treatise was not itself so original. In both its appeal to an optical argument, and its inference from the absence of real celestial spheres to the possibility of supralunary comets that were themselves optical phenomena, Rothmann's arguments were anticipated by Jean Pena in his preface to the *Optica et catoptrica* of Euclid published at Paris in 1557.[204] The degree of similarity is such that it is difficult to conclude that Rothmann's debt to Pena was anything other than direct.[205] More generally, the notion that there was no barrier between the sub- and supra-lunary realms and that a rarefied fluid, variously referred to as air, fire, or ether, extended continuously without interruption from the Earth to the stars, was an ancient cosmological doctrine most strongly identified with the Stoic school of philosophy. The Stoics, who promoted an animistic conception of the heavens, claimed that the planets and stars were nourished by the vapours arising from the Earth and propelled themselves through what, by implication at least, was a fluid celestial substance.[206] Aided by the presence of Stoic cosmological doctrines in Cicero's *De natura deorum* and, to a lesser extent, Seneca's *Quaestiones naturales*, these notions continued to find expression

---

[202] Snellius 1619, 105: 'Non igitur refractio illa accidit propter diversa aetheris & aeris diaphana, sed propter diversa diaphana vaporum & aeris, qui ultra illud spatium, in quo vapores & nubes continentur, semper purissimus est.'

[203] Snellius 1619, 134: 'Statuimus igitur materiam cometarum esse halitus ad superiora elevatos et perpurgatos, atque ita a DEO Opt. Max. qui pro sua sapientia et providentia cometas in certos, ut postea dicetur, usus hominibus ostendit, in corpus coactos.'

[204] Pena 1557, aa2r–bb5v, esp. aa2v–bb2v; Aiton 1981, 101.

[205] Barker and Goldstein 1995, 390–391; Granada 1999, 115–136; Granada 2004.

[206] Lapidge 1978. See also Barker 1985, 1991; Barker and Goldstein 1984.

in certain natural philosophical works of the later Middle Ages.[207] Thus, although they fell out of favour following the entrenchment of Aristotelian cosmology in university curricula, they remained sufficiently well known to be condemned in orthodox treatments. The claim that planets moved 'like birds in the air, or fish in the sea' was a formulation that neatly encapsulated the ideas of fluidity and self-propulsion, and as such it was explicitly rejected in at least two important textbooks: Melanchthon's *Initia doctrinae physicae* of 1549 and Reinhold's commentary on Peurbach's *Theoricae novae planetarum* of 1542.[208] It is highly unlikely that Tycho was unacquainted with both of these works. While a knowledge of such texts would help to explain Tycho's early resistance to the doctrine that the heavens were fluid more readily than his later acceptance, it could be argued that both these texts, and the second especially, would have primed Tycho to recognise the significance for the reality of the spheres of ethereal comets with a motion very different from that of the planets.

By the 1580s a number of sixteenth-century authors had advanced fluid-heaven cosmologies indebted, to a greater or lesser degree, to the ancient Stoic teachings.[209] In his later letter to Rothmann of 17 August 1588, Tycho noted the existence of two authors who appeared to share Rothmann's opinion that the heavenly substance was nothing other than air.[210] One of these was Giordano Bruno, whose cosmology was most famously set out in his *La Cena de le Ceneri* of 1584.[211] As far as can be determined, Tycho's knowledge of the Brunonian doctrine was derived solely from his *Camoeracensis Acrotismus* of 1588; Tycho therefore became acquainted with it after his reading of Rothmann's cometary treatise.[212] The other author whom Tycho recognised to have something in common with Rothmann was none other than Pena. It is unlikely that Tycho realised the full extent of the similarity, however, since he had not seen Pena's work, and was still attempting to obtain a copy of it as late as November 1590.[213] His knowledge of Pena appears to have derived solely from Johannes Praetorius' *De Cometis* of 1578, which Tycho read at some point during his preparation of the *De*

---

[207] Cicero, *De Natura Deorum* II.15; Lapidge 1988, 81–84, 99–112; Burnett 1998, 222–225; Ronca and Curr 1997, 42.

[208] As noted by Aiton 1981, 99–100. The idea that the planets moved in a way analogous to birds through the air was also expressed by Ptolemy; see Taub 1993, 117. For some other users and critics of the simile, see Grant 1994, 274 n. 16.

[209] Barker 1991; Blackwell 1991, 40–45; Lattis 1994, 94–102.   [210] *TBOO* VI, 135.39–136.1.

[211] See Jaki 1975, esp. 145–149; also Michel 1973; Gatti 1999, esp. 43–85.

[212] See Fiorentino and Tocco 1879–1891, I.1, 54–190, esp. 176–177, 183; also Granada 1996, 15–30. Tycho obtained a copy of this text as a gift from Bruno; see Westman 1980a, 96; Sturlese 1985, 314.

[213] *TBOO* VII, 212.37–42.

recentioribus phaenomenis and the *Progymnasmata*.[214] However, this reading cannot certainly be dated prior to that of Rothmann's manuscript treatise. It remains impossible to demonstrate, therefore, that Tycho was acquainted with positive references to fluid-heaven cosmologies, or took comets to be evidence in their favour, before his reading of Rothmann's work on the comet. Equally, however, as he was clearly thinking deeply about both comets and cosmology before 1585, and since he disagreed with many of the premises underlying Rothmann's own 'dissolution' of the spheres, it cannot be proved that Tycho's conversion to the doctrine of fluidity was not, as he indicated, entirely independent.

Whether Tycho was indeed effacing a debt to Rothmann in his letter of 20 January 1587, was simply obscuring the chronological proximity of his own conversion to a doctrine of fluidity, or was telling the truth when he said that he had independently dispensed with real celestial spheres, a motivation for making this claim needs to be established. As was the case with Wittich and the transfer of the Uraniborg instrument designs to Kassel, the issue of credit seems an obvious cause for concern. It is less clear, however, that such a concern would have been associated directly with the demonstration that the heavens were fluid. Tycho did not, as has sometimes been asserted, state in the *De recentioribus phaenomenis* that he had first shown that the planetary spheres were not real.[215] And had he wished to be credited with doing so, he would, I am inclined to believe, have offered a set of arguments for their imaginary status comparable in exhaustiveness and detail to those with which he attempted to prove the ethereality of the comet of 1577, and would have striven to do so before Rothmann's treatise on the comet of 1585 could be completed and published. In the absence of any rival publication by Rothmann, Tycho was, it is true, credited with having provided this demonstration in his cometary work by certain other astronomers.[216] But there is little evidence to suggest that this recognition was something he particularly sought to secure. He tended himself to attribute the demonstration of fluidity, in the

---

[214] *TBOO* III, 155.36–38; *TBOO* IV, 356–358. Granada 2002, 128–130; Granada 2004, 244–245.

[215] Grant 1994, 345, offers a misleading translation of *TBOO* IV, 159.1–10, to this effect. What he construes as Tycho's claim that he had *first* demonstrated the nonreality of the spheres on the basis of the cometary paths is in fact part of Tycho's deferral of a detailed account of his new world-system to a later volume, when or where it will *previously* have been established that comets' courses are incompatible with spheres.

[216] Kepler, in particular, seems to have done so, and was followed by later writers, while Heydon attributed the demonstration to comets without mentioning Tycho individually. See *KGW* VII, 260–261; Heydon 1603, 302; Riccioli 1651, II, 242–243; Grant 1994, 356; Jardine 1988a, 207, 230; Wallis 1995, 16–17. The matter was more complex than Kepler's remarks would suggest, for the reasons discussed at length in Donahue 1981.

De recentioribus phaenomenis and elsewhere, to the phenomena themselves, 'carefully observed by us', and may even have looked to Rothmann to provide some of the evidence that would persuade his contemporaries.[217]

With respect to his new planetary system, however, it was highly likely that Tycho was keen to protect his claim to have evolved it independently. Certainly, it was his later concern to secure credit for his geoheliocentric hypotheses, in the face of the publication of a similar scheme by his rival and antagonist Ursus, that led him to elaborate the chronology for his world-system's development which placed so much emphasis on his recent cometary analyses. This chronology, according to which Tycho arrived at his final geoheliocentric outline before coming to accept heavenly fluidity, cannot wholly be trusted: as we have seen, although Tycho was confident enough in January 1587 about the form his world-system would take to drop a hint to Wilhelm about it, he did not achieve the observational confirmation he would have needed to publish it until later that year, some time after he had committed to fluidity in his letter to Rothmann. In most respects, however, Tycho's disingenuous account could turn out to be true. In the *De recentioribus phaenomenis* itself, Tycho claimed that his thoughts concerning the true system of the world dated back to 'more than four years ago', and this does seem to be plausible. If Tycho truly believed for a while, as per his letter to Brucaeus of 1584, that the failure to observe Martian parallax in 1582 empirically disproved the theory of Copernicus, and yet he considered the Ptolemaic models objectionable for the reason stated in *De recentioribus phaenomenis*, that they depended on circular motions that were not properly uniform, then he would indeed have been led to consider the possibility of alternative hypotheses.[218] And that Tycho was already groping towards his world-system no later than 1584 finds a measure of support in his accounts of Ursus' visit to Hven in September of that year. These elaborate stories of how Ursus came into the possession of a 'faulty' geoheliocentric diagram that failed to show the intersection of the orbits of Mars and the Sun have every indication of being desperately improvised.[219] But if they contained absolutely nothing truthful in them at all, it is difficult to see why Tycho would bother to lie, as early as the printed pages of *De recentioribus phaenomenis*, about the

---

[217] Thus he cited Rothmann's observations of the 1585 comet favourably in *De recentioribus phaenomenis*, as evidence of the supralunarity of comets and fluidity of the heavens, and he encouraged Rothmann to complete and publish his account. See *TBOO* IV, 223.28–224.3, 350.28–38; *TBOO* VI, 147.17–21.

[218] *TBOO* IV, 156.3–11. Tycho's objection to the Ptolemaic system, as expressed here, is fundamentally the same as Copernicus'; see Evans 1998, 419–420.

[219] Rosen 1986, 12–14, 30–50; Jardine 1988a, 10–11, 29–34.

timescale of his world-system's development. Concern that Ursus or others might also lay claim to a geoheliocentric scheme best explains, I think, both Tycho's wish to publish his world-system once it was empirically verified, and his desire to show that the critical step of accepting the fluidity of the heavens had been taken independently.

Tycho chose not to conceal in his letter to Rothmann of January 1587 the fact that the notion of a fluid heaven was crucially related to his evolution of a new system of the world. Indeed, that he did so can again be attributed to his desire to secure credit for this invention, rather than for anything else. What he had to say, although it continued to obscure the chronology of the evolution of his thought, is revealing in that it exhibited his dependence on a crucial intellectual resource, the notion that the order of the cosmos was divinely ensured. Thus, immediately after his claim that he had, 'not, for many years, been of the opinion that some orbs really exist in heaven', he stated that he had:

always judged natural knowledge of motion to be congenital to each planet, or rather divinely imposed on it. By this they are constrained to observe the rule of their course in the most fluid and most rarefied ether, most regularly and uniformly, having need of no support or mover; just as the Earth itself is suspended in the fluid air, stabilised by no pillar, pressed down on all sides on its own single centre. But that I should indeed think so, a certain theory concerning the arrangement of the heavenly revolutions other than the Ptolemaic or Copernican, far more agreeable than these, and recently ascertained by me, informed by experience itself – about which this is not the place to be speaking – also very much confirms.[220]

Dispensing with the celestial spheres made Tycho's world-system possible – or, as he chose to put it here, the evolution of his world-system gave him good reason to support the notion of an entirely fluid heaven. At the same time, however, it generated a natural philosophical problem: how then were the heavenly bodies supposed to be moved? To attribute their regular motion to divine action was again an epistemologically minimalist position, one that did not commit Tycho to making a clear choice between God acting immediately or through a secondary cause. Indeed, it is possible to believe Tycho's claim to have 'always' believed planetary motion to be 'congenital' or 'divinely imposed', only to the extent that this description was entirely compatible with motion supplied by real celestial orbs. Similarly, Rothmann's assertion in his manuscript treatise that comets were terrestrial vapours raised into the heavens and 'compressed into certain bodies by God the Almighty', can be seen to have employed divine agency in order

---

[220] *TBOO* VI, 88.15–25. Cf. *TBOO* IV, 223.38–224.3.

to explain what could otherwise have seemed entirely mysterious.[221] Faith plugged the gaps left in accounts of the universe when mathematics was used to dispense with received natural philosophical truths without substituting new ones.[222] Yet Tycho and Rothmann's recourse to divine agency reflects more than just its status as the ultimate explanatory tool. It also reveals their joint indebtedness to a particular intellectual milieu.

Earlier I noted that the enterprises pursued at Uraniborg and Kassel depended on a belief in the possibility of a universal astronomy. Since antiquity, this belief had been underpinned in multiple ways, even if – perhaps for this very reason – it was rarely articulated in full. It was implicit in the technical mathematical literature of astronomy and it seemed to be confirmed by the experience of those who undertook to observe the heavens on a regular basis. It drew support from basic cosmological doctrines that were themselves empirically and mathematically informed, such as the claim that both the Earth and the heavens were spherical. It could be justified and explained by more sophisticated views concerning the heavens and their motions, including the attribution to them of such characteristics as mathematical perfection, a simple nature, and lack of susceptibility to generation or corruption. It might be sustained by the view that, for one or more of these reasons, the heavens were closer to the divine than anything sublunar. And it could be harnessed to more explicitly theological claims concerning divine agency and the role of the heavens in the overall scheme of the cosmos.[223] Since none of these individual foundations for astronomy was wholly independent of the others, the idea of a stable cosmos amenable to mathematical study came to be underwritten not only by particular experiences, modes of reasoning, and doctrines associated with the fields of mathematics, natural philosophy, and theology, but also by an understanding of the ways in which these disciplines were interrelated.[224]

Precise construals of that relationship varied from one scholar to another. This was true even within the overarching framework that, having been instituted in the curricula of the medieval universities, persisted to shape much, if not all, of sixteenth-century astronomy.[225] That framework, according to which mathematics was subordinated to an Aristotelian natural philosophy that had itself been made subordinate to revealed Christian

---

[221] Snellius 1619, 134, as cited in n. 203 above.   [222] Cf. Methuen 1997, esp. 512.
[223] See, *inter alia*, Plato's *Timaeus* 27C–31A, 40A–B, 47A–C, as discussed in Cornford 1937; Aristotle's *De caelo*, I.2–3; Ptolemy's *Syntaxis* I.1; Cicero's *De Natura Deorum* II.15–16, II.20–21.
[224] See Weisheipl 1978; Jardine 1988a, 225–257; Taub 1993, 19–38, 135–154; Grant 1994, esp. 11–59; Methuen 1998, 107–158; Martens 2000, 11–38; Howell 2002, 1–12.
[225] Grant 1994, esp. 19–23.

knowledge and theology, is often held to have constrained mathematicians in their ability to make claims about nature.[226] It is certainly true that mathematical demonstrations were not generally considered to be a legitimate route to *physical* knowledge, in the Aristotelian sense. Nevertheless, throughout the later Middle Ages the standing of mathematical astronomy *vis-à-vis* cosmology remained open to question. Along with optics, harmonics, and mechanics, astronomy was often accorded, as a subject concerned with the quantitative study of physical reality, a special status intermediate between mathematics and philosophy.[227] In practice, this could mean merely that the technical work of mathematicians was to be guided by certain principles – such as the circularity and uniformity of celestial motion – handed down to them by philosophers. That was seemingly enough in itself, however, to motivate serious attempts to respond to the perceived conflict between Aristotelian cosmology and Ptolemaic astronomy in ways that preserved the merits of both: most notably by adopting, in the form of what has been termed the *Theorica* compromise, a physical interpretation of the mathematical planetary models that could be considered an elaboration, rather than a contradiction, of the concentric nested orbs of Aristotle's *Metaphysics* and *De caelo*.[228] Dissatisfaction with that approach in the fifteenth and sixteenth centuries led both to the promulgation of rival non-Aristotelian cosmologies – including the ones indebted to the Stoics – and to attempts to construct astronomical systems that, being constructed of homocentric orbs only, would satisfy Aristotelian natural philosophy on the strictest of readings.[229] It also, partly in response to the emergence of the Copernican alternative to Ptolemaic astronomy, prompted some scholars to distinguish still more clearly different roles for the natural philosopher and the astronomer in studying the heavens, along the lines of the argument recorded by Simplicius (*c*.490–*c*.560) in his commentary on Aristotle's *Physics*.[230] Maintaining this crisp distinction between mathematical astronomy and physics often went hand-in-hand with adoption of a sceptical position regarding the reality of astronomical hypotheses, the geometrical devices used by mathematicians to model and predict the heavenly motions – or at least about the capacity of men to distinguish between competing sets of hypotheses that would, if they existed, give rise to the same visible phenomena.[231] Occasionally, a deeper form of doubt was expressed,

[226] E.g. Westman 1980c, 107–109; Grant 1994, 36–39.   [227] Jardine 1988a, 229–230.
[228] Tredennick 1933–1935, II, 152–163; Guthrie 1939; Grant 1978, 280–284; Jardine 1988a, 230.
[229] Pedersen 1978a, 320–322; Jardine 1988a, 231–235; Lattis 1994, 87–94; Swerdlow 1999.
[230] See Fleet 1989, 45–48; Grant 1994, 36–37.
[231] Jardine 1979, 1987, 1988a, 237–238, 1988b, 697–702; Barker and Goldstein 1998; Methuen 1998, 159–204; Barker 2000, 72–82.

one that questioned whether mankind was intellectually equipped to construct hypotheses complex enough to mimic the heavens with any degree of precision. Yet, except as part of attacks on the soundness of *all* forms of rational knowledge, scepticism about mathematical astronomy rarely extended to doubt concerning the validity of its fundamental tools, sense-perception and the demonstrations of geometry.[232] Increasingly, indeed, the certainty of mathematics was contrasted favourably with that enjoyed by other forms of reasoning.[233] Hence, although mathematicians were frequently thought unable to supply an understanding of the true *causes* of the celestial phenomena, the scholastic tradition within which they worked helped to indemnify the enterprise of seeking a stable and exact account of events in the heavens. And this scholastic framework ultimately assisted the transformation of cosmology through mathematics in the later sixteenth century since, as we have seen, faith in a divinely underwritten cosmological order could protect the project of understanding the heavens via observation and geometry even as particular tenets of Aristotelian natural philosophy were called into question.[234] This was especially true in the case of those individuals who, like Tycho and Rothmann, were trained within the universities but pursued their studies elsewhere. Unlike many of their contemporaries, they experienced few *external* constraints on their freedom of thought and expression as a consequence of the institutional separation of disciplines.[235]

The particular scholastic tradition to which both Tycho and Rothmann were indebted was the one that had been established by Philipp Melanchthon (1497–1560) at the University of Wittenberg and was subsequently exported to other institutions in both the Empire and Denmark. Melanchthon had encouraged the study of the heavens for two related reasons: because the movements of the stars and planets provided clear evidence of God's providential construction of the cosmos, and indeed were active elements within the divine system of governance; and because the heavens were a means by which God communicated with mankind, providing a crucial arena for the display of divine portents and signs. For Melanchthon, therefore, study of the heavens involved both astronomy and a species of astrology, and although he affirmed the inseparability of these disciplines, the latter was arguably the more important to him of the two. This was because, in Melanchthon's conception, astrology was primarily

[232] Jardine 1979, 1987, 1988b; Dear 1988, 23–47; Popkin 1988, 678–684.
[233] Rose 1969; Dear 1988, 48–79, 1995, 32–62; Jardine 1988b, 693–697.
[234] See also Methuen 1998; Barker 2000, 82–88; Martens 2000; Barker and Goldstein 2001.
[235] See Westman 1980c, esp. 121–123; Jardine 1998.

a study of the causal relationships operating between the heavens and the Earth, and such causal study not only licensed the inclusion of these traditionally mathematical subjects under the category of natural philosophy, it also led the student to knowledge of the first cause, God. Astronomy was an essential prerequisite for the study of astrology, and geometry and arithmetic were prerequisites for the study of astronomy. Study of all of these branches of mathematics was therefore mandated in Melanchthon's curriculum.[236]

Tycho and Rothmann both expressed Philippist notions when they interpreted novel celestial phenomena in terms of their status as providential signs.[237] Rothmann, for example, did not simply say comets were divinely caused; they were formed by 'God the Almighty who, according to his wisdom and providence, displays comets for man's use'.[238] Tycho described the nova of 1572 as 'a wonderful prodigy of God, creator of the whole world machine, fashioned by Him outside every order of nature at the beginning, and now shown to the eventide world'; comets, in his opinion, were 'unnatural wonders of God, by which he means to signify something other than what the natural courses' of the heavens normally revealed.[239] The similarity of these views is unsurprising given the nature of their authors' respective educations: Rothmann was a graduate of Wittenberg; Tycho had also spent some time there, and had received most of his education at the Philippist strongholds of Copenhagen and Rostock.[240] But while Melanchthon's views were extremely influential in the astronomy of the later sixteenth century, they were also rather fissile. Various inflections of Melanchthonian teachings were developed and articulated, partly as a consequence of the combination of Melanchthon's views with non-scholastic philosophies, and partly because of the evolution of different positions on the relative authority of scripture, natural philosophical doctrine, and empirical evidence.[241] Tycho himself displayed his clearest commitment

---

[236] For Philippist cosmology and natural philosophy, see Kusukawa 1993, 1995, esp. 124–173; Methuen 1996b, 1998, esp. 61–106; Barker 2000.

[237] Wilhelm may also have been indebted to Philippist conceptions of providence; see Moran 1978, 52–54, 125–126; Methuen 1997, 508–509.

[238] Snellius 1619, 134, as cited in n. 203 above.

[239] *TBOO* I, 19.19–22; *TBOO* IV, 390.13–38; Christianson 1979, 137. Cf. the views of other commentators on the nova in Methuen 1997.

[240] Granada, Hamel and von Mackensen 2003, 10; Thoren 1990, 9–39. On the Philippist climate of sixteenth-century Denmark, see Shackelford 1989, 232–238; Lyby and Grell 1995.

[241] See, for some examples, Methuen 1998, 107–204. Besides Rothmann and Tycho, significant astronomers working within the Philippist tradition included Maestlin, Kepler, Praetorius, and Helisaeus Röslin, none of whom shared *precisely* the same epistemologies and cosmologies. See Westman 1975b, 289–305; Granada 1996, 109–160; Methuen 1998, 205–224; Barker and Goldstein 2001.

to the Philippist tradition in the oration he delivered at the University of Copenhagen in 1574.[242] On that same occasion, however, he spoke in terms that indicated an equal debt to a non-Aristotelian, alchemical philosophy. Whereas Melanchthon affirmed the role of the heavenly bodies as causes of change in the sublunary realm, including changes within the human body, Tycho described those causes in astro-alchemical terms of a semi-Paracelsian sort.[243] Thus it is not so surprising that Rothmann and Tycho, despite their common Philippist heritage, were unable to agree in every respect about matters of astronomy and cosmology, and construed the relationship between mathematical, natural philosophical, and exegetical techniques to be employed in arriving at an understanding of the cosmos in rather different ways. Whereas Rothmann prioritised mathematics, Tycho displayed greater commitment to other sources of knowledge of the universe.

The correspondence between Kassel and Hven rendered these differences visible. It did so as the process of epistolary calibration proceeded and expanded, developing from Tycho's attempts to align Rothmann's observational practices and results with his own into a campaign to justify and win assent to his new system of the world and broader cosmology. To some extent this development was a natural consequence of the connections that had already been made in Rothmann's cometary treatise and Tycho's response to it. The continuing debate about refraction, for example, was in part simply a discussion about an empirical phenomenon that was highly significant for the practice of observational astronomy. As such, it is likely that it would have been pursued for its own sake. Yet this phenomenon had also been shown to be highly consequential for cosmology; it was relevant, one might say, to the task of putting philosophical flesh on the bare bones of Tycho's new planetary system. This meant that settling the question of who was right with respect to refraction's cause and extent became, from Tycho's point of view, even more urgent.[244]

In his 1587 letter to Rothmann, therefore, Tycho asserted that the *mathematicus* was wrong to suppose, as implied by his treatise on the comet of 1585, that refraction affected the apparent positions of celestial bodies uniformly. At an altitude of 30° above the horizon, refraction only led to an error of half a minute of arc; whereas near the horizon the apparent position was altered by twenty-eight minutes.[245] Observations of the

---

[242] *TBOO* I, 143–173; Dreyer 1890, 76–77; Thoren 1990, 81–84, 86.
[243] *TBOO* I, 157.5–158.4. See also Shackelford 2002.
[244] For a brief summary of the debate, see Moran 1982, 101–107.
[245] *TBOO* VI, 91.40–92.5. Cf. Snellius 1619, 102–111.

fixed stars were less susceptible to the problem, although the Hessen data that Tycho had received indicated that Rothmann derived a lower component of refraction than he did.[246] For this reason, he instructed the *mathematicus* to:

investigate this more precisely, when you can, with instruments well corrected, and you will see that the whole matter is as I say. Unless, perchance (which, however, I would not easily believe), the different nature of the horizons causes a greater intrusion of refraction here than where you are. From which it would follow that it occurs to a greater degree in places further north than in the south, and this even if the air is wholly pure and calm near the horizon in both cases – as is necessarily required in this observation above all others.[247]

Tycho was convinced that, meteorological disturbances aside, there should not exist any local variation in atmospheric refraction, not even a variation that was predictable and consistently related to latitude. He therefore challenged the Kassel observers to revise their evaluation of the extent of refraction in accordance with his own. At the same time, he objected on empirical grounds to Rothmann's explanation of the phenomenon as due to terrestrial vapours. Such exhalations could not explain solar refraction at an altitude of 30°, as he had found through solstitial observations, particularly when Venus, a humidifying planet according to astrological lore, was not above the horizon.[248] Hence Rothmann was wrong to deny the existence of a distinction in kind and transparency between sub-lunary and supra-lunary matter. Detectable refraction was, as the medieval authorities Alhazen (960–c.1040) and Witelo (c.1230–c.1280) had maintained, the consequence of light striking the interface between the air and the ether.[249] But that the magnitude of refraction was much smaller than these authorities had indicated could be attributed to their false belief in the solidity of celestial matter, and consequent failure to realise that elemental air, becoming more attenuated as it ascended, had quite similar properties to the celestial ether at their mutual boundary.[250]

In letters of 11 and 21 September respectively, Wilhelm and Rothmann responded to Tycho's analysis of refraction. As no detailed explanation had been addressed to Wilhelm, he did not touch on the dispute over the

---

[246] *TBOO* VI, 92.8–10.   [247] *TBOO* VI, 92.30–37.
[248] *TBOO* VI, 92.39–93.11. See, on Venus, Ptolemy's *Tetrabiblos* I.4.
[249] The *Optica* of Alhazen (Ibn al-Haytham) was transmitted to the West in the thirteenth century as the *Perspectiva* or *De aspectibus*, and was used by Witelo, along with other texts, in the composition of his *Perspectiva*, also called the *Opticae libros decem*. Both works were published in Risner 1572. On these authors and their works, see *DSB* VI, 189–210, XIV, 457–462.
[250] *TBOO* VI, 93.11–23.

cause of the phenomenon. But commenting on the 5° difference in stellar longitudes between the two observatories, and stating that he did not intend to defer to Tycho's values, the Landgrave referred to refraction as one source of error at Uraniborg. Having 'spoken seriously to our *mathematicus*' about the discrepancy, Wilhelm had learnt that Rothmann had performed fresh observations of the stellar longitudes with the newly corrected quadrant; even, as was Tycho's preferred method, employing the diurnal appearances of Venus.[251] However:

Venus would give no other place for the fixed stars. He supposes that you attribute somewhat too much to the refractions. He also supposes that you obtained the time throughout with the equatorial armillaries, but the refraction of the Sun occurs in a vertical circle, and not following the longitude of the ecliptic, [so] it could easily be that you assumed a time somewhat too short; which could not happen if you obtained the time through the azimuth of the Sun.[252]

Rather than accommodate Tycho's attempts to align the Hessen observations precisely with his own values, Wilhelm and Rothmann were utilising one of the points of contention, the degree of observed refraction, to challenge Uraniborg's claim to astronomical authority. 'It is not possible to really know on which side the error lies', wrote the Landgrave, 'but however it is, we are unable to stand back from our observations, which we established with no less care than you did yours'.[253] And he went on to assert that a publication of the Hessen stellar positions, which he had commanded Rothmann to prepare, would neither mention the five-minute discrepancy, nor remove it; with the result that sets of tables produced by the two noble astronomers would be accredited with more trust because, 'each will stay with his observations and no one shall change anything for the sake of the other'.[254] Rothmann made the same claim in his own letter, pointing out that such a small difference would offend no one, 'provided that they understand the subtlety of this matter'.[255]

Rothmann also discussed Tycho's explanation of refraction, defending his claim that there was no observational evidence in favour of a distinction between ether and air. If the apparent displacement were due to a boundary of two transparent fluids with different optical properties, then it would, he asserted, apply detectably to all celestial objects except those directly above the observer.[256] Since such refraction was not observed, the phenomenon provided evidence for Rothmann's own view that between the Earth and

[251] *TBOO* VI, 107.11–18.   [252] *TBOO* VI, 107.19–26.   [253] *TBOO* VI, 107.26–30.
[254] *TBOO* VI, 107.41–108.3.   [255] *TBOO* VI, 117.21–24.   [256] *TBOO* VI, 111.31–33.

the fixed stars there was 'nothing other than air moving around the seven wandering stars'.[257] However, close to the Earth there was a band of denser air, thickened by terrestrial vapours, as shown by the lightening of the sky during the hours of twilight.[258] It was this band of air that caused refraction; not, however, simply in virtue of refraction at the pure, dense air boundary, since this possibility was ruled out by the very argument Rothmann had used against Tycho. Rather, Rothmann ascribed the effect, somewhat obscurely, to the increase of the span traversed by rays of light as the originating objects moved closer to the horizon.[259] He concluded that:

it is most evident, that neither are there different transparencies of the ether and the air, nor can refractions be caused otherwise than by dense and vaporous air. But from this very fact, if nothing else, refractions would certainly show differences in different places. For it is very true that refractions are not so great here as they are with you.[260]

Whereas Tycho interpreted the smaller allowance for refraction utilised by the Kassel observatory as evidence of erroneous practice, Rothmann viewed the same data as proof that 'the ancients' had been correct to assert that atmospheric refraction varied according to the latitude of the observer.[261] While taking this view was not to deny the possibility of a universal astronomy, it clearly had different implications for observational practice, as well as for the ease with which true celestial positions in astronomical tables or ephemerides could be transformed to give the apparent positions of objects above the local horizon.

In his reply of 16 August 1588, addressed to the Landgrave, Tycho appeared to concede the points made by Wilhelm and Rothmann about the inappropriateness of merely adjusting the Kassel stellar longitudes so as to agree with Uraniborg's.[262] But he did not refrain from continuing to identify possible causes for their persistent 'error'. That the correction of the Kassel quadrant had not removed the discrepancy could be attributed to the intrinsic unreliability of the method they employed, using timed meridian

---

[257] *TBOO* VI, 112.11–15.
[258] *TBOO* VI, 112.15–29. Rothmann refers here, in addition to Alhazen and Witelo, to the *De crepusculis* (1542) of Pedro Nuñez (1502–1578). Smith 2003, identifies the *De crepusculis et nubium ascensionibus* of Ibn Mu'adh as a potentially important source for the understanding of the relationship between atmospheric refraction, twilight, and terrestrial vapours employed by Rothmann. It is perhaps helpful to point out, as Smith does not, that the treatise in question was first published in conjunction with that of Nuñez, whose own work was likely indebted to it, and republished in Risner 1572. In both cases, it was attributed to Alhazen. See Sabra 1967, 78 n. 1–2.
[259] *TBOO* VI, 112.29–113.15.   [260] *TBOO* VI, 114.16–21.   [261] *TBOO* VI, 114.34–37.
[262] *TBOO* VI, 130.28–32. See also 131.20–25, where Tycho acknowledges that greater trust will follow from not altering these values.

transits.[263] But it was refraction that misled Rothmann into believing that the observations made by means of Venus and the Sun confirmed his earlier values; at the time that he had employed the method, around the winter solstice, both were too low in the sky to be safely observed.[264] And it was this refraction that had led Rothmann to assert the existence of a slight error in the solar ephemerides Tycho had sent to the Landgrave.[265] Moreover:

what the *mathematicus* of Your Highness suspects from the fact that my longitudes differ from yours by five minutes, that I attribute too much to refraction, he would himself acknowledge to be far from the truth, if he had understood sufficiently how I investigated the extent of refraction some years ago, for both the Sun and the stars, with exquisite care and by various means, through a variety of instruments. For I do not hold that this is to be assumed as one wills, but as the very experience of heaven provides.[266]

Notwithstanding the fact that, in his own letter, Rothmann had indicated a belief in the variability of atmospheric refraction with terrestrial latitude, Tycho felt it necessary to strenuously rebut the suggestion that his own practices were erroneous. Possible sources of error alluded to by Wilhelm and Rothmann were discussed and rejected. Thus, he asserted that *his* use of clocks did not endanger the accuracy of his observations, because unlike the Moon, neither the Sun nor Venus moved sufficiently quickly across the sky for deviations from precise time determinations to affect the final outcome.[267] As for his use of equatorial armillaries, far from being ignorant of the fact that refraction applied vertically above the horizon, and not following the line of the ecliptic, he had produced geometrical tables to make determining the necessary correction exceedingly simple. Furthermore, when he employed these instruments to determine sidereal time, he selected stars crossing the meridian that were near the celestial equator: a procedure that maximised the accuracy of the determination, and admitted no error of refraction.[268]

From the agreement of his repeated investigations into the matter, Tycho was convinced that his evaluations of refraction were free from all error.[269] He wrote:

But in case it is required in order to oppose these many doubts about our observations, I shall add only one thing that is indisputably beyond all contention. If any flaw in my places of the fixed stars, even the least, occurs, from whatsoever causes ultimately, either those or others, it would certainly have revealed itself openly in

[263] *TBOO* VI, 124.31–41. [264] *TBOO* VI, 125.20–26. [265] *TBOO* VI, 127.2–11.
[266] *TBOO* VI, 127.32–39. [267] *TBOO* VI, 128.7–15. [268] *TBOO* VI, 128.21–33.
[269] *TBOO* VI, 127.42–128.7, 129.9–12, 129.34–130.6.

the reciprocal observing; [for] I indicated in the last letter to Your Highness, how I corrected the places of the fixed stars very carefully, not only many times through Venus, when she appeared occidental to the Sun, and very high and clearly in the year '82, but also through the same planet as she appeared in the east, around the middle of September in the year '85, with the mediating place of the Sun precisely ascertained in both cases; and from an exact examination I discovered that these were correct, without any detectable error.[270]

As this statement indicates, Tycho understood that correct evaluation of the extent of atmospheric refraction was critical to the accurate determination of stellar and planetary positions, and hence to his entire project of restoring astronomy.[271] If any error were admitted in this matter, it could undermine all of his observational labour.

In his letter to Rothmann of 17 August 1588, Tycho returned to both the astronomical and cosmological issues of refraction. He raised new objections to Rothmann's belief that the supra- and sub-lunary fluids were identical. What would the relative proportionality of the elements be, he asked, if air filled the immense space from the Earth to the fixed stars?[272] The force of this question derived from Tycho's particular conception of the symmetry and order underlying creation, although its full significance was not to be disclosed until later.[273] At a physical level, he objected that, without some distinction of matter, men would be deafened by the sounds accompanying the swift motions of the stars and planets.[274] In any case, the latter would be worn down and degraded by elemental air; so that to place this in the space above the moon would be for astronomers to 'strew heaven with defects'.[275] And addressing Rothmann's geometrical demonstrations, he elaborated once again his own beliefs about the cause of refraction, and its variation with altitude.[276] Then, having already asserted that, as a divine creation, Heaven was 'something abstract, and like something immaterial, escaping our comprehension',[277] he employed the distinction between philosophy and astronomy to downplay the significance of his debate with Rothmann over causal explanations:

But we dispute so much about whence the refractions occur, whether only from the intervening vapours, as it seems to you, or whether not only from these, but also to some extent from the difference in the transparencies of the ether and the air, in some combination, as seems right to me, to no purpose; since the chief

---

[270] *TBOO* VI, 128.33–129.4.   [271] *TBOO* VI, 127.41–42.
[272] *TBOO* VI, 135.17–21.   [273] See the discussion in Chapter 3.
[274] *TBOO* VI, 135.21–24. For the origins of this argument, see Aristotle's objections to the Pythagorean concept of the harmony of the spheres in *De Caelo* II.9, in Guthrie 1939, 190–197.
[275] *TBOO* VI, 135.14–26.   [276] *TBOO* VI, 136.18–139.26.   [277] *TBOO* VI, 136.12–14.

issue concerns the re-establishing of the motion of the heavenly bodies, and the restoring of astronomy, and the purging it of faults. And although knowledge of refractions conveys aid necessary to this objective, because without their prevention the very places of the stars cannot be derived from heaven, since it suffices for the astronomer to know how high the impediments of refraction reach, and how much they are at certain altitudes, especially when the same untroubled clearness of the air appears, whence these refractions proceed, whether even from one or two causes, it is therefore not of much importance to debate subtly and ambiguously. Just as also the issue of the very substance of heaven does not properly belong to the astronomer to decide. For he struggles to investigate, by certain observations, not what heaven is, and from what the shining bodies of it are made, but only how all these move. For the rest is left to be disputed by theologians and natural philosophers; among whom even up to now this thing remains insufficiently explained.[278]

Such was the impenetrable mystery of the subject, and the feebleness of the human mind, that there was little hope of determining anything of the nature of the heavenly substance, *except* that it in no way partook of the nature of anything elemental.[279]

Clearly, Tycho was not actually as epistemologically modest as this attempt to close down the debate might seem to imply. Although he strove to distinguish mathematical from philosophical or theological analysis, and to suggest that the latter ought to be reserved for others to perform, he was not truly an advocate of the sceptical position that physical knowledge of the heavens' true nature was unobtainable by man. Indeed, his reintroduction of the very bone of contention, the ethereal nature of the heavens, as an axiomatic truth, reveals his prior commitment to certain cosmological principles that prevented him from acceding to Rothmann's view that the ether was a fictional substance, or even conceding that different opinions of the substance of the heavens were equally plausible. And since, notwithstanding his own attempt to terminate the discussion, he could not quite resist continuing the attack, something of the origin of these notions emerged at this time: it was, at least in part, from his alchemical studies and philosophy that Tycho derived his commitment to the existence of a quintessential substance from which the heavens were formed.

Alchemy, or as he preferred to call to it, the spagyric or pyronomic art, and sometimes 'terrestrial astronomy',[280] was not a subject about which Tycho liked to correspond openly. On several occasions he expressed his concern

---

[278] *TBOO* VI, 139.26–140.3   [279] *TBOO* VI, 140.3–12.
[280] Christianson 2000, 90–91; *TBOO* VI 144.39–145.4, 146.24–27; *TBOO* VII, 238.1–5. On the term 'terrestrial' or 'inferior' astronomy, see Crosland 1962, 5–6.

about doing so, and once he indicated that it would only be acceptable if a code or cipher were employed.[281] However, there was both an occasion and an obligation to say something on this topic in the Kassel correspondence, because Rothmann had enquired about a pair of emblems that Tycho had published on the reverse of a poem printed at Uraniborg and addressed to his kinsman Falk Gøye (*d.* 1623). These images represented astronomy and the spagyric art and hinted at the relationship between them. Tycho explained that, as Rothmann had surmised, they were intended to suggest the intimate connection between the heavens and the earth: 'the superior are connected to the inferior through a wonderful resemblance, in such a way, that neither can truly be perceived nor deeply explored without the other'.[282] For the benefit of his correspondent, who had confessed to possessing little knowledge of alchemical philosophy, but expressed a desire to know more, he quickly sketched some of the correspondences that existed between the heavenly bodies, the metals, and human bodily parts, such as the identification of the Sun with both gold and the heart, and Saturn with lead and the spleen.[283] And he mentioned some of the principal authors and texts that Rothmann might consult for further information: Hermes Trismegistus, Geber, John of Rupecissa, Arnold of Villanova, Ramón Lull, Thomas Aquinas, Roger Bacon, the *Turba Philosophorum*, Albertus Magnus, Isaac Hollandus, and Theophrastus Paracelsus.[284] He also affirmed the truth of the latter's assertion that direct experience of alchemical operations was more valuable and instructive than book-learning alone.[285] Adjacent to these remarks, in the margin of his letter, Tycho added:

What if, from this very spagyric art applied to things which the Earth fosters and contains (which partake not only of the nature of the elements, but also of that of the ether combined with them), it could be visibly demonstrated that they are allotted a subtler and far more excellent, and more perfect substance (that which is celestial in them) than the elemental, whether it was airy, aqueous, earthy or even fiery; and this was set loose by howsoever much activity and succession of separations? Whence it could be inferred quite plainly, and concluded with

---

[281] *TBOO* VII, 5.2–18, 195.30–34, 310.7–13. It should be noted, however, that Tycho was wary about the means of communication, not about the idea of sharing alchemical knowledge; see *TBOO* VII, 42.34–36, 94.39–95.42; *TBOO* VIII, 372.18–373.13. Tycho's attitude to alchemical secrecy has been debated by Hannaway 1986; Shackelford 1993. On this topic, more generally, see Crosland 1962, 35–42, 51–52; Principe 1992.

[282] *TBOO* VI, 117.41–118.18, 144.39–145.4 *cont.* On this aspect of Tycho's cosmology, see Christianson 1968; Segonds 1993; Shackelford 1991, 95–105. For Tycho's alchemy more generally, see Figala 1972; Hannaway 1986; Shackelford 1989, 187–229, 1993, 2002; Keil 1992.

[283] *TBOO* VI, 145.4–35. Cf. *TBOO* I, 157.5–158.4.   [284] *TBOO* VI, 146.9–15.

[285] *TBOO* VI, 146.15–21.

sufficient certainty, that also in the whole macrocosm, its ethereal part consists of a far more perfect and excellent nature than that which could deservedly be equated to any of the elements, even to the air. Would not in this way your arguments about the airy substance of heaven seem satisfied, even by the very inferior and terrestrial astronomy?[286]

In other words, alchemical decomposition of terrestrial substances that produced pure and clear distillates was to be taken, in virtue of the relationship of resemblance between the heavens and the Earth, as evidence of the existence of quintessential matter.[287]

At the level of data-interpretation, Tycho informed Rothmann that it was simply not true that there was any real difference in the extent of atmospheric refraction in solar observations determined at the two observatories. The conclusion that there was had been derived from false comparisons: the meridian altitudes of the Sun at the winter solstice, for example, were not the same for the two different locations. But if the refraction incurred at the meridian altitude of the solstitial Sun at Kassel was compared with that found at the equivalent meridian altitude of the Sun at Uraniborg (that is, at around 5° into Aquarius or twenty-five into Scorpio in the zodiacal calendar), then there was no significant discrepancy.[288] From this, however, Tycho deduced that there was an error in the Kassel determinations of *stellar* refraction; one that was not to be explained away by recourse to differences in the vapours or climates of the two different latitudes.[289] Thus:

From all these things it is also sufficiently gathered, that those things which are openly stated by you, and which you are ingeniously and splendidly trying to convince me of, concerning the nature of the vapours, and the theory of refractions derived from them only, and even, after that, the substance of the whole heaven, declared to be airy and elemental, do not rest upon secure foundations.[290]

Rothmann's reply to Tycho was written on 19 September 1588. He rejected Tycho's attempts to assign responsibility for determining the nature of celestial matter to others. 'Unless this question is decided by us', he asserted, 'it shall never be decided by anyone, whether a theologian or a natural philosopher'.[291] In rejecting theological approaches to the question, he deployed

---

[286] *TBOO* VI, 359, note to 145.42. Rothmann resisted this interpretation, arguing that Tycho's arguments provided evidence of the action of a solar influence on terrestrial generations, rather than of a distinction between celestial and terrestrial matter. See *TBOO* VI, 153.25–30.

[287] On the identification of alcohol with quintessential ether, see Multhauf 1954.

[288] *TBOO* VI, 140.28–141.7.   [289] *TBOO* VI, 141.7–30.

[290] *TBOO* VI, 141.30–34.   [291] *TBOO* VI, 149.16–17.

what has become known as the 'accommodationist' approach to the relationship between the truths of scripture and those of nature: having no bearing on salvation, the topic of the celestial substance was not adequately dealt with in the sacred scriptures, for these had been written for, or accommodated to, the common man's comprehension.[292] And with respect to the potential contribution to be made by philosophers, he claimed that the only route to knowledge of these matters was the mathematical one, as provided by the application of trigonometrical and optical demonstrations to the heavens.[293] The traditional doctrine advanced by philosophers, that there was nothing elemental, corruptible, and changeable in heaven, was just as easily denied as asserted.[294] Rothmann attempted to defuse Tycho's *a priori* objections by pointing out that he had only stated that the fluid surrounding the celestial bodies was elemental and changeable; the existence of comets in the supralunary regions demonstrated that this had to be admitted. He had not, therefore, 'strewn heaven with defects', but had liberated it from 'a pointless substance, uselessly invented against all reason and experience' and 'substituted the true judgement'.[295] As for the claim that the planetary motions would produce an audible sound, Rothmann rejected Tycho's inference from common experience, on the basis that air at ground-level was qualitatively different from air in the celestial regions. He argued that air resisted motion, and hence was audible when moved, only because of its saturation with incompressible water.[296] At higher altitudes, it was reasonable to suppose that the thinner air, devoid of water, flowed freely without resistance, and hence would neither produce any sound, nor erode the bodies in the heavens. And if, as Tycho claimed, the motion of the air would wear the planets away, would not the Earth itself have been worn away long ago?[297] Also, he joked:

if so great a sound could be produced by that pure air, as could strike our hearing, I should have thought that all the Gods would long ago have been choked by excessive fear. For the sound of this dense air produced by the most violent winds would be perceived far more by the Gods, than it would be by us. However I recall that once, at a site five miles away, there was a most violent wind, so that buildings and trees were overturned; of the sound of which, however, I heard nothing here.[298]

In opposition to Tycho's arguments about the proportionality of the elements, Rothmann cited Melanchthon, who had 'rebuked the Peripatetics,

---

[292] *TBOO* VI, 149.16–23. See Williams 1948, 176–177; Hooykaas 1977, 114–124, 1984, 33–34; Harrison 1998, 129–138; Granada 2002, 67–113; Howell 2002, 3–12.
[293] *TBOO* VI, 149.24–27. [294] *TBOO* VI, 149.31–33. [295] *TBOO* VI, 150.5–16.
[296] *TBOO* VI, 150.20–23. [297] *TBOO* VI, 150.28–35. [298] *TBOO* VI, 150.35–41.

and taught the true opinion of Aristotle'.[299] On Tycho's elemental model
the proportionality of air to water and earth combined was 1: 14 608.
What kind of ratio was that?[300] With reference to Tycho's assessment of
Rothmann's optics, the *mathematicus* pointed out that he had not asserted
that atmospheric refraction ceased at an altitude of 30° above the horizon;
he had merely assumed this for the sake of his argument that *if* there was
a limiting point, the span of dense air that light rays penetrated before
they were refracted was readily deducible.[301] It was the fact that refraction
did not extend to the vertex that supported his case for the identity of the
sub- and supra-lunary fluid, since whenever there was refraction away from
the perpendicular, there had to be a boundary between two transparent
substances.[302] But in making this point, Rothmann failed to appreciate the
distinction Tycho was employing between refraction detectable in principle, and refraction detectable in practice. He stated:

I don't know how you adhere so to your opinion, that you even allow, contrary
to all experience, that refraction persists right up to the vertex, when, however,
you have never detected this with your instruments. Certainly on this basis it is
possible to fabricate anything whatsoever.[303]

On the evidence of this remark, it could be argued that Rothmann was
less sophisticated than Tycho in his understanding of the limits of the
capacities of instruments. On the other hand, since he did not share his
Danish correspondent's commitment to the existence of an ether revealed
by alchemical theory and experiment, he had no reason to interpret his
failure to detect refraction at high altitudes any differently. Both Tycho
and Rothmann were arguing from the evidence, but what they took to be
evidence relevant to the issue at hand differed significantly.

Rothmann's letter of 19 September 1588 also contained his response to
Tycho's *De recentioribus phaenomenis*, which had by then been brought to
Kassel from Uraniborg by Gellius Sascerides. He considered both Tycho's
discussion of the comet of 1577, and the new world-system described in
the volume's eighth chapter. The former, he suggested, lacked a satisfactory
account of the material nature of the phenomena. He could not determine

[299] *TBOO* VI, 150.42–151.3. Rothmann's reference is to the *Initia doctrinae physicae*, first published
       at Wittenberg in 1549, the final section of which is entitled *de Elementis, et eorum qualitatibus,
       et alterationum et mixtionum causis*. Melanchthon does not appear to address the question of
       proportionality directly; Rothmann may be thinking of his emphasis on the possibility of the
       mutation of one element into another. See Melanchthon 1563, 168v–194v, esp. 169r.
[300] *TBOO* VI, 151.4–9.   [301] *TBOO* VI, 151.10–21.
[302] *TBOO* VI, 151.28–36.   [303] *TBOO* VI, 152.13–16.

whether Tycho held that cometary tails were produced by the refraction of light through the comet, as he had written in his own cometary treatise, or whether they were material, as he was now firmly persuaded.[304] As for the Tychonic system, he both challenged the originality of the scheme, and objected to Tycho's claim that it was manifestly superior to the Ptolemaic and the Copernican schemes. If the geoheliocentric model were simply the 'inversion' of the hypotheses of Copernicus, that is the heliocentric system adapted geometrically to the immobility of the Earth, then it and the Copernican scheme would be empirically indistinguishable. And unless there was an observable difference, the Copernican system was to be preferred, because it did not attribute retrograde motions to celestial objects and therefore conceptually simplified the observable phenomena.[305] Nor, Rothmann claimed, was acceptance of the Copernican account prevented by physical reasoning or biblical evidence. Copernicus himself had refuted the natural philosophical arguments more than sufficiently, and the accommodationist principle dispensed with any scriptural obstacles. Mathematical demonstrations were the proper route to knowledge of the question;[306] although this was not to say that notions of divinity were not involved in determining the issue. God was 'the author not of confusion but of order', and the nature he created was not, therefore, unnecessarily complex.[307] Unsurprisingly, perhaps, as a product of an education that had stressed the connections between astronomy, natural philosophy, and theology, Rothmann was resistant to being forced into the role of a *mathematicus* constrained, rather than aided, by these disciplinary boundaries.

Rothmann's arguments naturally prompted vigorous replies from Tycho, with the consequence that subsequent letters of the correspondence continued to be characterised by lengthy debates on these topics.[308] Thus, while Tycho persisted, in his letter to Wilhelm of 21 February 1589, to prosecute the notion that Kassel's stellar longitudes were erroneous because of errors in assessing refraction,[309] in his letter to Rothmann of the same date he maintained his attack on a broad series of fronts. To begin with, he renounced, with conspicuous insincerity, any further attempt to force Rothmann to change his mind about the substance of the heavens.

---

[304] *TBOO* VI, 155.9–156.30. See Chapter 3.   [305] *TBOO* VI, 156.38–158.18.
[306] *TBOO* VI, 159.18–160.6.   [307] *TBOO* VI, 158.35–39.
[308] See also, on the cosmological debates, Blair 1990; Granada 1996, 61–76, 2002, 87–110; Howell 2002, 73–108.
[309] *TBOO* IV, 156.3–157.5.

Rothmann might stick to his opinion that the heavens were airy. Likewise, anyone who wished to profess the view that the heavens were watery, fiery, earthy, or some combination of the four elements might do so freely.[310] If forced to concede that there was something elemental in the supralunary realm, then he would side with the Paracelsian doctrine, that the celestial bodies were fiery. But he did not concede it. He considered that the heavens were ethereal, and claimed that they enjoyed the same relationship to the Earth as the soul (*anima*) to the body – a guarded remark, not fully explained by Tycho elsewhere in his writings, but again strongly suggestive of the role that astro-alchemical doctrines played in his cosmology.[311] Turning to the related question of refraction, he introduced another piece of evidence derived from his alchemical studies. Distillation of water and of wine could produce two very pure liquids with undetectable differences in their refractive properties, and yet these two substances would exhibit very different characteristics in other respects. Thus, even if there were no apparent refraction due to a boundary between aether and air, the existence of two distinct materials was not disproved by this evidence.[312]

With respect to the question of world-systems, Tycho flatly refused to accept that either Copernicus or Rothmann had explained away what he took to be the manifest physical absurdities of positing a triple motion of the Earth.[313] He also upbraided the *mathematicus* for advancing the view that scriptural exegesis had nothing to contribute to cosmology.[314] Tycho repeatedly claimed, in these letters and elsewhere, that scripture was a key reason to favour his world-system over that of Copernicus.[315] Thus, although he conceded that the Bible was not, primarily, a work of natural philosophy, he took the view that it did not contain anything contrary to the truth. Moses, in Genesis, may not have revealed the 'innermost secrets' of astronomy, but neither had he had said anything to which astronomers themselves could not also assent. Nor were the statements of other biblical authors contrary to truths established through physics.[316] Tycho was of course not alone in claiming that the Copernican account was ruled out by scripture, properly understood; Melanchthon had also adopted this position, and the Catholic Church was officially to do so in the seventeenth century.[317] But it is difficult to know *precisely* what part Tycho's understanding of scripture played in persuading him to reject the Copernican world-system, since he never

---

[310] *TBOO* VI, 166.30–167.6.   [311] *TBOO* VI, 167.6–167.24.
[312] *TBOO* VI, 168.8–36.   [313] *TBOO* VI, 177.6–10.
[314] *TBOO* VI, 177.10–178.7.   [315] *TBOO* VI, 177.6–12, 186.14–20; *TBOO* IV, 156.3–157.5.
[316] *TBOO* VI, 177.14–19, 177.41–178.2.   [317] Blackwell 1991, 111–134; Howell 2002, 53–57.

cited specific biblical passages as evidence of the Earth's immobility. It is not clear, therefore, whether his conviction that natural philosophical reasoning showed it to be impossible was primary and informed his scriptural exegesis, or whether his reading of scripture was the more fundamental.[318] However, on the evidence of the later face-to-face conclusion of his debate with Rothmann (on which see Chapter 3), natural philosophical objections derived both from Aristotelian terrestrial physics and his alchemical cosmology were paramount in his conscious rejection of the mobility of the Earth. His chief arguments were theological, in the loose sense that they were shaped by his understanding of the providential construction of the cosmos, but they were not exegetical.

In the last of the substantial letters that Rothmann sent Tycho from Kassel, dated 22 August 1589, Rothmann responded by defending his own sophisticated and epistemological principles. Claiming that Tycho's reply was tantamount to an accusation of impiety, he pointed out that his exegetical technique was that employed by St Augustine of Hippo; and he wondered archly whether the Danish astronomer judged 'that when two men say the same thing, it is not the same'.[319] He attacked Tycho for the unsound use of authorities, and continued to argue, for his part, from personal experience and mathematical demonstration.[320] Thus, while both Tycho and Rothmann believed in an underlying cosmological order, one that legitimated an inferential move from observation and mathematics to philosophy, there was a significant difference in their implementation of this concept. Tycho adhered to certain *a priori* physical and metaphysical claims about the cosmos, which invalidated some possible outcomes, whereas Rothmann ostensibly rejected cosmological claims as truth-criteria, rather than products, of mathematical analysis.[321]

From Tycho's perspective, his correspondence with Kassel can be read as an attempt to force a convergence of observational data, instrumental technology, cosmology, and method. In each of these areas, Tycho endeavoured to persuade Wilhelm, and more particularly Wilhelm's *mathematicus*, to adopt the position that he favoured himself. Tycho also attempted to convince the Kassel practitioners to match the scope of his own astronomical enquiry, by utilising their fixed-star determinations to investigate the

---

[318] Cf. Howell 2002, 107–108.

[319] *TBOO* VI, 181.26–31. On Augustine in relation to the biblical exegesis of Copernicans, see Reeves 1991; Howell 2002, 28–32, 59–67; and the literature cited in n. 292.

[320] *TBOO* VI, 182.12–17.

[321] Of course, Rothmann retained certain fundamental beliefs that did constrain his cosmology: for example that the world was created by God, and that the heavenly motions were circular.

motions of the planets.[322] But the correspondence did not lead to the outcomes he hoped for. In 1590, Rothmann visited Uraniborg, and his debates
with Tycho continued in person. Thus, Tycho's epistolary calibration of
the Kassel observatory was ultimately frustrated: he failed to force, through
his letters, a reassessment of the Kassel observations that eliminated the
few minutes of arc difference in the stellar longitudes; he failed to restrict
the Kassel observing practices and data-handling techniques; he failed to
impose his understanding of refraction, or his distinction between the sub-
and supra-lunary; he failed to win support for his world-system; and he
failed to convince Rothmann that his combination of empirical, metaphysical, and theological argument, was indeed a valid one. This failure
was not predetermined, for Tycho could have proved more persuasive, or
the *mathematicus* less stubborn. But we might be tempted to conclude that,
at times, Tycho must have regretted the time and effort he expended on
corresponding with the Landgrave and Rothmann.

On the other hand, we have seen that with the letters flowed a great deal
of information, including observational data that provided a high level of
corroboration for the work of each observatory. And partial appropriation of ideas and techniques can be demonstrated: Tycho was prompted
to investigate the latitudinal alterations in stellar positions by Wilhelm
and Rothmann, and the latter were persuaded to try Tycho's method of
observing with Venus as a mediator. There is also circumstantial evidence
to suggest that Rothmann's cometary treatise assisted Tycho's development
and articulation of his new system of the world. Thus, while Tycho was
denied the gratification of seeing his project exported to the Hessen observatory *in toto*, an outcome that would have been the ultimate corroboration
of his techniques and agenda, he nevertheless profited from his exchanges
with the Kassel practitioners.

In addition, there was a value in these exchanges that was only partially
dependent upon the successful elimination of practical, theoretical, and
epistemological differences. Relating as it does to the reputation of the
participants, and their entitlement to credit for various inventions and
discoveries, this value is something of which we have already seen glimpses.
There is more to be said, however, about the way in which Tycho, in
particular, sought to capitalise on the time and effort he invested in his
communications with Kassel, and what role considerations of this kind
may have played in sustaining the correspondence. This is now what we
shall move on to consider.

---

[322] E.g. *TBOO* VI, 143.3–6, 143.10–13.

V INTELLECTUAL PROPERTY, CREDIT, AND THE
EXCHANGE OF GIFTS

We are also sending you the following: observations of some distinguished stars, as they were observed by our *mathematicus* both on the meridian and also through a sextant; but in complete confidence, where you should also keep them, and not disseminate them yet, but only compare them with your observations. (Wilhelm to Tycho, 14 April 1586)[323]

The observations you sent here we guard very carefully, and we do not show for inspection by others; and you should not fear their publication. (Rothmann to Tycho, 21 September 1587)[324]

Although a great deal of astronomical material was exchanged between Hven and Kassel, both Tycho and Wilhelm expressed a desire to retain control over that which issued from their respective observatories. On Tycho's part, this secrecy might seem to be in tension with his attempts to transform Kassel into a semi-autonomous centre of Tychonic astronomy. It is worth considering, therefore, what the reasons for requesting secrecy were, and how such pleas modify or complicate our understanding of the correspondence as a tool of calibration.

The requests of Tycho and Wilhelm, that material they exchanged should not be distributed or published, were made necessary by the prevalent scholarly habit of circulating copies of letters and other documents in manuscript. Given the inevitable delays in conducting a long-distance correspondence, it was only prudent – when time and resources allowed – to retain a copy of any letter sent to another, for one's own future reference. These copies could be used when missives went astray; when Rothmann visited Hven in 1590, he brought with him a letter to Tycho that had never reached the island.[325] But they could also be shown and sent to individuals other than their designated recipients, as happened both when Rantzau forwarded to Tycho a copy of the letter he had received from Wilhelm, and when Rothmann did the same with the Landgrave's letter urging him to make use of the opportunity for communication provided by Frischlin. In these instances, enclosing the letter merely added emphasis to the point made by the correspondents in their own epistles: namely that Wilhelm very much desired to hear again from Tycho. At other times, however, letters were copied expressly in order to make their contents more widely available.

On 4 November 1588, for example, Tycho wrote to Heinrich Brucaeus, responding to a letter in which he had thanked the Danish astronomer

---

[323] *TBOO* VI, 50.19–23.   [324] *TBOO* VI, 117.31–33.   [325] *TBOO* VI, 200.18–20.

for sending a copy of the *De recentioribus phaenomenis*, and had given his reactions to the new world-system it described.[326] With his reply, Tycho sent copies, or partial copies, of two other letters: one that Caspar Peucer, a former professor at Wittenberg, had written to Heinrich Rantzau after seeing the diagram of the Tychonic system contained in the book, and one that he himself had written to Peucer in response, dealing with both his comments and those made 'by a certain other astronomer on the same subject'.[327] Having already been informed that the geoheliocentric world-system was being made known to others by 'a certain runaway employee' of his,[328] Tycho took the trouble in this lengthy letter to Peucer to establish a detailed chronology for the genesis of his new arrangement of the planets: the very chronology, in fact, which emphasised the importance of Tycho's observations of Martian parallax in 1582 for the development of the geo-heliocentric world-system, and the importance of the study of cometary courses for his realisation that the heavens were fluid.[329] This account was something that Tycho was clearly keen to disseminate. A second copy of the letter was sent by him to Rantzau, and another found its way to at least one other member of the wider astronomical community, Michael Maestlin in Tübingen.[330] It is rather telling that Tycho chose to provide the fullest account of his astronomical programme, including his world-system, in a letter to Peucer, even if the occasion to do so was fortuitous. Peucer was not only the doyen of the Wittenberg astronomical tradition; he was also the son-in-law of Melanchthon, the defender of his posthumous reputation, and a prominent casualty of the ongoing dispute between the Philippists and the hard-line Gnesio-Lutherans.[331] By writing to Peucer, Tycho was once again signalling his debt to Philippist conceptions of the relationship between theology, cosmology, and mathematics.

In replying to Peucer, Tycho had attached copies of letters exchanged between himself, Wilhelm, and Rothmann, 'in which are relayed very many things pertaining more than a little to the whole topic of astronomy and the better institution of observations and the very restoration of this

[326] *TBOO* VII, 141–144, especially 142.4–13. For Brucaeus' letter, see *TBOO* VII, 147–164.

[327] *TBOO* VII, 148.20–30. Peucer's letter had of course been forwarded to Tycho by Rantzau; see *TBOO* VII, 127.23–34. The 'certain other astronomer' was Georg Rollenhagen (1542–1609), rector of the *gymnasium* at Magdeburg. See Rosen 1986, 20.

[328] *TBOO* VII, 135.38–42. This news had come via Georg Rollenhagen. The 'employee' was later to be identified as Ursus.

[329] *TBOO* VII, 127.34–131.13.

[330] For Rantzau, see *TBOO* VII, 126.6–9. Maestlin's acquaintance with the letter is noted by Gingerich and Voelkel 1998, 4–5, 28–29. The letter itself is discussed in part by Blair 1990, Howell 2002, 84–86.

[331] Westman 1975a; Kolb 1977; Kühne 1983.

art'.[332] Since Tycho pointed out his debates with Rothmann on refraction, and their relevance to determination of the true heavenly motions, as a noteworthy feature of the exchanges, we can be certain that these copies included the letters of January 1587, in which Tycho first mentioned his new world-system and the fluidity of the heavens.[333] Tycho exhibited a certain desire for discretion on the part of his new correspondent; he asked that Peucer read the letters through, consider them carefully, and then return them to Uraniborg without letting them be seen by anyone else.[334] Yet, contrary to the impression such a request might create, Peucer was only one of several of Tycho's contacts to be made a party to the Hven–Kassel exchanges. Rantzau, for example, was sent copies of the letters from as early as March 1586.[335] And whilst acting as Tycho's intermediary, he received some telling instructions regarding the parcel of letters that passed through his hands *en route* to Peucer in late 1588: he was to open it up and read, or copy, any items with which he was not already acquainted.[336] Hagecius, in Prague, was sent copies of the Hven–Kassel exchanges from July 1586 onwards, while Brucaeus thanked Tycho for sharing the correspondence with him in a letter of 28 October 1587.[337] In the case of both men, the letters to Wilhelm and Rothmann of January 1587 were amongst the materials that Tycho was anxious to communicate, so this wider correspondence clearly establishes a context for Tycho's careful treatment of his views on the nature of the heavens in the Kassel exchanges.[338] Long before they were published – indeed while they were still part of an ongoing correspondence – the Hven–Kassel letters were being used by Tycho to shape his reputation as an astronomer and establish his claim to certain discoveries.

Rothmann was not informed in advance that his letters to Tycho were being shared with an audience. He might never have learnt that the correspondence had been seen by Peucer if his brother had not visited Tycho at Uraniborg. 'I know also from the letters of my brother Johannes', he wrote on 27 July 1589, 'that you have sent our disputations to Master Lord Peucer; and that Master Peucer advances different conclusions, not wishing to support either of us'.[339] And he was seemingly unaware of Tycho's wider

---

[332] *TBOO* VII, 140.1–9. [333] *TBOO* VII, 140.9–12. [334] *TBOO* VII, 140.12–16.

[335] *TBOO* VII, 101.37–102.7. [336] *TBOO* VII, 126.14–19.

[337] *TBOO* VII, 108.15–23, 113.22–40, 115.4–23.

[338] Tycho indicated in his letter to Hagecius of 25 January 1587 that only the lack of time prevented him from enclosing a copy of his most recent letter to Rothmann along with the one sent to the Landgrave. That Brucaeus received the January 1587 letters is evident from his comments on the discussion of the altered obliquity of the ecliptic. See *TBOO* VII, 113.33–40, 115.4–10.

[339] *TBOO* VI, 201.39–202.1. On Johannes Rothmann, see *TBOO* VIII, 462; Leopold 1986, 23. His visit to Hven in 1589 is recorded in Tycho's weather diary. See *TBOO* IX, 74.

distribution of the Hven–Kassel exchanges prior to being apprised of the decision to put the correspondence into print.[340] But although he expressed disquiet at the thought of his letters being published, and asked Tycho not to discuss his cometary treatise in print until it had itself been finished and printed, he does not seem to have reproached his Danish correspondent for circulating material in manuscript.[341] His objections to Peucer's involvement seem to have derived from a concern about being judged for his opinion on matters concerning which he was not finally decided, or on theological and philosophical grounds in matters that were, as he saw it, a question of mathematics – rather than from any perception of unfairness in Tycho's epistolary practice.[342]

In fact, Tycho's behaviour with respect to the Hven–Kassel correspondence and the exchanges with Peucer was conventional in two senses, even if the reasons for distributing these letters were rather particular. First, their circulation was quite consistent with Tycho's practice on other occasions. Between 1589 and 1594, for example, Tycho was involved in a bitter epistolary dispute with John Craig, who disagreed with his supralunary placement of comets, that saw him send copies of his and Craig's exchanges to correspondents such as Brucaeus and Hagecius.[343] The purpose of doing so was – just as Rothmann suspected to be Tycho's motivation for forwarding the Hven–Kassel exchanges to Peucer – in order to enlist other scholars as arbiters and allies in what had become an increasingly intractable debate.[344] More generally, Tycho's circulation of letters was consistent with the values and practices of his scholarly contemporaries. Letters were not considered intrinsically private communications in the early modern period, and it was not assumed that the author retained legal or moral rights over the dissemination of their contents.[345] Thus Tycho was also the recipient of a number of manuscript letters treating astronomical matters, forwarded to him by individuals who rightly supposed that he might find them of interest.[346]

---

[340] *TBOO* VI, 190.22–37.   [341] *TBOO* VI, 28.6–12, 182.19–30.   [342] *TBOO* VI, 182.19–30, 202.1–4.

[343] Mosley 2002b.

[344] *TBOO* VI, 193.25–194.5. Some other examples of Tycho's use of letters are discussed in Jardine 1988a, 23–25.

[345] On the status of letters and the problem of authorial control over the circulation of material in manuscript, see Root 1913, 420–421; Saunders 1951, 515–516, 518–520; Braunmuller 1981; Love 1998, 43. The emergence of the modern legal notion of a letter, according to which the document is the property of the recipient, but the copyright in the text belongs to the sender, is discussed by Ransom 1951.

[346] E.g. the exchanges between Peucer and Wilhelm reproduced in the *Progymnasmata*; see *TBOO* III, 114–115, 120–123, 127–129.

While this explains why it was necessary to specify when, for whatever reason, one did not want material included in a letter to be passed on to others, it is only a partial explanation for the behaviour of Tycho and his Kassel correspondents. Why did they not want their astronomical work to circulate freely? The Landgrave's concern over the Kassel data can perhaps be attributed to a desire not to be responsible for inaccurate or unreliable observations. The correspondence with Tycho was, after all, partly concerned with establishing the validity of his observatory's findings. But his avowed intention was to publish a star catalogue as soon as Rothmann had completed the necessary calculations. Thus, like the disquiet Rothmann exhibited at the possible publication of a Tychonic critique of his manuscript cometary treatise, Wilhelm's concern was clearly with *premature* distribution of the stellar data, not with secrecy for its own sake. A concomitant cause of anxiety may have been the thought, clearly expressed by Tycho, if not by the Landgrave, that material of the sort transmitted through their epistolary exchanges could be appropriated by others. Thus, as he put it in his letter of 18 January 1587:

I ask of your Highness, by whatever entreaties I can, that he should not wish to share these observations of ours with anyone, and to take care that they are not made public, especially as regards the longitudes and latitudes of the fixed stars; which, for this reason, I have had written down separately on their own sheet of paper. For your Highness knows as well as anyone how much time, effort, care and expense is required, before it is permitted to attain accurate knowledge in this matter. And I have certainly shared these findings of mine with no one else in the whole of Germany, or elsewhere, of whatever status, and I have suppressed these up to now, although urged by many for this reason. For since many men do not wish to produce anything honourable in these matters by their own efforts, they do not blush to honour themselves by the findings and efforts of others, and to hawk them as their own, with serious injury to the author. Indeed, I do not doubt that your Highness will keep this to himself alone; and in the same way I shall make no one privy to the observations your Highness is going to communicate to me.[347]

By copying the observations separately, Tycho provided for the possibility that Wilhelm would want to circulate the text of his letter, without at the same time exposing his data to the eyes of the unscrupulous. He thereby sought to retain some control over what he quite clearly perceived to be his intellectual property.

It is noteworthy that Tycho did not merely explain his concerns, and express the hope that Wilhelm would respect and understand them. He also

[347] *TBOO* VI, 74.17–31.

represented it as an act of courtesy to allow an astronomer to retain control over his own productions, and suggested that an intimate bond had been recognised by his exchange of observations with Wilhelm. This approach was one characteristic of patronage relationships in the early modern period. Renaissance princes and nobles maintained relations with their clients by bestowing gifts and favours. While these supplemented a regular salary, or substituted for one, they also invited reciprocation, driving a cycle of mutual obligation in which each party rewarded the other in a way commensurate with their abilities, seeking to discharge, partially if not absolutely, the debts that had accrued. The unwritten contract of patronage was thus constituted by the exchange of services and gifts; indeed, a patron–client relationship might well begin with the patron accepting a gift that he or she had not solicited. And once established, the negotiation and maintenance of the relationship proceeded through certain linguistic conventions, including an exaggerated courtesy and, over time, declarations of friendship and intimacy.[348] This is the pattern that can be seen in the exchanges between the Landgrave and Tycho.

Patronage operated in a similar manner in scholarly contexts. This is unsurprising, perhaps, given that many scholars – along with poets, artists, and artisans – received essential support from nobles and princes. Such individuals shared an ability to render their patrons famous, and collectively they promoted the exhibition of princely virtue through direct and indirect cultivation of scholarship, the arts, and civic beneficence, thereby supplementing and to some extent supplanting traditional celebrations of martial prowess and conquest.[349] Indeed, emphasis on peaceful routes to the immortality of fame provided scholars themselves with a way to disengage from courts, and to relate to one another as clients and patrons.[350] But successful propagation of these ideals also allowed princes and nobles, such as Wilhelm and Tycho, to fashion themselves as aristocratic participants in the pursuit of knowledge, and to claim an enhanced reputation as their reward. To the extent that this prospect motivated astronomical investigations, it encouraged practitioners to ensure that their work was publicly visible. Secrecy, therefore, could only ever be a temporary measure.

Tycho and Wilhelm were each engaged in multiple patronage relationships. Wilhelm, as a ruling German prince, was both a client of the Holy

[348] See Kettering 1988; Neuschel 1989, 1–37, 69–102. For the anthropology of gift-exchange, see Mauss 1990.
[349] The relationships are complex, and the literature large. For a sample, see Anglo 1990; Hale 1994, 392–413; Dewald 1996, 33–40; Jardine 1996, 183–221; Raggio and Wilmering 1996, 5–35.
[350] Findlen 1994, 365–376.

Roman Emperor, and his own domain's chief patron. He exploited existing patronage ties to pursue his scholarly interests, which included natural
history and alchemy as well as astronomy, but he also constructed them
de novo.[351] Tycho was himself a patron of those students and scholars who
worked at Uraniborg.[352] Yet he was also a client of the Danish crown, both
as a member of the hereditary nobility, and as the fief-holder of Hven,
granted to him particularly so that he could pursue his investigation of the
heavens.[353] In return, he was expected to provide his patron with certain
astronomical productions, including horoscopes and astrological interpretations of celestial portents.[354] Tycho's obligations to the Danish crown
indicate that astronomical documents could be appropriate elements in
the management of a patron–client relationship. But whereas Frederick II,
and later his son Christian IV, would not have attributed value to every
form of astronomical material, Wilhelm's own interest in the art of astronomy allowed him to receive a wider range of Tychonic productions with
favour. Thus we can consider the observations and other documents transferred between Hven and Kassel in the course of the correspondence, even
the letters themselves, as gifts: tokens by which the relationship between
these two aristocratic scholars was constructed and mediated.[355]

The gift-value of astronomical work has implications for the analysis of
those very situations in which Tycho's authority over his astronomical productions was challenged and contested. The incident of alleged plagiarism
involving Wittich and the design of his instruments, and perhaps also that
of Ursus and the geoheliocentric world-system, may have contributed to
the fear, expressed in his 1587 letter to Wilhelm, that observations might
fall into the hands of unscrupulous men who would not hesitate to 'hawk
them' improperly.[356] Elsewhere, he certainly suggested that Wittich was
motivated by greed, writing that:

he imitated, although imperfectly, certain features of the *rimulae* of the *pinnacidia*,
especially suited to the nocturnal observations of the stars (which invention, when
he first saw it here, he declared that he had not come to Denmark in vain, and
thought better than a great deal of gold) . . . A certain member of my household,

[351] Moran 1978, 76–97, 111–128.
[352] Dreyer 1890, 117–127; Thoren 1985; Shackelford 1991, 102–105.
[353] Dreyer 1890, 85–87, 109–113; Thoren 1990, 103–104; Christianson 2000, *passim*.
[354] Christianson 1979.
[355] For gift-giving in relation to the sciences, see Findlen 1991, 1994, 346–392; Biagioli 1993, 36–73; Jardine 1998, 53–58. On letters as gifts, see Ransom 1951, 122; Constable 1976, 16; Jardine 1996, 238.
[356] Tycho apparently first wrote of Wittich retailing his secrets to others in 1581, shortly after he had left Hven, but many years before his fears were confirmed by seeing the letter of Wilhelm to Rantzau. See *TBOO* VII, 62.35–63.8.

who was here at that time, reports that he heard him say, that when he imitated the construction of our instruments, he would accomplish the same thing with fewer *rimulae*, and he would keep this saving to himself.[357]

As Tycho had learnt from Rothmann and Jacobi, it was this 'saving' of labour with respect to the sighting slits that had necessitated the reconfiguration of the Kassel instruments after Wittich's botched attempt at their improvement. The import of the statement, therefore, was that the pursuit of riches, rather than renown, led directly to bad practice. Tycho's own relationship with money was troubled; precisely, indeed, because he depended upon patronage to supplement his inherited income, and seemed unwilling, perhaps for reasons of honour, to realise the commercial value of his alchemical and astronomical activities.[358] For someone to appropriate part of his astronomical work, and to sell or give it to a prince like Wilhelm, was to deprive him not only of public recognition, but also of a potentially valuable gift to a patron. For this reason, Tycho probably viewed such plagiarism as the theft of his property in a very concrete sense. While the damage could be undone, and the public acknowledgement of his *inventiones* and their gift-value recovered, this in itself required the expenditure of significant effort. In the episodes involving Wittich and Ursus, much of that labour was devoted to the writing of letters.

This account best explains Tycho's apparently contradictory, but, I believe, sincere, combination of displeasure with Wittich for conveying his design principles to Kassel, and delight that Rothmann and Wilhelm had access to instruments as accurate as those he employed at Uraniborg. And it is supported by certain features of the later correspondence. Tycho had received no letters from Wilhelm for almost three years by the time that he finally wrote on 15 May 1590. In that letter, the Landgrave asked Tycho to allow Rothmann to visit Hven, and inspect his instruments. But he also raised non-astronomical concerns: 'we are waiting to learn as well, how things stand just now, for you, on the one hand, and also for your gracious master, the young King, and the whole government of the Danish Kingdom'.[359] Similar requests for political news were made in subsequent letters of 3 February and 31 August 1591.[360] In the meantime, Rothmann had come to Uraniborg in August 1590; and, having departed after a stay

---

[357] *TBOO* VI, 90.1–7, 90.10–13. Cf. Tycho's remarks in the *Mechanica*, *TBOO* V, 25.32–39.

[358] In general, a tension existed between the courtly and the commercial spheres in early modern Europe. However, local circumstances varied and the extent to which aristocrats engaged with the market changed over time. See Kellenbenz 1954; Petersen 1968; Davis 1961; Midelfort 1992; Dewald 1996, 93–97; Kamen 2000, 77–81.

[359] *TBOO* VI, 213.1–4.   [360] *TBOO* VI, 225.18–23, 241.2–3.

of one month, had failed to return to his post in Kassel. In fact, although he informed neither Tycho nor Wilhelm of his decision, Rothmann, suffering from ill-health due to his contraction of syphilis, had decided to give up astronomy, and instead had made his way back to his home town of Bernburg.[361] As a consequence of his disappearance, the astronomical content of the letters quickly diminished. Tycho made one last great effort in September 1591 when, realising that Rothmann's failure to return to Kassel meant that Wilhelm still lacked a full description of his observatory, he had an amanuensis compile the *Synopsis*.[362] This consisted of a detailed account of his instruments, including their decorations and poetical adornments, plus a description and plans of Uraniborg, and a map of his island.[363] Much of the material was later repackaged as the *Mechanica* of 1598, a book that Tycho employed as a gift during the patronage crisis precipitated by his departure from Denmark. But while they may have been well received by the Landgrave, the descriptions sent to Wilhelm did not stimulate further astronomical exchanges.

Near the end of Wilhelm's life, Tycho also began to write to Moritz, Wilhelm's son and expected successor. In his first letter to the future Landgrave, he requested a commendatory poem for his forthcoming book, explaining that:

I consider that this will by no means be dishonourable for your gracious Prince; for I have decided to print it on the second page of my first book (three copies of the first quire of which I have attached here). If, therefore, your gracious Prince will deign to write the said poem on the page that is blank, next to the Imperial Privilege, with his own hand, and send one of them back to me, then I could soon have it printed following the other one (which I likewise await daily from the most serene King of Scotland, who is also a very good poet; and which I shall place on the first page, following the portrait, since he is a King, and joined with our Royal Family by marriage).[364]

Citing Moritz's great reputation as a poet, Tycho stated that, given his father's approval of Tycho's endeavours and own efforts to restore the arts, it would be highly appropriate for him to write a verse on the subject of astronomy.[365] He sent a poem of his own, in praise of Moritz, to supplement his letter and add weight to his appeal. The young prince modestly replied that his poetic talents had been exaggerated, but nevertheless composed some verses for Tycho.[366] The astronomer gratefully commented, in his return letter, that:

[361] Granada, Hamel and von Mackensen 2003, 13–14; Barker 2004.   [362] *TBOO* VI, 246.12–23.
[363] *TBOO* VI, 250–295.   [364] *TBOO* VI, 307.32–308.3.   [365] *TBOO* VI, 308.5–9.
[366] *TBOO* VI, 311.14–17, 312.26–313.12.

Those verses will certainly show a most glorious Τεκμηριον [token] of the favour and goodwill of your Highness towards such sublime arts and also to me, consecrating an everlasting memorial to Posterity at the beginning of my astronomical works: but for my part it will offer μνημοσυνον [a memorial] of my devoted regard for your Highness, that will never perish through forgetfulness.[367]

What is clearly revealed in this passage is Tycho's attempt to establish a new patronage relationship with the Hessen court. Both the status differential, and the nature of the transaction are evident, unobscured by the intellectual excitement of the astronomical discussions. The developing relationship is mediated by an exchange not of observations, but of verses. Tycho asks to be allowed to add the endorsement of the prince to the publication of his work; and in return offers, at one and the same time, to display his patron's discrimination and virtue for all of posterity. So that Moritz can be sure that he *is* being discriminating, he holds out the prospect of proximity to an Imperial privilege and another poem, one written by King James VI of Scotland. The exchange allows both patron and client to accumulate status, and to mark that increase with tangible tokens: letters and handwritten poems, in the first instance, with the promise of a printed text to follow.

A bulkier, but otherwise similar item in the exchanges between Hven and Kassel was referred to in one of the last of the letters. Writing on 21 September 1592, Tycho described to Moritz how he had been attempting for some time to acquire certain animals for the zoological collection of Wilhelm's *Lustgarten*.[368] Wilhelm had first sought assistance from Tycho regarding a Norwegian animal that he believed could be found in the royal zoological gardens: called a *rix*, it was described as taller than a stag, with slender horns, but Wilhelm could find no account of it in any of the *historia animalium* with which he was familiar. Accordingly, he asked in February 1591 that Tycho supply him with a picture and a natural history of the beast.[369] This request caused some confusion at Uraniborg: Tycho could identify no creature by that name held either in Copenhagen or the gardens at Fredericksborg, but there had been some Norwegian reindeer in the city the previous year, and Tycho thought it was likely that these were the beasts of which Wilhelm had heard.[370] Aided by his Norwegian assistant Aslachus, Tycho sent Wilhelm a description of reindeer and their character in April 1591, and promised to let him know if he learnt anything more. At the same time, he mentioned that he himself possessed an elk, which he would happily send to the Landgrave if he desired it.[371]

[367] *TBOO* VI, 313.26–31.   [368] *TBOO* VI, 337.18–28.   [369] *TBOO* VI, 225.23–33.
[370] *TBOO* VI, 230.12–28.   [371] *TBOO* VI, 230.28–40.

It was elk that found a place in the last of the Hven–Kassel exchanges, although there was some further discussion of both reindeer and *rix* in the preceding correspondence. The name *rix* was eventually associated by the two correspondents with the oryx, an animal that was depicted and described in the *Historia animalium* of Conrad Gesner (1516–1565), if not very helpfully; although an African animal, Wilhelm still wished to know of Tycho whether he had seen one or knew where one could be found.[372] Wilhelm had already seen and kept reindeer, but had discovered that they could not survive the summer's heat of so southerly a latitude.[373] And he also owned an elk; an animal that 'prospers, runs about, dances, and is of good cheer, and especially when we ride around our garden of animals in our green carriage, runs as close as it can to us, and attends us just like a dog'.[374] But the Landgrave was keen to obtain a companion for this animal, particularly a breeding partner, and therefore accepted the offer that Tycho had made to send him another.[375] As it turned out, however, the elk that Tycho owned died on its way to Uraniborg from his Scanian estates; at the castle at Landskrona, it imbibed a quantity of beer, became inebriated, and fell down some steps, fatally breaking a leg.[376] The process of obtaining more elk was protracted and difficult, so that by the time a Hessen retainer sent to Tycho to accomplish the task had returned from Norway with a pair of the animals, the astronomer had received news of the Landgrave's demise.[377] Tycho decided nevertheless to send the beasts to Moritz, in memory of his father.[378]

Comprised as it is of apparent absurdities – the companionable elk, the drunken one, and the elk despatched *in memoriam* – it is tempting to see the saga of these animals as an amusing but essentially irrelevant episode in the history of the relations between Tycho and the Landgraves of Kassel. Nevertheless, the story adds something of weight to arguments about the nature of the Hven–Kassel exchanges. Elk were considered to be of particular significance in the courtly culture of gift-giving. Thus Edward Topsell, in his *Four-footed Beasts* of 1607, wrote that:

In *Swedia* and *Riga* they are tamed and put into Coaches or Charriottes to draw men through great snowes, and upon the yce in the winter time they are also most swifte, and will run more miles in one day, then a Horsse can at three. They were wont to be presents for princes, because of their singular strength and swiftnes, for which cause *Alciatius* relateth in an emblemme the answer of Alexander to one

---

[372] *TBOO* VI, 240.22–28, 244.36–41. See Gesner 1563, 68v.   [373] *TBOO* VI, 231.32–38.
[374] *TBOO* VI, 231.38–232.4.   [375] *TBOO* VI, 232.7–11 and 240.6–17.   [376] *TBOO* VI, 238.22–30.
[377] *TBOO* VI, 244.15–35, 296.11–19, 303.27–304.9, 305.7–306.25, 337.18–338.25.
[378] *TBOO* VI, 338.2–4 and 339.17–25.

that asked him a question about celerity; whether hast doth not alway make wast: which *Alexander* denied by the example of the Elke . . .[379]

Topsell's work was based principally on Gesner's *Historiae animalium*, the work that both Wilhelm and Tycho had studied (in German editions) in order to identify, and distinguish between, elk, reindeer, and oryxes.[380] His description of the animal closely followed Gesner's, which also made reference to the custom of making elks into gifts and to the Alciati emblem regarding their swiftness.[381] Both Wilhelm and Tycho likely perceived, therefore, that to succeed in furnishing the Landgrave with an elk would mark Tycho out as a man of great service to his patron. It was an act that, even after Wilhelm's death, had the potential to increase the astronomer's standing.

Tycho's printing of the correspondence in 1596 can be read, at one level, as an attempt to realise the value of this, and other services to Wilhelm, through their publicisation. The strategy unifying all pursuit of posterior fame, whether as scholar, client, or patron, was that of display.[382] For without embodiment, whether in text, artefact, or specimen, knowledge and virtue eluded recognition. Both Wilhelm and Tycho were aiming to produce lasting monuments through their life's work. But in the process of achieving this objective, one form of display could be traded off against another. Thus, while some gifts had value because of their intrinsic material and artisanal nature, some because they manifested the intangible, and some because they could contribute to future productions of both kinds, all embodied the act of giving, and the act of receiving. This was an aspect of a gift that might be displayed when the given thing was itself to be kept hidden, or when it no longer existed to be shown to an audience. A public value could thereby be imparted to the exchange of observations that were not themselves to be made public. Perhaps contrary to expectation, therefore, the culture of early modern patronage emerges from this analysis as a sophisticated instrument for licensing, and controlling, the distribution of property, intellectual and material.

---

[379] Topsell 1607, 213. For the emblem referred to, see Henkel and Schöne 1996, 466–467.

[380] *TBOO* VI, 231.24–30, 234.40–235.2, 244.36–41. Wilhelm specified that he was referring to the German Gesner published at Zurich in 1563; Tycho divulged that his edition was German but not identical to Wilhelm's.

[381] Gesner 1551–1558, I, 3.33–37: 'Alces vivae captae principibus aliquando inter munera mittuntur. Elegans & argutum est Andreae Alciati epigramma . . .'; Cf. Gesner 1563, 40r: 'verschenckt mans gmeinlich grossen herren an die hoef als ein sondere vereerung.' On the difference between Topsell's work and Gesner's, see Ashworth 1990, 316.

[382] For some pertinent remarks on this topic, see Eamon 1991.

These claims should not be understood to indicate that scholarly activity was necessarily parasitic on a culture of courtly display and princely extravagance. Patronage was essentially a strategy of management, and the accumulation of honour was pursued alongside other objectives. A court like Wilhelm's was a centre for the construction of a conspicuous demonstration of princely virtue that, because of his interest in natural philosophical, medical, and astronomical endeavours, depended upon a network of clients and transactions capable of mobilising specimens, theories, artefacts, and data. Practical benefits were concomitant with this acquisition of objects, and with the construction of the paths of communication that enabled the accumulation of them. A herbarium or laboratory could be a source of medicaments and commodities;[383] a correspondent who provided specimens or scholarly expertise might also supply political intelligence.[384] Furthermore, interpreting the Hven–Kassel correspondence as exchanges constitutive of a patron–client relationship in no way contradicts the earlier reading of them as a means of overcoming the practical difficulties of constructing and validating universal astronomy from a single location. Indeed, it helps to explain how Tycho could attempt to reproduce his entire astronomy at the Landgrave's court; and anticipate that, in doing so, he would achieve a conspicuous form of valuable corroboration. For, as objects legitimated within the culture of display, letters that were a necessary adjunct to astronomical observation also served to represent one's own achievements and virtues. In both these ways, epistolary communication was an important part of the process through which Tycho and Wilhelm achieved their shared goal of a contemporary fame that would endure for many generations.

VI CONCLUSION

But a little while after I had finally returned to my homeland, at about the age of 28, I quietly made preparations for another longer journey. For I had made up my mind to settle in the town of Basle or thereabouts . . . For I liked that place more

---

[383] Moran 1978, 72–74, 115–124.

[384] And vice versa: Moran 1978, 76–102, demonstrates how specimens were obtained from contacts whose primary role was political, as well as from correspondents cultivated for this purpose. Diplomacy, trade, and scholarship all benefited from the construction and maintenance of correspondence networks; the practical difficulties of epistolary communication acted as a disincentive, certainly for a prince like Wilhelm, to develop lines of communication that were not exploited in whatever ways possible. At many courts across Europe, diplomats were used in the collection of art; and the use of *Kunst-* and *Wunderkammern* as arenas for the display of both artefacts and natural specimens suggests that it is likely other objects were moved through such channels. See Ami, de Michele and Morandotti 1985, 25; Gutfleisch and Menzhausen 1989; Fucikova 1997a, 13, 23–24.

than other regions of Germany, partly on account of the famous University and the remarkably learned men there, partly because of the healthy climate and the agreeable living; and because Basle is located as if at the point where Italy, France, and Germany, the three biggest countries in Europe meet; so that it would be possible by correspondence to form friendships with many illustrious and learned men in different places, and circulate my findings for public use that much more widely. (Tycho Brahe, *Mechanica* (1598))[385]

With the death of Wilhelm in 1592, Tycho's correspondence with Kassel all but ended completely. Despite Tycho's attempts to encourage him, Moritz was much more interested in pursuing alchemical investigations than observing the heavens, and seems not to have seen Tycho as a useful contact for this purpose.[386] A single letter-exchange occurred between Tycho and Rothmann at the end of 1594 and the beginning of 1595, when the latter, to Tycho's surprise, announced his continued existence, and enquired after the progress of Tycho's intended publications.[387] But Rothmann was writing from his new home in Bernburg; and almost nothing was subsequently heard from him again.[388] Of course, Tycho also possessed other correspondents, and wrote many more letters before his life ended. Nevertheless, when he published his *Mechanica*, and made reference to the desirability of epistolary communications, it seems clear that he was thinking, in part at least, of the letters he had exchanged with the Kassel astronomers. Indeed, whether Tycho had truly appreciated the value of corresponding with other scholars in his youth remains somewhat uncertain: there are few very letters extant from before the 1580s. But the letters that I have considered in this chapter, and which Tycho had himself thought worthy of publication only two years prior to the *Mechanica*, amply illustrate the complex role that correspondence could play in the communication of early modern astronomy.

What I have presented in this chapter is a reading that may provide a model for investigating the other letters sent and received by Tycho, and indeed the correspondences of Clavius, Kepler, Galileo, and other practitioners, and of interested patrons of astronomy such as Wilhelm and Rantzau. As I have argued, the conduct of correspondence was not unconstrained; nor was communicating by letter a casual exercise. In addition to facing considerable practical difficulties in communicating at a

---

[385] *TBOO* V, 108.26–36. On Basle as an intellectual centre and focal point of scholarly exchange, see Bietenholtz 1971.

[386] See Moran 1985.   [387] *TBOO* VI, 314.11–315.26, esp. 314.27–30, 315.337.

[388] For evidence that Tycho later heard of him indirectly, via Rollenhagen, see Chapter 3. Rothmann was in contact with Moritz in March 1597; see Granada 2002, 311–312.

distance, individuals were sensitive to epistolary conventions learnt during their early education. And because letters, even those exchanged between scholars, were seen as flexible and sophisticated communications whose circulation could contribute to the fashioning of the public-self of their senders and recipients, they were often composed with great deliberation and care.

Epistolary culture presented certain difficulties to individuals who sought to maintain control over the distribution of their findings and inventions. Yet letters were also a principal means of asserting one's rights to an item of intellectual property, and thereby of defining one's place for posterity. Furthermore, the fact that letters had the potential to lead to unintentional disclosures is merely the obverse of their ability to mediate the transfer of objects of value: astronomical observations, instrument descriptions, maps, pictures, poems, and even zoological specimens. In the case studied, these objects were considered worth exchanging and controlling, not so much for immediate financial gain, but because of the courtly culture of gift-giving and display that helped to license the sending of letters, and even the support of astronomy. To underestimate the significance of attributions of virtue and fame in the everyday pursuit and representation of astronomical endeavours is to risk mistaking the desire to publish in appropriate form, and at an appropriate juncture, for the sign of an intrinsically secretive nature. Moreover, courtly culture was itself multi-valent, being as much an institutionalised form of behaviour for the management of practical and political objectives as pomp and affectation: each, one could say, was a by-product of the other. Social structures not only dictated certain features of astronomical correspondence; they were also, through imbuing such communications with additional significance, enabling of this particular epistolary practice.

In discussing the contents of the letters exchanged by Tycho, Wilhelm and Rothmann, I have attempted to make clear two points in particular. The first is that, following from their use as a means of communication, these letters are a valuable source of information about the practice of astronomy and the development of theory at both Kassel and Hven. To this, I add only that the more we understand of early modern epistolary culture, the better equipped we are to recover and interpret such material. My second claim, however, is that the correspondence must be considered as something more than a diary, or set of diaries, in the form of a sequence of letters. The letters are not incidentally instructive about the astronomical activities and cosmological beliefs of Rothmann and Tycho, but are actually constitutive of a form of astronomical practice. Communication by letter was one way for astronomers to overcome the contingent obstacles that prevented observation of phenomena at one particular location. Moreover, for individuals pursuing the project of restoring or correcting the astronomy of the ancients, the comparison of observations produced at different sites provided the principal means of confirming the absence of local impediments to a universal science. And indeed, Tycho, Rothmann, and Wilhelm found, in the close agreement of their cometary observations, welcome evidence that their instruments, observing techniques, and reduction methods, produced results of a much higher accuracy than those obtained by their predecessors and contemporaries.

As we have seen, the exchanges between the two sites were partly collaborative, partly adversarial. Both sides were eager to facilitate the movement of observations in order to accommodate the comparative evaluation of their data. But the small discrepancies that were found called into question, over a period of time, instrument-construction and alignment, observing methods, techniques of recording and retrieving data, and corrections due to atmospheric refraction. Tycho criticised each of these areas of the Hessen practice, pressing Rothmann to make alterations that would lead to the duplication, within the limits of the human senses, of his own determinations. And, as the debate over data developed, his efforts to calibrate the Kassel observatory to his own standard extended to the determination of theories of refraction, the conception of celestial matter, the true system of the world, the principles of scriptural exegesis, and indeed the scope of the Hessen observational project. Whether this is to be taken as evidence of supreme confidence about his own labours or, on the contrary, a fundamental insecurity, will probably depend upon each individual's intuitions about Tycho's psychology. It is plausible, however, that Tycho consciously recognised that a slight discrepancy provided more of a challenge to his claims to have founded a new world-system on precise observations than data whose clear divergence from his own would have allowed it to be dispensed with more readily. And that issues of credit conditioned Tycho's conduct of the correspondence from the very earliest stages has, I think, been clearly established.

Rothmann's resistance to Tycho's authoritarian attempts at calibration was clearly a source of frustration. It stemmed, in part, from different epistemological priorities: both Wilhelm and Rothmann expressed more concern with the credibility of their star catalogue, which they held to be improved by the presence of slight differences from Tycho's, than with the convergence of the two observatories on a single set of values. But Rothmann's stance also indicates the depth of his own natural philosophical thought, and his commitment to a form of Copernicanism. By not acquiescing, Rothmann engaged Tycho in a debate about technical, physical, and theological issues in which both were stimulated to articulate their opinions and develop their arguments. Although both drew on materials prepared for publication, they were not, I think, simply rehearsing claims made on other occasions. One consequence of this is that the collected correspondence is somewhat reminiscent of a didactic dialogue or debate, in which Wilhelm features as a somewhat compromised and generally remote arbiter of the proceedings.[389] I shall argue, in the next chapter, that this resemblance was not lost on Tycho himself. Letters were, after all, considered to be, 'conversations between absent friends'. As such, they might profitably be 'overheard' by hitherto unknown members, as well as existing acquaintances, of the wider international astronomical community.

---

[389] From time to time Tycho appealed to Wilhelm as a judge, although only indirectly in his disputes with Rothmann. See, for example, *TBOO* VI, 130.15–24. On princely disdain for taking a position on epistemological matters see Biagioli 1993, 159–209, esp. 208–209. Such disinterestedness, however, was perhaps less a problem with Wilhelm than with his heir Moritz, to whom both Tycho and Ursus appealed by means of dedications in their printed volumes; see, on this, Chapter 3.

CHAPTER 3

Books and the heavens

After the birth of printing, books become widespread. Hence everyone
throughout Europe devoted himself to the study of literature ... Every
year, especially since 1563, the number of writings published in every
field is greater than all those produced in the past thousand years.
Through them there has today been created a new theology and a
new jurisprudence; the Paracelsians have created medicine anew and
the Copernicans have created astronomy anew.

Johannes Kepler, *De stella nova* (1606)[1]

The importance of manuscript letters to the practice of sixteenth-century
astronomy may have eluded the notice of some historians of science, but the
same can hardly be said with respect to texts that were printed. Explanations
for this difference are not difficult to find. Professionally obliged to publish
themselves, historians have long been inclined to view a printed page as the
natural outcome of intellectual activity, and hence as a privileged medium
for the articulation and dissemination of ideas. And even though the history of science has increasingly moved away from its origins in the history
of great thinkers and great thoughts, the effect of recent historiographical
currents has been to multiply the number of ways in which printed matter
is studied: there has therefore been a net gain for the history of the book,
despite the decline in the attention paid to standard editions of canonical
works.[2] Of course, it would be easy to exaggerate the extent to which the
adoption of new methodologies and new objectives has been responsible for
shaping the priorities of individual historians; it is likely, after all, that prag-
matic choices, made in response to the constraints imposed by institutional
structures and local resources, have proved just as important. And books are,
after all, typically easier to work with than manuscripts: both easier to locate,
and easier to read. Yet however one accounts for it, the outcome is the same:

[1] *KGW* I, 330–332, as translated in Jardine 1988a, 277–278.
[2] See Chartier 1988, 1989; Tanselle 1995; Johns 1998a; Frasca-Spada and Jardine 2000, 1–12.

anyone who sets out to explore the role of books in the culture and communication of early modern science seems less likely to find new land, than to end up crossing and recrossing territory that has already been charted.

Unsurprisingly, given its long-standing prominence in the currently unfashionable grand narrative of the Scientific Revolution, this applies at least as much to the subject of astronomy as it does to any other branch of early modern learning. Contributions to a history of early modern astronomical books have ranged from the broad coverage supplied by national and international bibliographic surveys,[3] to detailed analyses of the composition and reception of those texts (most notably Copernicus' *De revolutionibus*) that are especially celebrated.[4] Some works have received the honour of a facsimile edition, occasionally one produced from the extant copy of a notable owner;[5] or else, a greater blessing in the eyes of many, an annotated translation.[6] Others, and certain outstanding collections, have been displayed in exhibitions and their accompanying catalogues, thereby making evident the great variety in the form and function of such texts.[7] And with the aid of contemporary inventories, and scattered survivals, some scholars have made considerable progress in reconstructing the reading practices of astronomers, and establishing the contents of their personal libraries.[8] In these ways and others, therefore, evidence regarding the overlap between the history of the book and that of astronomy has been brought into view. The cumulative effect of this labour is impressive enough that it would not seem unreasonable to proceed by calling for more of the same kinds of scholarship.

Over and above the consideration granted to individual books, readers of books, and collections of books, however, historians of astronomy have also been invited to give particular thought to the mechanisms by which they were produced. Certain scholars have advanced the claim that the technology of the printing press enacted a profound transformation on the early modern world; that it, indeed, was instrumental in bringing about both the Scientific Revolution and the preceding Reformation.[9] With respect to

[3] E.g. Houzeau and Lancaster 1964; Zinner 1964.
[4] E.g. Gingerich 1973a, 1973b, 2002; Swerdlow 1974; Westman 1975a, 1990; Neugebauer and Swerdlow 1984; Verdet 1989; Zinner 1988.
[5] E.g. Kepler's *De revolutionibus* and Tycho's copy of Apian's *Astronomicum Caesareum*.
[6] On the drawbacks of working with such editions and translations, however, see Grafton 1989.
[7] E.g. Swerdlow 1993b; Bennett and Bertoloni Meli 1994; Macdonald and Morrison-Low 1994.
[8] E.g. Prandtl 1932; Nørlind 1970, 336–366; Czartoryski 1978; Grafton 1992, 1997a, 185–224; Müller 1993, 1995.
[9] Most notably Eisenstein 1979. But see also, on the sciences, Drake 1970; Eamon 1985. More recent appraisals of the relationship between print-culture and the Reformation can be found in Edwards 1994; Gilmont 1998; Pettegree and Hall 2004; Pettegree 2005, esp. 128–124.

the study of the heavens, the import of this claim is that the printing of texts was of such significance in shaping astronomical practice, disseminating theories, establishing new modes of thought, and enabling particular forms of collaboration, that it was a *sine qua non* for the new astronomy that emerged in the seventeenth century.

Written by one of the chief architects of the reconstituted discipline, by the author, indeed, of an *Astronomia nova* (1609), the statement opening this chapter might be thought to be a strong endorsement of this thesis. In addressing the impact on human affairs of great planetary conjunctions (one such having occurred in 1563), Kepler showed himself to be both a sophisticated student of history, and an insightful commentator on changes within his own culture and time.[10] Just as more recent scholars have done, he asserted that the development of printing was implicated not only in the transformation of astronomy, but also in the religious schisms, educational reforms, and maturation of philological expertise with which it coincided. But there are several subtleties to Kepler's account, and some omissions as well, which diminish its authority as a witness-statement that the printing press was the great early modern agent of change. Kepler acknowledged that church reform, pedagogic innovation, and humanist concerns were interrelated phenomena, for example; but not that each of these was also important with respect to developments in astronomical practice. He failed to mention that the emergence of the reading strategies deployed in the transformation of astronomy preceded Gutenberg's invention; or that, except in the case of the mixed blessing of the presses' fecundity, claims about the distinctiveness of print and scribal culture are easy to make, but difficult to sustain in the face of the evidence.[11] And he did not disclose what contemporaries would have known already: namely, that his appropriation of the press as an ally of Copernicanism, occurring as it did at a time when few authors openly supported the heliocentric theory (and indeed, when much of the vast output of astronomical works said little or nothing about changes in cosmology) was both bold and polemical. Indeed, astronomy's transformation was itself more thoroughgoing than Kepler acknowledged; involving not just the elaboration and adoption of any one new theory or practice, but a renegotiation of the relationship between the different forms of knowledge of the heavens that accompanied and licensed all such innovation. And so the attempt, whether Kepler's or any other scholar's, to reduce one complex

---

[10] See the full passage and commentary in Jardine 1988a, 277–286; Grafton 1991, 196–197.
[11] See Love 1987, 1998, 35–89; d'Amico 1988a, 23–24; Sherman 1995, 117.

historical event, the astronomical revolution, to an epiphenomenon of another, a revolution in communications, must inevitably look suspect.

Reservations such as these have now been expressed by a number of scholars, on a number of occasions.[12] I do not mean to suggest, therefore, that by arguing for the need to adopt a more nuanced view of the importance of printing and printed books to astronomy, I am making an entirely original point. But by exploring the production and use of books with respect to Tycho, as I propose to do here, I hope to correct a significant number of misapprehensions. Tycho's mastery of a press, which quickly became a part of his mythology, has contributed to his characterisation as an exemplary figure of the new age of the book: a hero not just of the Scientific Revolution, but of the printing revolution as well. As one recent commentator has pointed out, however, claims about his importance and success as an astronomical publisher are undoubtedly overstated. His chief works were not widely distributed during his lifetime, and he encountered considerable problems in managing their production.[13] Yet in the rush to qualify our understanding of the role of print in the transformation of early modern astronomy, there also is some danger of marginalising the Dane, portraying him as a more singular individual than he actually was.

This chapter cannot accommodate a comprehensive study of Tycho Brahe as publisher and author. The focus here will therefore be on the one Tychonic publication, the *Epistolae astronomicae*, with whose contents we are, to some extent, already familiar. His edition of the Hven–Kassel correspondence, arguably the most singular of all of Tycho's books, shall be considered not only with respect to its composition and purpose, but also with regard to its distribution and use by a number of readers. And as we move towards an understanding of the character and history of this particular work, it will be possible to shed some light on Tycho and his contemporaries as readers, and critics, of a wider range of texts.

I FROM MANUSCRIPT TO PRINT

(i) To please all men of learning and goodwill?

There is an important difference between a book and a letter, in that the latter must be adapted as far as possible to the immediate occasion, and to contemporary topics and individuals, whereas a book, intended as it is for general consumption, must

---

[12] See, *inter alia*, Grafton 1980; Johns 1998b, 6–57. [13] Johns 1998b, 16–17.

be contrived to please all men of learning and good will. (Erasmus, *De conscribendis epistolis* (1522))[14]

In setting out what distinguished a letter from a book, Erasmus was probably thinking of the sort of work for which he became famous: a text that, published by Froben of Basle, or Martens in Louvain, would be distributed and sold across the whole expanse of Europe.[15] His description may reinforce the impression that in the sixteenth century, as today, a work issued in print was always intended for release to a general reading public. But this was not necessarily the case; while the presses could, and typically did produce numerous copies of any one text, not all of these were destined to be commercially distributed. Many authors participated in a form of private distribution, one that the business practices of early modern publishers themselves helped to foster. Rarely paid for assigning their copy to a printer, scholars who were might be rewarded with complimentary texts rather than money. An arrangement whereby authors secured publication of their works by contracting to purchase a substantial portion of the finished edition was probably as common.[16] Individuals who found themselves in either situation quite naturally sought an advantageous disposition of the copies they obtained. But in addition to authorial distributions of this sort, which occurred alongside, and as a consequence of, the familiar commercial transactions, there were others that constituted the entire method of dispersion.[17] Books, like letters, could be important tokens in the early modern economy of credit and patronage, and as such their value did not always depend on the breadth of their audience.[18]

Printed at Tycho's own press in the grounds of Uraniborg,[19] the *Epistolae astronomicae* might be thought one example of a work never intended for commercial distribution. Such is certainly the impression conveyed by recent statements on the nature of the astronomer's publishing enterprise. His printing office, it has been argued, was 'as geographically isolated on the island of Hven as it was socially isolated from the companies of the European book-trade', and his intention was 'to bypass the structures of

---

[14] Translated in *CWE* XXV, 14.

[15] On the enormous popularity of Erasmus' *Adages* and *Colloquies*, see Febvre and Martin 1984, 274.

[16] E.g. Kepler, in relation to his *Mysterium Cosmographicum*; see Caspar 1993, 66, 75. The issue is discussed by Zeeberg 2004, 40, 45.

[17] As noted by Rosen 1986, 88, the distribution of Ursus' *Chronotheatrum* was of this sort.

[18] For a useful discussion of book distribution and patronage in relation to one of Tycho's close contacts, Heinrich Rantzau, see Zeeberg 2004, 29–50.

[19] Strictly speaking, one should refer to the presses of Tycho's printing office; as he stated in his letter to Thaddaeus Hagecius of 25 January 1590, he was then operating two, rather than the one he owned previously. See *TBOO* VII, 222.33–40.

the international book-trade altogether'.[20] And it is quite true that, when he himself paused to explain his acquisition of the technology of print, Tycho gave not the slightest hint of any commercial imperative.[21] Indeed, as an aristocrat, for Tycho to have embarked on publishing as an enterprise of profit would perhaps have been seen by him and his contemporaries as somewhat improper.[22] Yet if Tycho did not envisage the distribution of his works by sale, then the size of the print-run he adopted as a standard for his works seems rather large.[23] The evidence indicates that he produced, or aimed to produce, 1,500 copies of each of his books – a quantity greater than the first and second printings of *De revolutionibus* combined, and equal, at the time, to the print-run of many a single-edition bestseller.[24]

In fact, when it came to establishing a broad audience for his work, Tycho was clearly not above allying himself with representatives of the mercantile classes: two instances of such collaboration, both involving the commercial production of Tychonic celestial globes, will be considered in the following chapter. And Tycho's own printing office was not, in any case, so isolated from the structures of the European book-trade that he would have found it difficult to distribute his productions commercially had he chosen to do so. Precisely because of his physical location, Tycho found it advantageous to maintain good relations with nearby booksellers and printers: partly in order to facilitate the carriage of his correspondence, as discussed in Chapter 2, but also to obtain the paper, type, ink and

[20] Johns 1998b, 14.

[21] Writing to Brucaeus in 1584, Tycho stated that, although persuaded by the exhortations of friends to publish on the subject of recent celestial phenomena, an outbreak of plague had made it too dangerous to contemplate producing his book in Copenhagen. Later, however, when the disturbance was over, the printers of the city were already occupied with other projects; and so, in order that his own publication might be timely, it seemed appropriate to establish a printing office on Hven. See *TBOO* VII, 81.6–26.

[22] Both for the reasons given in the last chapter, and because of the long-standing sentiment against the commodification of knowledge described by Davis 1983.

[23] It is unfortunate, therefore, that in using the case of Tycho to challenge the arguments of Eisenstein, Johns 1998b, 6–20, takes as his measure of the size of the astronomer's editions, and indeed his ambitions, the hastily produced, and anomalous, *Mechanica*.

[24] Thoren 1990, 366, notes that 1,500 copies was the standard print-run for Tycho's works, but does not cite the evidence that justifies this statement. On 1 November 1589, Tycho wrote to Hagecius that he had only distributed a few copies of the *De recentioribus phaenomenis* from a total of 1,500; on 7 January 1600, he sought to enlist the assistance of Bartholomew Scultetus in having 1,500 copies printed of all the works he had yet to complete; and a few days later he wrote to Jacob Monaw for the same reason, again citing the figure of 1,500 as his target. See *TBOO* VII, 213.13–21; *TBOO* VIII, 235.9–34, 237.21–29. For the estimate that the 1543 and 1566 editions of *De revolutionibus* each had a print-run of 400–500, see Gingerich 2002a, XIV. On edition sizes of this period in general, see Febvre and Martin 1984, 216–220.

skilled labour required to maintain his printing operation.[25] There is clear evidence, moreover, that he entertained the notion of establishing links with commercial publishers at other locations. Thus, having at one time volunteered his services as a printer to certain correspondents, and finding that this offer was one on which he felt he had to renege, he was apparently quick to pursue alternative arrangements with commercial operations. In July 1586, for example, he told Hagecius that he intended to enquire whether the famous Christophe Plantin of Antwerp would publish the physician's manuscript of George of Trebizond's commentary on Ptolemy's *Syntaxis*.[26] He later wrote that he would make enquiries in Frankfurt for another man willing to print it.[27] These are not the statements of an individual who faced insurmountable difficulties, practical or ideological, in dealing with the commercial world of the book. Indeed, far from it being his aim to bypass the international book-trade, Tycho repeatedly expressed his intention to exploit it. In August 1588, he informed Thaddaeus Hagecius that one work, the *De recentioribus phaenomenis*, would be available for sale at the next-but-one fair at Frankfurt, at that time the greatest of the trade's great emporia.[28] He was later to ask Bartholomaeus Scultetus, based in Görlitz, and other correspondents, in Prague and Vienna, to enquire whether local booksellers would be interested in purchasing this and other texts from him at a reasonable price.[29] The commercial distribution of texts that took place after the astronomer's death was not, therefore, a betrayal of Tycho's legacy on the part of his heirs.[30] Instead it was the fulfilment of an ambition that the astronomer had deferred too long to accomplish himself.

To be sure, the astronomer's continued failure to send the *De recentioribus phaenomenis* to Frankfurt demonstrates that his motives were very different from those of the publisher for profit. Few businessmen, if any, could or would have held back from seeking to realise their investment in a work once it was printed, yet this work awaited its general release for about fifteen years.[31] This interval is long enough, indeed, that it might be considered

---

[25] The evidence pertaining to this in Tycho's correspondence is extensive. See, in the first instance, Tycho's procurement of type from the Rostock printer Myliander, as revealed in a letter from Brucaeus of 10 December 1584, *TBOO* VII, 88.13–17.

[26] *TBOO* VII, 105.27–30. See, on this text, Swerdlow 1993b, 149.

[27] *TBOO* VII, 112.29–41. As noted by Monfasani 1984, 672, Tycho's enthusiasm for seeing the text printed waned, and the manuscript was eventually returned to Hagecius.

[28] Tycho meant the completed work of three volumes; see *TBOO* VII, 122.23–28. On the Frankfurt fairs, see Thompson 1911; Febvre and Martin 1984, 228–232.

[29] *TBOO* VII, 327.16–21, 329.35–40, 332.31–35.   [30] *Contra* Johns 1998b, 18.

[31] The volume was issued in 1603 from Prague, by S. Schumann, with additions by Tycho's son-in-law Franz Tengnagel. See *TBOO* IV, 491–497; Zinner 1964, no. 3930. Christianson 2000, 124, claims

evidence for Tycho having thought better of a commercial release. It seems to have resulted, however, for more interesting reasons.

Tycho authored, and had printed, five major works.[32] The first of these, *De nova stella* (1573), contained his astronomical and astrological analyses of the nova of 1572, and was produced in the shop of the Copenhagen printer Lorentz Benedicht. But, Tycho later complained, 'few copies of it were printed, and (because of a certain niggardliness of the publisher) far fewer were conveyed to those outside the country'.[33] The later texts were all begun at Uraniborg; and two of them, the 1588 *De recentioribus phaenomenis*, and the *Epistolae astronomicae* of 1596, bear on their title-pages the observatory's imprint. The fourth work, the *Mechanica* of 1598, an illustrated account of Hven and Tycho's instruments, was finished in a very small edition at Wandsbek, near Hamburg.[34] And the fifth, the *Progymnasmata*, published posthumously in Prague in 1602, contained Tycho's catalogue of stars, his models of lunar and solar motion, and a revised study of the 1572 nova. This is a respectable print legacy, one might think; and it was one, moreover, that was supplemented by the appearance in 1627 of the *Tabulae Rudolphinae*, the final fruit of Tycho's observations, as compiled and edited by Kepler.[35] But it represents, nevertheless, only the partial realisation of a much grander ambition.

Evidence for the true scope of Tycho's publishing programme can be found in letters that he wrote in the 1580s and 1590s. In his lengthy 1588 epistle to Caspar Peucer, for example, he stated that in addition to a three-part work on the recent phenomena of the celestial realm, of which the completed *De recentioribus phaenomenis* was only volume two,[36] he intended to produce, in five or six years' time, an *Opus astronomicum* containing seven

that some copies of the *De recentioribus phaenomenis* were sold at the Frankfurt fairs of 1588, but offers no evidence in support of this claim. The work was certainly not listed in the Willer catalogues for either the spring or autumn fairs of this year, as reproduced in Fabian 1972–2001, IV. In the absence of evidence to the contrary, I stand by the claim that this work of 1588 only received a much later commercial release. Dreyer was of the same opinion; see *TBOO* IV, 492–493.

[32] The Uraniborg press also produced a *Diarium astrologicum et metheorologicum* for the year 1586, issued under the name of Tycho's assistant, Elias Olsen Morsing, that contained an appendix written by Tycho. See *TBOO* IV, 399–414, 512–513. Tycho later contributed a preface to the *Astrologia* of his assistant Jacobi, printed in 1591; see Christianson 1968. But I do not think that either of these works, or the poetry issued from the Uraniborg press, should be regarded among Tycho's major publications.

[33] This comment introduces and justifies the partial reprint of the work in the *Progymnasmata*; see *TBOO* III, 96.36–38. The *De nova stella* was not advertised in the Willer Frankfurt catalogues of 1573 or 1574, a fact which bears out Tycho's complaint; see Fabian 1972–2001, I-II.

[34] Wandsbek was one of the estates of Heinrich Rantzau; see Christianson 2000, 213–214.

[35] See Gingerich 1971b.

[36] The first volume was to treat anew the nova of 1572; the third, comets other than the one of 1577. See *TBOO* VII, 131.35–42.

parts.[37] The later sections of the work would treat the positions of the fixed stars, and the motions of the Sun, Moon and planets.[38] But the first and second books would be devoted to a discussion of instruments and mathematical techniques; in order, as Tycho put it, that these topics would not cause him the trouble, in the later parts of the work, that they had occasioned his predecessors.[39] Tycho's *Opus*, this comment suggests, was conceived by him as an answer to Copernicus' *De revolutionibus* and Ptolemy's *Syntaxis*, and it was intended to be superior to both in its organisation as well as its contents. Since, however, what Tycho termed the 'first and mechanical part of astronomy' was issued as the *Mechanica*,[40] and the *Progymnasmata* combined material on the nova with other sections of his *Opus*, what he had originally planned to issue as one work was separately produced, and what he had intended to keep distinct he was forced to amalgamate. The chief causes of this reorganisation of material seem to have been two: the extent of the astronomical work and intellectual effort his projects required, and the practical difficulty of maintaining a printing operation in the face of shortages of labour and material. But without doubt his patronage-crisis of the late 1590s was an exacerbating factor.[41] When all the problems he faced are considered together, it becomes unsurprising that other planned publications, including those of his correspondents, had to be abandoned.[42] Circumstances conspired to ensure that Tycho's ownership of a press made his experience of getting work into print no less vexing and time-consuming than other astronomers of the sixteenth and seventeenth centuries.[43]

The reason, then, that the *De recentioribus phaenomenis* failed to appear at Frankfurt year after year, is that from the day it was finished until the day that he died, Tycho saw it as just one part of an incomplete series of works.[44] Initially designated as the second volume of his trilogy on heavenly

---

[37] *TBOO* VII, 132.8–9, 132.32–133.3.   [38] *TBOO* VII, 132.21–32.

[39] *TBOO* VII, 132.9–21.   [40] E.g. in his letter to Hagecius of 25 January 1587; *TBOO* VII, 111.38–112.2.

[41] See the discussion in Thoren 1990, 312–316.

[42] The failure to print Hagecius' manuscript copy of George of Trebizond's commentary on Ptolemy has been mentioned above. In addition, Hagecius sent Tycho a corrected version of his *Dialexis* (1573) on the new star of 1572, hoping that Tycho would print it, or else incorporate selections into his own treatment of the phenomenon. Tycho did discuss the revised text in the *Progymnasmata*, but he did not publish the whole work. See *TBOO* III, 19–43; *TBOO* VII, 99.5–28. A list of the works that Tycho published, or proposed at one time or another to publish, was drawn up by Kepler after his death. See *KGW* XX.1, 91–95; Grafton 1997a, 195.

[43] Notable examples include Kepler, whose difficulties in publishing a range of texts, but particularly the *Astronomia Nova* and *Tabulae Rudolphinae*, are well documented; Rothmann, who drew up a list of astronomical publications that, comprising twelve separate works, was near Tychonic in scale, but did not see even one of them in print; and Flamsteed, whose problems with the *Historia Coelestis* stemmed from his difficult relationship with Newton and Halley. See Caspar 1993, esp. 308–318; Hamel 1998, 77–84; Johns 1998b, 543–621; Voelkel 1999.

[44] Tycho was explicit on this point. See, for example, his letter to Hagecius of 25 January 1590, *TBOO* VII, 224.5–6. In addition, the completeness of the 1588 text was itself compromised after Tycho's

phenomena, it was subsequently intended to be a companion to the *Progymnasmata*; a book that, mostly ready by 1592, was itself destined to remain unfinished for an entire decade longer. Both works, therefore, and the *Mechanica* as well, were distributed only to a privileged few: the *Progymnasmata* in the smallest number of pre-publication copies sent to selected correspondents;[45] the *De recentioribus phaenomenis* a little more widely; and the *Mechanica* to a number of princes and courtiers, whom Tycho thought could assist him in obtaining and consolidating Imperial patronage, as well as to fellow scholars with an interest in astronomy (see Appendix). Similar reasoning might have induced Tycho to withhold the *Epistolae astronomicae* from the general reading public; in his letter to Peucer of 1588, he spoke of publishing his correspondence as an appendix to the series on celestial phenomena.[46] And as its full title makes clear, the volume of Kassel correspondence was supposed to be followed by other books of letters, at least one of which Tycho actually began to compile, edit, and print.[47] After his death, however, the astronomer's heirs were able to despatch only '820 gutte exemplaria' of the *Epistolae astronomicae*, plus an unspecified number of mediocre copies, to the Nuremberg bookseller Levinus Hulsius, who sold them on commercially under his own imprint.[48] The number of poorly printed texts may indeed have been substantial, for Uraniborg sheets of low quality formed the core of the 1610 edition ostensibly produced by the Frankfurt printer Godefridus Tampachius.[49] But it is difficult to believe that these actually amounted to the 650-or-so copies required to produce the total of 1,500 previously mentioned. The shortfall can be attributed to the fact that, unlike the other works printed on Hven, the letter-book *was* commercially distributed within Tycho's lifetime. It was sold at Frankfurt in the spring of 1597, and it was advertised accordingly in the book-fair's catalogue.[50]

Part of the importance of the *Epistolae astronomicae*, therefore, is that it was the first text indicating the full scope of the Tychonic astronomical

---

decision to supplement it with a refutation of criticisms by John Craig, one of the recipients of the privately circulated copies. This text is to be found in *TBOO* IV, 417–476; for the decision to append it to *De recentioribus phaenomenis*, see *TBOO* VII, 225.27–34.

[45] Tycho sent an incomplete copy of the *Progymnasmata* to Hagecius and Kurtz in 1593; see his letter to Kurtz of 19 April, *TBOO* VII, 349.27–32.

[46] *TBOO* VII, 140.22–30.

[47] In his letter to Scultetus of 7 January 1600, Tycho indicated that he had begun preparing the second volume of letters before he left Denmark; and indeed, had printed half an Alphabet, i.e. about twelve sheets, with two more Alphabets more to go. See *TBOO* VIII, 235.33–37, and, for this terminology, Schröter 1998, 24. Some of the explanatory remarks that Tycho proposed to add to his edition are reproduced in *TBOO* VII.

[48] See Dreyer, in *TBOO* VI, 343. Hulsius also sold the 1602 imprint of the *Mechanica*, produced using Tycho's engravings. See Zinner 1964, no. 3929.

[49] As is evident from inspection. See also Zinner 1964, no. 4262.

[50] See Tycho's remark to Moritz of 17 July 1597, *TBOO* XIV, 107.48–108.1; Fabian 1972–2001, V, 373.

project to receive any form of general distribution. Indeed, the role of the work in filling in the gaps in public awareness caused by the delays dogging Tycho's other publications was something that he himself publicly recognised.[51] He shaped the book accordingly, framing the Hven–Kassel correspondence with a quantity of additional material that refined its capacity to inform readers about Tycho himself, and about his astronomy. A portrait of the Dane, surrounded by the escutcheons of his distinguished ancestors, and displaying his personal motto, 'Not to seem, but to be',[52] was placed immediately after the title-page (Fig. 3.1).[53] Opposite was positioned a verse composed by Albertus Voitus, professor of poetry at Wittenberg,[54] in which Tycho was lauded as the Titan Atlas, whom 'Credulous Antiquity' wrongly thought had disappeared, changed into a mountain by Perseus.[55] Five further poems in a similar vein were added; and after them came a dedication, to Wilhelm's son Moritz.[56] Next was inserted a *praefatio generalis*, which advertised Tycho's hope of publishing further volumes of letters; and then another preface summarising the content specific to this, the *liber primus* of the series. And just before the Hven–Kassel exchanges themselves appeared, punctuated by a handful of addresses *ad lectorem*, two other letters were inserted. One of these, from Heinrich Rantzau to Tycho, was labelled as having 'demanded and brought about the following publication'.[57] The other was the letter from the Landgrave to Rantzau, that had initially triggered the entire correspondence. At the end of the work, finally, Tycho placed a poem of his own composition, one that praised and memorialised Wilhelm.[58] As we shall see, it is to this extraneous material, which occupied perhaps a tenth of the book's pages, that we must turn if we are to answer a number of questions about this, on the face of it, rather curious work.

(ii) A dedicated gift . . . ?

And it seemed fitting to dedicate the learned letters of the Most Learned Prince to none other than his son, heir not only to the goods of the father, but also his virtue, and inheritor also of his learning . . . Therefore, with so many reasons running together, it seemed both unbecoming and ungrateful for us not to name you in this dedication, even though in books which we have published hitherto, or have decided, the heavenly Numen willing, to publish in the future, we have for

---

[51] See *TBOO* VI, 21.19–26.

[52] 'Non haberi sed esse.' On this motto, see Shackelford 1993, 211; Christianson 2000, 114.

[53] On the production of this portrait, see Christianson 2000, 113–118.

[54] *TBOO* VI, 5.   [55] *TBOO* VI, 5.5–6.

[56] *TBOO* VI, 10–18. The date of composition is given as the vernal equinox, 1596.

[57] *TBOO* VI, 29.1–7.   [58] *TBOO* VI, 340.

3.1 The portrait of Tycho placed at the beginning of the *Epistolae astronomicae* (Uraniborg, 1596). By permission of the Syndics of Cambridge University Library.

certain reasons refrained from some special dedication, customarily employed by others more expediently and more agreeably. (Tycho Brahe, *Epistolae astronomicae*, Dedication to Moritz)[59]

In the epistolary essay placed between the opening verses and the two *praefationes*, Tycho dedicated the *Epistolae astronomicae* to Landgrave Moritz of Hesse-Kassel, Wilhelm's son. Like the two prefaces, this letter, dated to the vernal equinox of 1596, served to introduce the work by summarising the contents of the correspondence, describing how the original exchange

[59] *TBOO* VI, 14.29–32, 15.24–29.

of letters had come about, and offering an explanation for their appearance in print. At the same time, however, this document flattered Moritz by enumerating the qualities that made him the book's worthy recipient. In the process, it spoke to a contemporary anxiety to justify the existence of the nobility with reference to the merits of its individual members, rather than the mere inheritability of their titles and lands.[60]

Such a dedication, which might be addressed to a prince, nobleman, cleric, or a well-placed fellow scholar, was a standard way of positioning a text within the patronage-economy of early modern Europe.[61] Within the culture of display, the printed dedication allowed an author or editor to exhibit his civility and erudition; and yet do so without letting fall a much-cherished but frequently threadbare garment, the cloak of authorial modesty.[62] Indeed, the way in which a well-crafted example of the genre induced admiration for both writer and recipient, helps to make explicit the reciprocity of the (largely unwritten) contracts that underpinned Renaissance patronage of scholarship and the arts. Whether seeking to establish a new relationship, or to reaffirm an existing one, the basic objective of a dedication was the generation, through mutual association, of some mutual benefit.

To dedicate a book is, of course, to make it into a gift. And in both origin and practice, this mode of giving was closely related to another of long standing, the presentation of a text. The medieval custom had been to signify publication of a work by presenting it to a higher authority; an act that signified the recension presented was definitive, and that might be construed as a transfer of rights over the text's future distribution.[63] On the occasion of such a presentation, some indication of the donor's motivations and aspirations might seem to be appropriate; the written dedication became a way of incorporating such a statement, or placing additional emphasis on it, within the work as it was presented. In principle, the existence of this strategy in the age of print eliminated the necessity of actually delivering a copy of the work into the hands of the dedicatee, whether in person or by proxy. But as a means of both enhancing and exploiting the gift-value of a book, the custom of presenting texts was retained in the sixteenth century, and even extended.[64] It was important to ensure that the party honoured by a printed dedication noticed one's

---

[60] Kristeller 1990, 48–49; Midelfort 1992; Dewald 1996, esp. 15–59.

[61] Febvre and Martin 1984, 160; Chartier 1995, 25–42.

[62] On dedications as a genre, see Schottenloher 1953. For the rhetoric of authorial modesty, see Janson 1964, 51–52.

[63] Root 1913, 419–420, 426, 428; Grafton 1980, 280.   [64] Davis 1983, 73–81.

efforts; managing a broader distribution of the work through presentations helped to secure the visibility of both the document and the gesture. There was, moreover, an element of promiscuity in sixteenth-century dedicatory practice: texts were often dedicated, and presented, to multiple recipients.[65] Most commonly, this involved supplementing printed dedications with briefer manuscript ones. But texts might also be sequentially dedicated to different individuals, as second and subsequent editions were produced, contain a number of component parts separately dedicated, or be furnished with a single dedication that varied from one presentation copy to another.

Many of the gift-copies of early modern books were distinguished in material ways from the texts destined for commercial distribution: by the hand-colouring of their illustrations, for example, or by the use of an extravagant binding.[66] Tycho issued at least one such presentation copy of the *Epistolae astronomicae*. Preserved in the Herzog August Bibliothek, Wolfenbüttel, this volume is bound in similar fashion to gift-copies of the later *Mechanica*: in apple-green silk with gold tooling, showing Tycho's portrait on the front cover and the Brahe arms on the back, and having silk ties and gilt edges (Fig. 3.2).[67] It is possible, therefore, that this example was prepared for presentation after the *Mechanica* had been printed in 1598. Yet the astronomer did send copies of the letter-book to his correspondents prior to this date, and a letter of June 1597 preserved in Marburg not only confirms that Tycho despatched a copy to Moritz, but also displays an anxiety on his part that it might not have arrived.[68] It is likely that Tycho took the trouble to have this gift-copy specially bound.

At the root of Tycho's assertion that Moritz was the most appropriate dedicatee of the *Epistolae astronomicae* was his representation of the work as a memorial to Wilhelm. Publication of the correspondence for just this purpose had been requested by Hieronymus Treutler, the Landgrave's funeral

---

[65] Tycho himself had something to say on the subject of promiscuous presentations. In a letter written to Holger Rosencrantz, he justified having given the *Mechanica* and other texts to Ernst, Archbishop and Elector of Cologne, Prince Maurice of Orange, Duke Ulrich of Mecklenburg and Duke Johann Adolf of Holstein-Gottorp (not to mention Joseph Scaliger), even though he had not yet presented them to Rudolf II, the dedicatee, with the phrase, 'As if it were not customary to give books dedicated to one king or prince to others!' *TBOO* VIII, 112.10–14.

[66] Frequently, presentation copies were also distinguished from commercial ones by the absence of any privilege; see Armstrong 1990, 160–164.

[67] I am indebted to Nicholas Pickwoad for first drawing my attention to this copy. There is no inscription to demonstrate that the text was a gift, but the front flyleaf, on which such might have appeared, is missing. On the bindings of the *Mechanica*, see Dreyer 1890, 260–261. As Duke Heinrich Julius of Braunschweig-Wolfenbüttel was presented with a copy of the 1598 text, it is possible that this copy of the letter-book was bound and presented at the same time.

[68] *TBOO* XIV, 107.44–108.1.

3.2 (a) The front cover of a presentation copy of the *Epistolae astronomicae* (Uraniborg, 1596), displaying Tycho's portrait.

orator, in a passage that Tycho quoted in his own dedication. 'If it should only happen', Treutler had stated:

that at some time those letters concerning these [astronomical] matters, which he frequently sent, especially in his later years, to the Danishman Tycho Brahe, most noble both by birth and in virtue of his mathematical knowledge, should

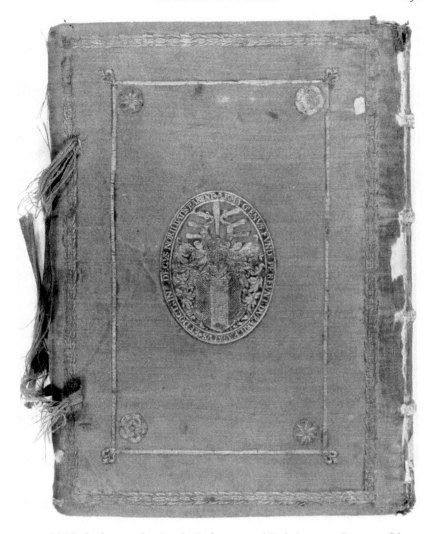

3.2 (b) The back cover, showing the Brahe arms and Tycho's motto. Courtesy of the Herzog August Bibliothek, Wolfenbüttel [Astronomica 8].

come to be published, then it would establish for posterity how diligent [and] how practised he was, and how much he was devoted to removing the faults of the tables of astronomy and to determining the positions of the stars more precisely through his laborious observations.[69]

---

[69] Treutlerus 1592, 82: 'Quod si eas tantum epistolas edi in publicum aliquando contingat, quas iis de rebus ad virum & genere scientia Mathematica nobilißimum Tychonem a Brahe Danum subinde

Posterity was a recurring theme not just in the introductory matter of the book, but also in the letters, where it featured as both the ultimate beneficiary of the work undertaken at Kassel and Hven, and as the custodian of the best sort of fame, the reputation that lasted. Tycho wrote of 'dedicating to posterity a universal astronomy, complete in all its numbers'; Rothmann stated that Wilhelm, 'wishes to leave to posterity a monument of his love towards astronomy'.[70] Aspirations of this sort were common enough among men of rank and education, but lasting recognition was not yet a prize that was devalued for being widely pursued. Thus, by suggesting that the *Epistolae astronomicae* would ensure that Wilhelm's name was known to later generations, Tycho's dedication invoked a powerful preoccupation of his age to render the work attractive to all those who would be pleased by the honour accruing to the Landgrave, and would share in the glory thus reflected. The chief of these, we must presume, as Tycho evidently did, would have been Moritz.

That it was indeed a point of some strategic importance, is indicated by Tycho's careful placement of his memorial poem on the work's closing pages. The elegy he penned addressed Wilhelm as a ruler so prudent, and so kind, that:

> all Germany perceived how wise
>   how zealous of peace you were.
> And troubled France realised it, and Britain,
>   and the people of the Low Countries acknowledged it no less.
> And Denmark, ever so far across the waves of the sea,
>   made a friendship adamantine-firm with you.
> Even Sweden also, which lies under the Bears of the North,
>   was not bereft of your fame and merit.
> Neighbouring Germany, Poland to the east
>   and Italy below the Alps knew you also.[71]

As suggested by the exchange of verses discussed in the previous chapter, poetry was held in high esteem in the Renaissance courts. At the same time, however, it was a mode of expression that, far from being the province of a talented few, all scholars with a humanist education were encouraged to make use of.[72] Indeed, with the exception of the *De recentioribus phaenomenis*, all of Tycho's publications contained a quantity of poetry of his and

---

extremis maxime annis misit, vel inde constabit posteritati, quam industrius, quam exercitatus fuerit: quantum ad tabularum astronomicarum tollendos errores, ad loca stellarum exactius cognoscenda laboriosis suis observationibus contulerit.' Cf. *TBOO* VI, 14.2–8.

[70] *TBOO* VI, 102.15–20, 117.14–16.   [71] *TBOO* VI, 340.18–27.

[72] Saunders 1951, esp. 509; Kristeller 1990, 12–13; Grafton 1991, 182–183.

others' composition, and a few elegiac verses were even issued separately from his printing office, at a time when 'he did not want the presses to lie idle'.[73] But although we must be careful not to exaggerate the specificity of poetry's use at the end of the letter-book, it seems clear that the content of the text on the work's final page was intended to be underscored by its format. The poem combined well with the dedication to frame the whole book as a monument to the Landgrave.

It is possible that, by taking his cue from Treutler, whom he was careful to describe as representative of the pre-eminent men of Moritz's authority,[74] Tycho sought to evade any doubts about the propriety of his publishing the letters. As noted previously, the documents that flowed between Hven and Kassel, and the objects that accompanied them, were tokens of friendship and favour, constitutive and representative of a form of client–patron relationship. The *Epistolae astronomicae* was, in a certain sense, a recapitulation of those exchanges; with gifts such as the elk, requested by Wilhelm but sent to Moritz *in memoriam* of his father, represented by the letters describing them. We do not know if the animals were still alive at this time; or if they were not whether, as seems likely, some record or remains of them had been deposited in the Hessen Landgrave's *Wunderkammer* or cabinet. But certainly in the epistolary form in which they reappeared in 1596, these gifts were still in the possession of Wilhelm's son and successor. And he, whether pleased or displeased with the final result, might have thought it appropriate to be consulted before they were published.

Yet the explicit connection which Tycho made between property and propriety, between Moritz's possession of the letters and his selection as the dedicatee of the text, was slightly different in nature than such considerations might suggest. The published letters, Tycho reasoned, 'will deserve a more certain trust, because I did not hesitate to present the father's letters to the son. And indeed, there is no doubt that copies of these are still preserved in your Chancery; from which anyone may readily infer that I have interpolated nothing of my own. Because that would have been a crime in itself.'[75] Whether Tycho did, in fact, commit a crime by these criteria is a question to which we shall have cause to return; but for now it is more important to note that, by re-presenting Moritz with objects that already belonged to him, the astronomer was able to invoke the Landgrave's authority in support of his claim that the *Epistolae astronomicae* was the faithful

---

[73] *TBOO* IX, 179.38. The line itself comes from one of the verses printed by Tycho in 1584. For more on Tycho's poetry, see Dreyer 1890, where his verses are indexed as 'poetical effusions'; Zeeberg 1994; Christianson 2000, esp. 45.

[74] *TBOO* VI, 14.13–16.   [75] *TBOO* VI, 14.24–29. Cf. *TBOO* VIII, 112.19–22.

reproduction of an actual correspondence. Clearly, an aura of credibility was one benefit that Tycho associated with his dedication of the work to Wilhelm's son and heir.

There may have been others as well, but if so what they were is more difficult to see. In a letter of September 1598, Tycho declared that he had neither anticipated nor received any reward from the Landgrave,[76] and despite his vast outlays in pursuit of his astronomical work, we can probably rule out the possibility that what he really hoped for was some cash reward, of the sort that Kepler was only too pleased to receive from the Styrian representatives to whom he dedicated his contemporaneously produced *Mysterium cosmographicum*.[77] It is more plausible that, if he held any hope at all that Moritz might enhance his fortunes, Tycho imagined the Landgrave interceding on his behalf with King Christian of Denmark. Such assistance from the Hessen prince would not be unprecedented, for according to an account found in the dedication (but not, interestingly, elsewhere) it was Wilhelm's intercession with Frederick II that had led to Tycho receiving the fiefdom of Hven, and the funds to pursue his astronomical work.[78] There is, moreover, an intriguing suggestion that the Dane solicited further aid when seeking to increase the extent of his support from the State during the early days of Christian's regency, in 1588.[79] But if Tycho did dedicate to Moritz with an eye to improving his domestic situation, it is entirely possible that he required no action from his princely ally for the work to serve this purpose. Simply by demonstrating how capably he courted a princely-patron, the *Epistolae astronomicae* might, in Tycho's opinion, have performed such a function.

In fact, Tycho did make one explicit request of Moritz in the dedication of the *Epistolae astronomicae*: he asked him to continue the astronomical work of his father and, in particular, to publish the catalogue of star positions and

---

[76] *TBOO* VIII, 112.19–24. [77] Caspar 1993, 66.

[78] *TBOO* VI, 11.3–21. See also Dreyer 1890, 84–85; Thoren 1990, 101–104. There is no mention of Wilhelm's representation to the Danish Crown in either Tycho's autobiographical comments in the *Mechanica*, or his letters dating from the period.

[79] In his letter to Rothmann 'around the summer solstice' of 1588, Tycho indicated that the bearer, Gellius Sascerides, would convey a private message to be relayed to the Landgrave. In a passage preserved in the copy Rothmann retained of his reply of 19 September, but not printed in the *Epistolae astronomicae*, Rothmann remarked that: 'Those things which Gellius Sascerides revealed to me about Nicholas Kaas, the Chancellor and regent of your most worthy kingdom, I reported to the Most Illustrious Prince, who also, as soon his affairs allow, will write to him about this matter.' See *TBOO* VI, 133.23–29; 360, n. to 161.38. The content of Tycho's message is unknown, but as described by Thoren 1990, 340–344, Tycho moved with almost indecent haste after the demise of Frederik II in April 1588 of that year to exploit his influence with the *Rigsraad* ruling on behalf of ten-year-old Christian, and was remarkably successful. An appeal to Wilhelm for further support could well have played a part in this strategy.

other observations that Wilhelm had sought to produce.[80] If Rothmann would not return to perform this role, and no one else in Germany could be persuaded to assume it, then, wrote Tycho, he would, 'willingly and with pleasure offer my service to you in this respect, and struggle with all diligence and zeal, sparing no expense, so that these deserving Atlas-like labours of your very great father on the subject of astronomy might be published as faultlessly as possible . . .'[81] This statement could be construed as indicating that Tycho was interested in the position of *mathematicus* to the Court of Hesse. But it seems unlikely that he would have coveted such a role for himself, rather than for one of his students. Although he was later to use presentation copies of his *Mechanica* as part of his determined bid to acquire a new patron for his astronomical work, at the time that he wrote the dedication to Moritz, he was only just beginning to experience the resistance to his demands on the resources of the Danish nation, the 'Machiavellian intrigues', that would induce him to leave his island and homeland.[82] Most probably, therefore, he envisaged being appointed to edit the observations whilst remaining at Uraniborg; a task that, even if it were never completed, would have extended his access to the data of Wilhelm and Rothmann. These data, as we saw in the previous chapter, had significant value to Tycho because of its power to corroborate and supplement his own observational work.[83]

Further perspective on the dedication of the *Epistolae astronomicae* may be gained by noting that dedications were much less decisive in securing long-term patronage than might be supposed. The two best-known cases of successful dedications by early modern astronomers, Galileo's *Sidereus Nuncius* (1610) and Tycho's own *Mechanica*, are not only anomalous in this respect, but also liable to a certain amount of misrepresentation themselves.

[80] *TBOO* VI, 15.33–38.   [81] *TBOO* VI, 16.5–12.

[82] For this description, see, *TBOO* VIII, 26.13–17, 49.20–25. It is noteworthy, however, that Tycho used the presentation of the letter-book to King Christian and his Chancellor Friis to press for further concessions from the Danish State. See *TBOO* XIV, 99–101; Dreyer 1890, 231–232; Thoren 1990, 368–375. Christianson 2000, 171–206, provides the richest account of the personal and political setbacks that Tycho suffered and that precipitated his departure.

[83] It is interesting to speculate whether, if Tycho had taken on the project, publishing the observations 'as faultlessly as possible' would have meant eradicating the five-minute discrepancy in the stellar longitudes. Perhaps it would: his pledge to Moritz was only that he would not abuse the material, nor appropriate anything contained therein, but would arrange everything to the honour of the House of Hesse. Moreover, as well as claiming that, during his visit to Hven, Rothmann had recanted his Copernicanism (on which, see below), Tycho indicated that the *mathematicus* came to accept that the discrepancy in the stellar longitudes was a product of an error at Kassel. See *TBOO* VI, 26.33–38. This might have served as a justification for altering the Kassel values as part of the process of publication. On the other hand, it is not clear whether, after the issue of the *Epistolae astronomicae*, he would have felt that such emendation was still necessary.

Galileo's success in winning Tuscan patronage derived not so much from the gift-value of the book itself, but rather from the carefully negotiated naming of the four newly discovered satellites of Jupiter as the Medicean stars, a gift to the House of Medici which the text then announced.[84] Likewise, Tycho's dedication of his *Mechanica* to Emperor Rudolf represented a gift much greater than the book: as underscored by his simultaneous presentation of two manuscript texts, containing his star catalogue and his work on the courses of the Sun and the Moon, what Tycho offered to Rudolf was the entirety of his astronomical project.[85] And whether this was decisive in securing Tycho Imperial patronage is, in any case, difficult to judge: Rudolf's decision to appoint him Imperial *mathematicus* appears to have preceded, not followed, his formal presentation of these books.[86] Often, however, dedications were acknowledgements of existing patron–client relationships rather than attempts to establish new ones. Many were made only with the hope of securing some immediate benefit, a small sum of money or a valuable item, rather than regular salaried employment.[87]

Occasionally dedicatees were chosen for other reasons entirely. In respect of Tycho's invitation to visit him at Hven, for example, Rothmann wrote that it was encouraging that Wilhelm had proposed the dedication of his work on the observations of the fixed stars to King Frederick II.

If he should persist in this judgement, I could come to you on that occasion. But if not, I wrote at Wittenberg, in the course of many sleepless nights, an *Organon Mathematicum* that contained the sexagesimal logistic, the doctrine of sines, the doctrine concerning the composition of ratios or the rule of six quantities, and the doctrine of triangles, both spherical and planar, clearly and briefly conveyed with many clever shortcuts, and very precious to me. I shall dedicate this to your Most Serene King so that an occasion for me to visit you is not wanting.[88]

Clearly, the results of the Kassel observing programme were to be in the gift of the Landgrave, and to the extent that he was happily maintained at the Hessen court, Rothmann's interest in the disposal of the work was actually quite minimal. A resulting nonchalance that might, in others, have been frivolously expressed,[89] emerged in Rothmann as a readiness to dedicate

---

[84] Drake 1957, van Helden 1989b; Biagioli 1990, 103–157; but see also Shank 1994; Biagioli 1996; Shank 1996.

[85] *TBOO* III, 335–389; *TBOO* V, 5–10, 165–189. [86] *TBOO* VIII, 163–166; Thoren 1990, 410–413.

[87] Zeeberg 2004, 45–50.

[88] *TBOO* VI, 118.19–29. This text was one of the twelve for which Rothmann obtained an Imperial privilege the following year. See Hamel 1998, 78.

[89] One example might be Heinrich Rantzau, who wrote or contributed to many treatises pseudonymously and, in at least one case, dedicated such a text to himself. See Mortensen 1994, 702; Zeeberg 2004, 14–15.

and present a book merely in order to provide a pretext for travelling to Denmark, and so to Uraniborg. His intention, expressed in the same letter, to address his cometary treatise to Frederick's chancellor, Nicholas Kaas, was perhaps similarly formed.[90] And to this suggestion Tycho gave a revealing response: despite not having the opportunity to discuss the topic with the Chancellor, he was sure that Kaas would find it an acceptable gift.[91] This illustrates the lack of any perceived impropriety in presenting a text to an individual with little personal interest or expertise in its subject. Princes and nobles often collected books as they collected works of art, curiosities, and specimens; to display their erudition (and their wealth), rather than to enlarge it.[92] Thus dedications were frequently addressed to an audience quite distinct from that to which the body of the text was directed, a practice that sometimes resulted in clear disparities in the characterisations and justifications of a work offered to the readers of its various parts.[93]

This was as true of the *Epistolae astronomicae* as it was of any other text of the period. Elsewhere in the prefatory material, as we shall shortly see, Tycho gave a quite different account of the genesis and purpose of the letter-book to that contained in the dedication to Moritz. And that it was not originally intended to be a memorial to Wilhelm, or a gift to his son, is made clear by consideration of the chronology of production. For as indicated in Chapter 2, Tycho's circulation of the letters in manuscript began almost as soon as the correspondence itself. And by February of 1590, more than two years before Wilhelm's death, Tycho was sending sheets of the partially printed text to individuals such as Hagecius, Peucer and Magini, and was proposing to conclude the work with the next pair of exchanges between Kassel and Hven.[94] So while it remains true to *one* authorial reading of the work to describe the text as a gift and memorial, this reading tells us little about Tycho's earliest motives for producing it. We have still to discover whom Tycho thought would read the body of the *Epistolae astronomicae*, and why he wished them to do so.

(iii) . . . or an astronomical text?

But indeed, to me personally this work did not seem as if it would be unprofitable or displeasing: since *Epistolae* contain many and various things, and are

---

[90] *TBOO* VI, 119.18–20.   [91] *TBOO* VI, 147.17–21.   [92] Garberson 1993; Jardine 1996, 190–201.

[93] Mortensen 1994; Westman 1990; Jardine 1991, 313–315.

[94] *TBOO* VII, 225.4–7, 225.41–226.4, 238.23–31. As Dreyer pointed out, *TBOO* VI, 343, Magini's receipt of pages 49 to 56 of the text (77 to 84 in his edition) was mediated by Gellius Sascerides. See *TBOO* VIII, 252.14–18.

eagerly read by very many people for this reason; and this with less boredom than other volumes treating almost the same subjects and discussing them at length. And indeed, the different opinions of others about the same things are more correctly recognised and evaluated when someone frankly and freely shares with a friend in letters that which he has ascertained and established on a topic he has studied. And so from this friendly and frank collation, truth is that much more readily and clearly extracted. And therefore many men renowned for their learning, both among the Ancients and the Moderns, have not unprofitably used this way of writing and conveying their findings to posterity. Wherefore, following in their footsteps, and gradually publishing this epistolary bundle, I also wanted to hold out an opportunity to other learned men, even among those who by chance are still unknown to me, so that concerning philosophical matters, above all indeed concerning those which deal with the *scientia* of the stars, they will not hold back from conversing in like manner with me through letters, or from helping me by supplying the material for this volume to be augmented by many books. In which case they would do something both pleasing to me and (I hope) not unuseful to others. (Tycho Brahe, General Preface to the *Epistolae astronomicae*)[95]

Tycho took the opportunity when introducing the *Epistolae astronomicae* to solicit letters for the further volumes of correspondence that he intended to publish. In doing so, he situated his *libri epistolarum* within a well-established tradition of letter publication for didactic purpose, one that encompassed not only the collections of ancient authors, such as Cicero, Pliny the Younger, and St Jerome, but also those of the humanists who followed their example. Collected letters were often, particularly in the schools, appreciated more for their literary exemplarity than for the nature of their contents. This was a natural consequence of the fact that the letter, considered a simpler exercise than the oration or the declamation, was typically used to introduce students to the arts of composition.[96] Yet even in the rhetorician's classroom, the detail of an epistle, its historical or political content, might be thought worthy of study alongside the author's diction and style.[97] In a professional discipline such as medicine, moreover, collections of letters could also serve as textbooks, and were frequently confected or compiled for this purpose.[98] Generally speaking, indeed, in the sixteenth century the letter held the place that would later be occupied by

---

[95] *TBOO* VI, 22.4–22.   [96] Henderson 1983a, 93.   [97] Grafton 1991, 23–26.

[98] *Epistolae medicinales* such as those of Pier Andrea Mattioli (1501–1577), Conrad Gesner (1516–1565), and Johannes Crato von Krafftheim (1519–1585) are likely to have come to Tycho's attention. Such collections were closely related in form and value to collections of medical *consilia*, physicians' written reports on the case and care of an individual patient, but in conformity with the recognised flexibility of the letter might accommodate a wide range of topics of medical interest and mediate disputes. See *DSB* IX, 178–190; Durling 1980; Siraisi 1991, 588–590; Agrimi and Crisciani 1994; Delisle 2004.

the essay and was, for this reason, often anthologised.[99] It would, therefore, have been tedious and unnecessary for Tycho to compile a list of the authors, both ancient and modern, who had, 'used this way of writing, and of conveying their findings to posterity', and just as tedious and unnecessary for his contemporaries to read it.

The visibility of the printed epistle was not, in any case, only due to its appearance in letter-collections. Even if the many prefaces, introductions, and addresses *ad lectorem* that adopted the epistolary form are discounted, one finds letters produced, reproduced, and excerpted within all manner of larger texts, and also, on occasion, published as pamphlets. In this fashion, many epistolary texts touching on astronomical matters preceded the Hven–Kassel correspondence through the presses. The *Narratio prima* (1540) of Georg Joachim Rheticus (1514–1574), the first published account of the Copernican world-system is perhaps the most famous example; others known to Tycho included the *Epistola ad Martinum Mylium* (1580) of his correspondent Hagecius, and the collection of letters published by Giovanni Battista Benedetti (1530–1590) in the fifth part of his *Diversarum speculationum liber* (1585).[100] And in the *Progymnasmata*, Tycho himself printed several letters discussing the 1572 nova, not all of which, to briefly return to an earlier theme, belonged to him as either addressee or author.[101] Thus the sentiment strongly expressed by Tycho, that printed letters might prove instructive for those with an interest in the heavens, was by no means unique to the *Epistolae astronomicae*.

Having already acquired a sense of the contents of the Hven–Kassel correspondence, it is tempting to suppose that this is all one requires in order to appreciate the astronomical value of Tycho's edition. Perhaps because he inclined to the same opinion, the second, general, preface to the book offered little more than a précis of the topics tackled within the exchanges: 'We discussed the construction of instruments, and their precise correction', he wrote, 'various astronomical observations with respect to the Sun, the Moon, and the rest of the stars; accurately determining the altitude of the Pole; discerning the greatest obliquity of the ecliptic more thoroughly than it was done before; astronomical hypotheses; the optical theory pertaining to refractions of the luminaries and the stars; and twilights'.[102] Elsewhere, however, he drew specific features of

[99] Clough 1976. Constable 1976, 39, notes that between 300 and 400 editions of letter-collections were published before 1580.

[100] Rosen 1959; Burmeister 1967–1968, I, 44–50. For Tycho's knowledge of Hagecius' text, see *TBOO* IV, 287.11–17, 507 n. to 287.12; for his use of Benedetti, see *TBOO* III, 251–253, and below.

[101] For example, the letter of Peucer to Hieronymus Wolf, 7 December 1572, *TBOO* III, 49.18–50.12.

[102] *TBOO* VI, 25.41–26.11.

the text to the reader's attention. Thus, in the first of his two longest interpolations *ad lectorem*, he accounted for the reproduction of astronomical data that, when it was originally sent to Wilhelm, he had asked be kept secret:

Although in the first volume of *De recentioribus phaenomenis* I have treated the correction of the places of the fixed stars more fully and, as it were, openly, nevertheless it is pleasing to add here those proceeding from our workshop which at that time I sent to the Most Illustrious Prince Wilhelm, Landgrave of Hesse, in accordance with his request, so that students of astronomy might see that the correction subsequently found, if any, was very slight and almost undetectable, and that our findings always remain highly consistent with one another; and so that they might also have here some intervals between the fixed stars, not written down in the earlier volume, with which they might exercise themselves, and test their instruments, if they happen to have any.[103]

Previously, Tycho had expressed the concern that his labours might be appropriated by unscrupulous rivals. But in the case of the stellar data, at least, it seemed that the very act of publishing his positions then made this consideration redundant. And by including this material, Tycho not only imparted to the letter-book the ability to function as confirmation of the reliability of his data; or rather, since the work that he named, volume one of the *De recentioribus phaenomenis*, was never completed, as a kind of promissory note for the data that would later appear in the *Progymnasmata*. He also enabled it to perform some of the same tasks as the original correspondence: facilitation of the comparison of the Hessen and Tychonic observations, and calibration of distant instruments to a high (Tychonic) standard.

That Tycho should have wanted the *Epistolae astronomicae* to be used in such a way fits well with his desire, as identified in the last chapter, to universalise his astronomical project. Such ambition could be taken as evidence of the domineering personality often ascribed to Tycho by his biographers and other historians, but it is difficult to see how he might have pursued the reform of astronomy without believing that what he found to be true should be more widely adopted. This was no less true of cosmology than it was of positional astronomy; and indeed, it was cosmology that exercised Tycho in the longest of his editorial interventions, made just after the appearance of the last-but-one letter from Rothmann. It was shortly after writing this letter that the *mathematicus* had visited Uraniborg, and

---

[103] *TBOO* VI, 75.37–76.6.

although in the course of the correspondence Rothmann had never wavered in his advocacy of Copernicanism, Tycho was here able to announce that:

while he was staying with me here for some weeks, I spoke with him frequently about this matter in person, and I destroyed his arguments, plausible enough in appearance, and I ultimately drove the man, otherwise persisting completely in his purpose, to the point where he at first hesitated somewhat, later more timidly and less certainly maintained these things, and finally seemed to wholly repudiate his former conceptions: to the extent that he affirmed that he had as yet published nothing on this subject, and would not subsequently do so, but had only declared these things to me for the sake of disputation.[104]

It is possible, given this account, Rothmann's subsequent departure from the astronomical scene, and the knowledge that he was suffering from a debilitating illness, to conjure a portrait of the *mathematicus* as a rather wretched figure, wracked by physical discomfort and harangued mercilessly by Tycho to the point of despair. Indeed, some commentators have connected Rothmann's illness and his reported recantation by suggesting that the trip to Denmark, and any cosmological concessions he made while visiting Tycho, were both motivated by a desire to obtain alchemical medicines from Uraniborg's laboratory.[105] Such an interpretation would have the merit of explaining the anomaly of Rothmann's sudden apostasy without recourse to the unsavoury hypothesis, that it was nothing more than a fabrication by Tycho. On the other hand, it would not exculpate Tycho entirely with respect to the charge of deception; in one of his letters to Wilhelm written after Rothmann's departure from Hven, Tycho indicated that the Hessen *mathematicus* had not divulged to him the nature of his illness, and consequently did not receive an appropriate remedy during the course of his stay. This remark is difficult to reconcile with the suggestion that this was precisely the purpose for which Rothmann made his way to Uraniborg.[106] But however one accounts for it, it was certainly extremely convenient for Tycho to be able to represent Rothmann as a reformed Copernican, a talented but misguided individual ultimately brought to his senses, since it gave him the opportunity to conclude the cosmological debates contained in the letters in a manner that he must have considered highly satisfactory. 'But so that to some extent it may be evident to others by what arguments to the contrary the objections he made can be overthrown', he wrote, 'I should like to add some of them here . . .'[107] Over the next five pages of the *Epistolae astronomicae*, he proceeded to do so.

---

[104] *TBOO* VI, 218.10–17. [105] Barker and Goldstein 1995, 400–401; also Barker, 2004.
[106] *TBOO* VI, 237.36–39. [107] *TBOO* VI, 218.19–23.

As the last word in the disputes discussed in the previous chapter, the arguments presented by Tycho are worth summarising here. In defending his belief in the Copernican hypotheses, Rothmann had asserted that the perpendicular descent of falling bodies was no obstacle to the attribution of a diurnal motion to the Earth; indeed, he argued that Copernicus had dealt with this criticism himself by pointing out that an object might possess a downward rectilinear motion and nevertheless share in the circular movement of a daily rotation.[108] Tycho responded by rejecting the principle of motion – inertia, as we have since come to think of it – underlying this statement. Participation in the diurnal motion would be possible, he conceded, if one supposed that the air rotated with the Earth and carried objects with it; but not in the case of solid bodies that were seen to cleave the air freely as they fell. And while some of these objects, made of gold or other metals, might indeed preserve in themselves the natural force producing the motion of the Earth, the womb from which they had been taken, it could not be conceded that others – such as stones or dry timbers – would similarly retain the vigour of the whole.[109] Consequently, one would expect the diurnal rotation, if any existed, to be clearly manifested in the trajectory of almost all falling bodies. But in the case of a shot fired from a cannon, where one could see clearly the interaction between the violent force imparted by the gunpowder, and the natural force of gravity, no such effect was apparent. For there was no detectable difference in the distance traversed by a cannon ball fired to the west, in accord with the Earth's postulated rotation, and one fired in the opposite direction. And yet, Tycho asserted, 'an iron ball fired at an angle from the greatest cannon, which they call a double Curtow, will scarcely reach the ground, its motion spent, within two minutes. In which time it should be, at the parallel of Germany, revolved twenty thousand paces more by the diurnal motion . . .'[110] In the absence of invincible arguments as to how the cannon shot could partake of a terrestrial motion without any effect on its trajectory, this was, Tycho thought, sufficient proof that the Earth underwent no daily rotation.[111] And in the analogous case, a missile fired vertically upwards on board a ship, it was clear (at least to him) that the position at which the object came to land on the deck varied according to whether the ship was moving or not, and the speed of its motion.[112]

---

[108] *TBOO* VI, 215.25–31. See Copernicus 1543, 6r–7v.   [109] *TBOO* VI, 218.48–219.9.

[110] *TBOO* VI, 219.24–46. The experiments had been performed, not by Tycho, but by Wilhelm, 'although not for this reason, but only to test the span and duration of the ball's forward motion'. See *TBOO* VI, 219.46–49. On 'double Curtow' as a type of cannon, see Blackmore 1976, 225. I am indebted to Ruth Rhynas Brown for this translation and reference.

[111] *TBOO* VI, 219.50–220.6.   [112] *TBOO* VI, 220.16–21.

Since the principles of motion implicit in these claims had already been rejected by Rothmann in his letters,[113] Tycho's response really amounted to no more than a refusal to subscribe to the terrestrial dynamics his opponent had adopted. However, in addition to claiming that the Copernican system led to physical absurdities, Tycho also argued that it improperly represented and interpreted Providence. Although it might seem that a diurnal rotation of the Earth would be more easily accomplished than a diurnal rotation of the whole sphere of the fixed stars:

> in this shines forth the inscrutable wisdom and power of GOD the illimitable Maker, who both wished and was able to impart to the bodies of heaven, so vast, a motion inconceivably swift, and at the same time uniform, divided and double. And who established the body of the Earth, inactive, heavy and unsuited to circular and perpetual motion, in its immobility, so as to be unmoved in perpetuity; whereby it is possible here, as if from the still centre of the heavenly bodies, which exist in a shining and fiery but inconsumable state, and which are wholly adapted to the swiftest motion, and freely pursue it, to behold their astonishing and unwearied courses, and to look upon the majesty of the divine Numen as if in a mirror.[114]

The belief that study of the heavens provided a privileged route to understanding of the Divine was one that, as we have seen, Tycho and Rothmann held in common. The Danish astronomer elected to differentiate himself, however, with regard to what could legitimately count as providentially ordered. Aristotle was not, Tycho claimed, incorrect to attribute a 'certain divinity' to the sky, even though it was finite in size and duration, because, as he had recognised, a 'certain infinity' was present in its motion. Movement, more spiritual and more worthy than stillness, had therefore to be ascribed to the heavens, not to the sublunary realm of the transient.[115]

Given the arrangement of his world-system, Tycho had no recourse to a mechanism of substantial rotating spheres to account for either the transmission of motion from the sphere of the fixed stars to the other celestial bodies, or the individual planetary courses. As discussed in Chapter 2, he came to espouse a doctrine of fluidity that, however he arrived at it, was far from unprecedented. It is not surprising, therefore, that in order to explain the regularity of the heavenly motions, he resorted to the analogy favoured by the ancient Stoic philosophers. 'It is clear', he wrote, 'that this

---

[113] *TBOO* VI, 215.25–217.28.
[114] *TBOO* VI, 220.48–221.8. The phrase 'established [the Earth] . . . in its immobility' (sua stabilitate fundavit) recalls several passages in Psalms, for example Psalm 23 in the Vulgate, 'Domini est terra et plenitudo eius / orbis et habitatores eius quia ipse super maria fundavit eum et super flumina stabilivit illum.'
[115] *TBOO* VI, 221.10–18.

[movement] occurs no differently, to some extent, than for the fish in the sea, or the birds in the air (if it be permissible to compare the small to the great)'.[116] As these animals might be moved by the ebb and flow of the tides, or by high winds, Tycho suggested, so the planets were subject to a movement of the ether; meaning, presumably, one generated by the rotation of the sphere of fixed stars. Yet creatures also move independently, through their own volition or 'natural desire', and planets do likewise – except that their motions are more enduring and more perfect than those of any terrestrial animal.[117] Tycho did not shy away from acknowledging the metaphysical implications of this picture: 'As that divine philosophy of the Platonists seems to have appropriately realised', he stated, 'heaven is animated, and the heavenly bodies are themselves animated, endowed with the living spirit of a particular heaven'.[118] Tycho's geoheliocentric world-system was clearly dependent on a vitalist conception of the cosmos.

Having considered, and dismissed, the attribution of a diurnal rotation to the Earth, Tycho moved onto the subject of the second terrestrial motion described by Copernicus, its annual orbit of the Sun. Towards the end of their correspondence, Tycho and Rothmann had clashed over what Tycho considered to be the manifest absurdities generated by positing such an annual motion in the absence of any visual consequences. Accommodating the lack of annual stellar parallax required a vast expansion of the apparently empty space between the fixed stars and Saturn, and a concomitant increase in the size of these distant stellar bodies. Stars of third magnitude would therefore have to equal the proposed terrestrial orbit in diameter in order to account for their appearance on the Earth.[119] Rothmann, in response, had argued that there was nothing self-evidently absurd about these particular consequences of the Copernican hypotheses, and suggested that to suppose the contrary was to set arbitrary limits on divine power and will.[120] Tycho's rebuttal of this objection rested upon the conviction that one could infer a great deal about the divine construction of the universe from study of its individual parts and from consideration of the relationship obtaining between them. Thus, that the created world displayed symmetry, harmony, and proportionality was demonstrated by anatomical as well as astronomical evidence: 'For the parts and limbs of

---

[116] *TBOO* VI, 221.32–42. 'If it is allowed to compare the small to the great' appears in Virgil's *Georgics*, IV.176; c.f. Ovid's *Metamorphoses*, V.416–417.

[117] *TBOO* VI, 221.42–49.   [118] *TBOO* VI, 221.45–49, *cont.*

[119] This objection was partly dissolved following the introduction of the telescope, which reduced the extent to which stellar images were affected, and hence enlarged, by scintillation or twinkling. See van Helden 1989a, 106–108.

[120] *TBOO* VI, 221.49–222.16.

all animals are ordered and arranged proportionally to one other, in such a way that any one of them holds an exact ratio to the whole, and to the other parts. Just as that exceptional painter DÜRER makes clear in those things which he demonstrated from the symmetries of the human body (which is a Microcosm) . . .'[121] Yet proportionality of this kind was lacking in the Copernican conception of the universe – the putative macrocosm – when the disparate sizes of the fixed stars and the Sun, or the ratio of the space occupied by planetary bodies to that left vacant, was studied and considered. Reflections such as these, Tycho claimed, finally helped to persuade Rothmann that heliocentric hypotheses resulted in an absurdity from which they could not be rescued by the doctrine of omnipotence. 'That ungeometric, and asymmetric, and disordered way of philosophising would produce something very foreign to divine wisdom and providence', he wrote, 'which even Rothmann himself, when he was with me, did not dare to continue to defend . . .'[122]

Tycho's explanation for the insertion of this lengthy addition to the Hven–Kassel exchanges, suggests that, just as the classroom role of letter-books was to provide students with models for engaging in epistolary conversations, the *Epistolae astronomicae* was partly intended to supply a model for participating in, and winning, cosmological debates. Like its inclusion of observational data and information about Uraniborg's instruments, this feature of the *Epistolae astronomicae* diminishes the contrast between it and the astronomer's other printed texts. And that contrast is further weakened by consideration of Tycho's conception of its readership, insofar as this is suggested by what is known, or may reasonably be inferred, about the book's distribution. Thus, although the *Epistolae astronomicae* was made available for purchase at Frankfurt in the spring of 1597, it was also sent or presented by Tycho as a gift to many individuals other than Moritz. Excluding the unusually large number of princes and chancellors to whom he presented his *Mechanica*, often in conjunction with a manuscript *Catalogus* of stars, after May 1598, it can be seen that there was a significant overlap between recipients of the *Epistolae astronomicae* and those of Tycho's other texts. More than three-quarters of the scholars known to have received a gift-copy of the *Epistolae astronomicae* were also given the *De recentioribus phaenomenis*, the later *Mechanica*, or both of these works. Many of these individuals were themselves authors of astronomical or astrological texts. (See the Appendix, I). On this evidence, therefore, Tycho was being entirely

---

[121] *TBOO* VI, 222.43–47. Tycho is perhaps thinking of Dürer's *Vier Bücher von Menschlicher Proportion* (1528). On this work, inspired by Dürer's reading of Vitruvius, see Panofsky 1955, 260–267.
[122] *TBOO* VI, 222.27–31.

sincere when he suggested that, except for its greater accessibility, there was little to distinguish his letter-book from other, more familiar, kinds of astronomical work.

Tycho's intended audience for the completed *Epistolae astronomicae* was not only signalled by his distribution of the text after the fact. Tycho also named certain of his readers in the general preface to the work, suggesting that it was these individuals, and not Hieronymus Treutler, who had provided the real motivation for the letter-book's appearance in print:

> while through so many passing years I have been examining the heavens with unflagging effort, many great men, distinguished and most illustrious in philosophical and, especially, in mathematical learning, have exchanged letters with me from diverse parts of Europe about this most exalted matter: which, although I had decided from the beginning to keep utterly to myself (as for a long time was the case), I was at length persuaded by the exhortations of friends and excellent men that I should not refuse to print and share with others to be read freely. But most of all these led me to it: the Most Noble and Renowned man, HEINRICH RANTZAU, Lord of Bredenburg, etc., the Splendid Vice-Regent of our Most Serene King in Holstein and adjoining dominions; then also that most renowned man, of great fame among the Germans, master CASPAR PEUCER, philosopher and most celebrated physician; and furthermore that most distinguished master, the Bohemian THADDAEUS HAGECIUS of Hajek, who is most skilful in medicine, which he practises, and equally in philosophy and above all in astronomical matters; and, which I consider to count for more, he is loving of truth and virtue, and is uncommonly furnished with purity and noble-mindedness of conduct.[123]

Having received manuscript copies of parts of the Hven–Kassel correspondence, Rantzau, Peucer, and Hagecius were certainly in a position to be able to recommend that Tycho print the exchanges. But the letter which Tycho published to illustrate the exhortations of these correspondents was ostensibly written by Rantzau on 17 January 1587.[124] This date is somewhat surprising, and has important implications. If correct, it would mean that the Holstein Governor asked the astronomer to see to the publication of 'the letters of the Most Illustrious Prince Landgrave, written to you, and yours to him, and also to his Most Excellent *Mathematicus*', as well as those Tycho had exchanged with other learned men on similar subjects, at a point in time when he is thought to have seen only a single item sent to Kassel from Hven.[125] And it would indicate that Tycho was urged to consider an

---

[123] *TBOO* VI, 21.26–22.1.   [124] *TBOO* VI, 29–30.
[125] *TBOO* VI, 30.14–21. Tycho sent Rantzau his first letter to Wilhelm on March 23rd 1586. He next sent Rantzau copies of his own letters and those of the Kassel astronomers on 19 October 1587. See *TBOO* VII, 101.37–102.7, 385.9–16.

edition of his correspondence with Rothmann and Wilhelm less than a year after these exchanges commenced. Assuming that he was not slow to appreciate the merits of this scheme, most of Tycho's own contributions to the correspondence would therefore have been written in full knowledge of the fact that they would later come to be printed.

There are, however, grounds for doubting the authenticity of Rantzau's letter of 17 January 1587. It is only slightly troubling that Rantzau mentioned Tycho's letters to Rothmann before he saw them – before, indeed, any existed. Although the first of these letters was not written until three days after Rantzau's exhortation was supposedly composed, the identity of the *mathematicus* was known to him from those parts of the Hven–Kassel correspondence that he had already seen, and the fact that Tycho would write to him individually could perhaps have been inferred.[126] It is even possible that Rantzau was independently in receipt of copies of the letters emanating from Kassel, and therefore had access to a greater proportion of the correspondence – including a letter from Rothmann to Tycho – than his Danish correspondent had supplied. But the fact that disavowals of authorial responsibility were a standard feature of the ancient prefaces considered exemplary by early modern scholars may give some cause for concern. By stating that a work had been produced in response to others' requests, classical authors, and those who emulated them in the sixteenth and seventeenth centuries, sought to avoid the charge of immodesty that might follow from appearing too eager to circulate their writings themselves.[127] And Tycho was certainly not unaware of this strategy. He employed it, indeed, when he had his *De nova stella* produced with a prefatory epistle exhorting him to publish that, as its author well knew, was actually written three weeks after printing of the text had already commenced.[128] It is not impossible, therefore, that Rantzau's letter was prompted by Tycho, and was perhaps devised and composed some time after the date under which it was printed. If this were the case, it would be rather significant, for Peucer and Hagecius certainly did not push for the printing of the correspondence, as Tycho suggested, but merely added their polite endorsement to his existing plan to do so when he raised it himself.[129] On the other hand, since the trope was so familiar and widespread, it also seems plausible that individuals anticipated their role by urging close acquaintances

---

[126] See *TBOO* VI, 32.34–37, 33.39–34.7.   [127] See Janson 1964, esp. 41, 54–55, 60–61, 116–117.

[128] Thoren 1990, 64.

[129] See, for example, Tycho's letter to Peucer of 13 September 1588, in which Tycho mentions the notion of publishing the letters in the very act of revealing that he is enclosing some copies – attributing the idea to Rantzau – and Peucer's predictably polite response; *TBOO* VII, 140.1–9, 140.22–30, 191.7–10.

to publish material with very little provocation.[130] We cannot be certain, therefore, to whom the idea of publishing the correspondence first occurred, Tycho or Rantzau. Consequently we cannot say precisely when this idea was conceived; even if, given what we know about the contents of Tycho's letters to Wilhelm and Rothmann of 18 and 20 January 1587, and his readiness to circulate them in manuscript, it seems far from implausible that publication was already in Tycho's mind by the time Rantzau's letter, purportedly composed just in advance of these two, would actually have reached him.

Yet if this remains ambiguous, one thing seems quite clear. Whatever other purpose it may have served, naming Rantzau, Hagecius, and Peucer in the general preface of the *Epistolae astronomicae* gave prospective readers another indication of the sort of person whom Tycho supposed would welcome, and benefit from, his edition of the letters. All three were eminent men, distinguished by past or present positions of responsibility and power, who had themselves been responsible for publications concerning the study of the heavens: works such as Rantzau's *Catalogus* of men who had promoted astrology, Peucer's *Elementa doctrinae de circulis coelestibus*, and Hagecius' *Dialexis* on the nova of 1572 and *Descriptio* of the comet of 1577. All of them, therefore, were recognisable as scholars who possessed more than a modicum of expertise and interest in astronomical matters. Even allowing for a desire to dignify his intended audience by naming the better qualified among its members, this again indicates that Tycho sought to address the *Epistolae astronomicae* not so much to *all* men of learning and goodwill, as to those who could read it, and appreciate it, as a work of astronomy.

Whether Tycho's readers did view the text in this way, however, is a different matter. The issue of how the *Epistolae astronomicae* was actually received is one that we shall come to eventually. Before we do, however, it seems appropriate to reflect on what it meant to be a reader of astronomical works in the late sixteenth century. How did one approach the texts important for the study of astronomy and cosmology, and how did one acquire them? These are questions that, keeping our focus still on Tycho, we shall now turn to consider.

---

[130] Arguably, moreover, Rantzau was precisely the sort of individual who would urge publication of the letters at the earliest opportunity. From 1593 onwards, he himself published several editions of *Epistolae consolatoriae*, i.e. letters of condolence sent to him by various correspondents after deaths in his family. See Zeeberg 2004, 18–21.

II ASTRONOMICAL READERS

(i) Textual astronomy

For if I am not mistaken, we are sinning when we obscure the opinions of noble authors by contaminating them with our own ignorance and infecting posterity with erroneous copies of books. For who does not realise that the admirable art of printing recently devised by our countrymen is as harmful to men if it multiplies erroneous works as it is useful when it publishes properly corrected editions. (Regiomontanus, *Disputationes contra Cremonensia in planetarum theoricas delyramenta* (*c*.1474))[131]

Representation of Tycho Brahe as an exemplary figure of the early modern 'print revolution' has rested not only on his ownership and use of a press but also on the nature of his introduction to the study of the heavens. As a young man learning astronomy against the wishes of his family, Tycho taught himself much of the art through the study of books. Indeed, as we have noted before, he himself explained that it was his youthful comparison of different tables and ephemerides, both with one another and with the phenomena visible in the heavens, that first alerted him to the 'intolerable error' in the texts, and thereby persuaded him of the need for the reform of astronomy.[132] This tale, certain historians have suggested, illustrates several of the key changes brought about by the invention of printing. It demonstrates how the increased availability of texts facilitated the acquisition and comparison of several works, where only one would have been used previously. It shows that the fidelity inherent in the new process of replication allowed individuals to see past incidental flaws, of the sort that had previously marred each copy of a work, to the more fundamental errors of the text as it had originally been written. And it indicates how elimination of the requirement for scholars to spend their time transcribing books they wished to study, enabled the unlikeliest of characters to make the transition from neophyte to expert with unprecedented speed.[133]

As indicated above, however, claims such as these about the distinct character of 'print-culture' have now been criticised by several experts in the field of book history.[134] Even so, the thesis that the printing press at least

---

[131] As translated by Pedersen 1978, 177.
[132] *TBOO* V, 107.22–25. On Tycho's early acquisition of astronomical texts, see also Thoren 1990, 10–11.
[133] I am paraphrasing Eisenstein 1979, esp. 596–602. For a largely favourable response by a historian of astronomy, see Westman 1980b.
[134] Grafton 1980; Hellinga 1983; Long 1997; Johns 1998b, 6–57; McKitterick 2003, *passim*.

contributed to, if it did not quite cause, the epistemological transformations of the later sixteenth century, retains a certain robust plausibility that many historians of science have found hard to resist.[135] It is easy to see why, with respect to astronomy and cosmology at least, this should be the case. The tradition of studying the heavens that extended back from the early modern period to antiquity always depended at least as much upon the transmission and interpretation of written sources as on direct inspection of the heavens. Indeed, important developments in late medieval astronomy in the Latin West were almost solely the result of the receipt and circulation of ideas and techniques originating in, or transmitted through, Islamic tables and texts.[136] And given the limited place of mathematical astronomy in the formal curricula of the Middle Ages, it seems likely that the importance of *loci* such as Oxford, Paris, and Vienna, as centres for its study was as much a consequence as a cause of the accumulation and availability of relevant manuscripts.[137] With texts so fundamental a part of astronomical practice, it might seem counter-intuitive to suppose that the effects on that practice of a profound alteration in the way of producing them would not be equally striking.

It is certainly true that much astronomical material underwent an early shift from manuscript to print. Gutenberg himself produced a table of new and full moons at Mainz in 1448.[138] More significantly, perhaps, Regiomontanus produced the first edition of Peurbach's *Theoricae novae planetarum* at his own printing office in Nuremberg, *c*.1472,[139] and advertised his intention to issue not only astronomical and mathematical works by authors such as Ptolemy, Euclid, Theon of Alexandria, and Proclus, but also his own innovative texts.[140] And although his untimely death prevented much of this plan being put into effect, Regiomontanus was soon followed by other astronomers with ready access to a press. Both Peter Apian (1495–1552), professor of mathematics at Ingolstadt, and Johannes Schöner (1495–1555), later a professor at Altdorf, managed a press of their own,[141] while Oronce Finé (1494–1555), Regius Professor at Paris, collaborated with several

[135] Jardine 1996, 135–180, 349–356; Christianson 2000, 103–104.
[136] For an introduction to this topic, see the essays in Walker 1996, particularly those by King, Pedersen, and Swerdlow; Montgomery 2000, 138–185.
[137] Lemay 1976; Kren 1989; Pedersen 1996, 179–180. For a good example of a medieval autodidact, see the work on the astrologer Pierre d'Ailly by Smoller 1994, esp. 43–60.
[138] Zinner 1964, no. 1; Pedersen 1978b, 157.
[139] On Peurbach and his text, see Aiton 1987; *DSB* XV, 473–479.
[140] Ehrman and Pollard 1965, 27–28; Pedersen 1978b, 157, 169–170; Swerdlow 1993a, 154–162.
[141] For Apian, see Gingerich 1971a, esp. 168; *DSB* I, 178–179. For Schöner, see Coote 1888; Schottenloher 1907; *DSB* XII, 199–200.

printing operations.[142] Certain publishing enterprises, including those of Erhard Ratdolt, Johannes Petreius, and Guillaume Cavellat, were particularly active in the production of mathematical and astronomical material.[143] And an increasing number of scholars who were not themselves printers, sent astronomical works to the press as the sixteenth century proceeded: not just treatises, tables, and instrument-books, but also the more ephemeral and popular broadside prognostications, verse texts, and almanacs.[144]

Yet despite, and to some extent because of, the proliferation of printed works, it is the textual element of astronomical practice itself that most constrains the claims that can be made about the transforming power of print. Astronomical books were not simply the products of astronomers' efforts, they were also an indispensable resource for the empirical reform of the discipline. Weather and location presented obstacles to the observation of contemporary phenomena that, as we have seen, encouraged individuals such as Tycho to seek data from others by means of correspondence. Manuscript treatises and books, equally, could grant access to the information that could be used to supplement or corroborate the products of his own observational labours. But while these contingent difficulties may have promoted the turn to others' texts, the need to take account of long-term celestial phenomena rendered it essential. As Tycho noted regarding the precession of the equinoxes, 'it is not consistent with the life-span of one man to obtain some detectable variation in this matter'.[145] And Rothmann expressed the same thought in a way that was even more explicit. Unlike the other liberal arts, he asserted, astronomy would be impossible to restore were its texts to be destroyed.[146] An astronomer had to inspect the heavens not just as they appeared in the skies, but also as they were represented in books.[147]

The reading of astronomical texts, however, was not always straightforward. Careful analysis might be required for the proper evaluation of the numbers to be found on the pages, just as much as for data obtained through observational means. A good example to be found within the

---

[142] On Finé, see *DSB* XV, 153–156; Hillard and Poulle 1971, esp. 335–349; Ross 1974, 1975.

[143] Shipman 1967; Risk 1982; Pantin 1988.

[144] The latter categories of text typically possessed a wider audience than the former, of course; but were by no means spurned by society's elite. See Capp 1979; Hammerstein 1986; Armstrong 1990, 16, 35–36; Niccoli 1990; Pantin 1995.

[145] *TBOO* VI, 73.26–27.

[146] The comment was made in ch. 18 of his unpublished *Observationum stellarum fixarum liber primus*. See Granada, Hamel and von Mackensen 2003, 174.15–19.

[147] As other historians have shown, astronomers also had to acquire certain abilities as readers in order to meet the expectations of broader scholarly communities and patrons. See Grafton 1991, 178–203, 1992 and 1997a, 185–224; daCosta Kaufmann 1993, 136–150.

Epistolae astronomicae concerns the shift in the latitude of the fixed stars, a phenomenon that was first brought to Tycho's attention when Wilhelm pointed out that Rothmann's reduced observations showed a difference of up to half a degree from the values to be found in earlier catalogues.[148] Rothmann himself thought that the long-postulated alteration in the obliquity of the ecliptic was the cause of the difference.[149] Since celestial latitude is measured from the ecliptic, any alteration in the obliquity would change this value for every celestial body, without implying any movement on the part of the stars. But the data did not wholly fit this explanation; a fact, the *mathematicus* wrote, 'that should be ascribed (I think) to the observations of the ancients, among whom sixths or even thirds of a degree were not of great importance; or also an error of this sort could have intruded in the copying . . .'[150] Tycho readily agreed with both of these suggestions. Indeed, the Danish astronomer went further than his correspondent, asserting that, like *scriptoria*, printing offices not only reproduced mistakes, they also created them.[151] With the century between Regiomontanus and Tycho having done little to diminish the ambivalence astronomers felt about the products of the printing press, individuals had to be prepared to read their books with considerable care.

When it came to tackling the written legacy of their predecessors, early modern astronomers deployed similar critical tools to those deployed by scholars working on other kinds of text.[152] Thus Tycho, investigating the shifting stellar latitudes in the pages of Ptolemy's *Syntaxis*, first looked to eliminate printing errors by an elementary procedure of collation: he compared his standard text with, 'a certain old copy of the *Almagest*, published nearly seventy years ago in Venice, which expresses the declinations not in numerical figures, as the Trebizond version does, but in letters to aliquot parts of a degree'.[153] It was a simple step, hardly comparable to the philological *tours de force* of some humanist scholars of the late sixteenth century, but

---

[148] *TBOO* VI, 49.23–29.

[149] *TBOO* VI, 57.13–18. The variation of this parameter with time had been identified by Copernicus, among others, and assigned a non-linear periodic quality in *De revolutionibus*. See Neugebauer and Swerdlow 1984, 129–148.

[150] *TBOO* VI, 57.18–25.   [151] *TBOO* VI, 71.38–72.5.

[152] For the humanist strategies of textual criticism and interpretation, see Kenney 1974; Grafton 1983–1993, 1991 and D'Amico 1988b. For some other examples of humanistic reading practices of astronomers/astrologers, see Grafton 1992 and 1999, 127–155.

[153] *TBOO* VI, 95.3–16. This was the 1515 edition by Petrus Liechtenstein of Gerard of Cremona's translation from the Arabic, and the first printed version of the *Syntaxis*. See Nørlind 1970, 358. There were reasons other than its use of figures to distrust George of Trebizond's work: Stephanus Gracilis, who in 1556 produced a translation of Book II from the Greek, indicated that he had begun by attempting to correct the existing version, but had found it so poor that it was better to begin again. See Pantin 1988, 244.

one that recognised an important fact: whereas a single miscast character in a word was usually obvious, and rarely destroyed the sense of the text, the same occurrence in a number could be as pernicious to the meaning as it was difficult to spot. In addition to a humanistic sensitivity to the normal processes of textual transmission and corruption, however, interpreters of astronomical texts had to apply expert knowledge of their own.[154] Tycho's decision to use the declinations of the stars in this analysis, rather than the latitudes given, was strongly informed by his astronomical knowledge. The ecliptic coordinates given in the Ptolemaic catalogue were not to be trusted, he thought, because in respect of the interval between the stars they not only failed to agree with his own observations, but were mutually inconsistent. For this reason, he only accepted the reported longitudes of a single star, Spica in the constellation of Virgo, which he thought Ptolemy and his predecessors must have located with a fair degree of precision. Applying his own values for the longitudinal interval between Spica and other stars, he arrived at the supposed antique stellar longitudes; he could then determine the co-latitudes from the spherical triangle these formed with the complement of the reported declinations.[155] This was a procedure, Tycho thought, that allowed him, 'wheresoever incoherence was clearly apparent', to 'bring the Ptolemaic numbers themselves into consideration'.[156] It enabled him, in other words, to test not only the hypothesis of the shifting ecliptic, but also the data-set on which this theory depended.

According to Tycho's analysis, the southern stars Castor and Pollux confirmed the supposition that the latitudinal shift was caused by the postulated change in the obliquity of the ecliptic.[157] But in the case of Capella, Ptolemy's declination for the star gave a latitude of $22^{5/6°}$, whereas he had stated a figure of $22° 30'$; only if an intermediate value of $22° 36'$ was adopted, did the change in latitude over time agree with the theory.[158] The star in the rear shoulder of Orion was consistent with the altered ecliptic provided that the declinations of Timocharis and Hipparchus were employed, as 'some error exists in the Ptolemaic declination for this place'.[159] The star in the forward shoulder of Orion, however, presented no problems for the

---

[154] For an example of how scholars without astronomical knowledge could go wrong, see Grafton 1983–1993, II, 203–205.

[155] *TBOO* VI, 93.40–94.10. Tycho gave as reasons for adopting Spica as his base-star that: (1) it had been adopted as such by Ptolemy himself, and (2) was located near the autumnal equinox, a fundamental point of the ecliptic system. Unfortunately, he was wrong to suppose that Ptolemy's longitude for this star was therefore more reliable than those for others. See Dreyer, in *TBOO* VI, 354, note to 96.27.

[156] *TBOO* VI, 95.3–16.   [157] *TBOO* VI, 95.17–31.

[158] *TBOO* VI, 95.31–37.   [159] *TBOO* VI, 95.38–41.

theory.[160] Cor Leonis, calculated from the declinations of both Hipparchus and Ptolemy, agreed tolerably well, although less so in the second case.[161] But Aldebaran was not consistent with the hypothesis, leading Tycho to conclude that there must be a significant error in both the latitude and the declination of this star in the Ptolemaic catalogue. And indeed, he argued, this could be demonstrated from the declinations assigned by Timocharis and Hipparchus, the putative change in declination between Timocharis and Ptolemy (which was too large), and the position Ptolemy assigned to the neighbouring star Oculus Tauri.[162] 'In this case, which seems to oppose our premise', he reasoned, 'faith is not warranted'.[163]

Tycho applied the same procedure to the stars in the northern part of the sky, with similar results.[164] But he completed his analysis with two of the sky's brightest stars, Sirius and Spica. The former was particularly problematic, because the declinations of Timocharis, Hipparchus, and Ptolemy produced latitudes of 39° 18′, 39° 10′, and 39° 10′ again, whereas Tycho's observations gave 39° 31′. But the alteration of the obliquity of the ecliptic should have led to a decrease in latitude with the passage of time.[165] This was, Tycho wrote, a tricky knot to disentangle; 'for the uniform consensus of all makes it difficult for some error to lie hidden in the observations of the ancients; and that they could have fantasised so detectably about such an illustrious star, which of all the fixed ones is the greatest and most conspicuous, does not seem credible'.[166] Yet an error in his own observations, or some exemption from the rules governing the other stars' motion, was equally implausible.[167] 'Wherefore', he wrote, 'by winnowing this matter variously, I am able to offer nothing except that the observations of the ancients of the declination of this star, although they were perhaps made well enough with respect to the sky, were transmitted to us less correctly through the fault of the copyists'.[168] With respect to his base-star Spica, however, Tycho's conclusion was almost the opposite. Although the values provided for all three ancient astronomers again conflicted with the theory, he decided that the quantities involved were small, and could be attributed to the fact that, 'those ancients had almost no concern for three or four minutes' of arc.[169] The ancients' tabulated data only gave ecliptic coordinates to a sixth of a degree, he pointed out, and their method of observation could easily have generated errors of this size, since they relied upon the Moon for determining the position of the stars, but lacked a lunar model

---

[160] *TBOO* VI, 95.41–96.3.   [161] *TBOO* VI, 96.3–9.   [162] *TBOO* VI, 96.9–97.4.
[163] *TBOO* VI, 97.4–7.   [164] *TBOO* VI, 97.2–98.27.   [165] *TBOO* VI, 98.31–99.2.
[166] *TBOO* VI, 99.2–6.   [167] *TBOO* VI, 99.6–11.   [168] *TBOO* VI, 99.11–15.
[169] *TBOO* VI, 99.34–37.

sufficiently adequate to take account of its parallax.[170] Consideration of a printed edition alone allowed the reader to distinguish between observational error, original to the text, and subsequent corruption. Or so, at least, Tycho thought.

Not to be outdone in explicating Ptolemy, Rothmann, though agreeing broadly with his correspondent, offered an even bolder hypothesis. 'Concerning the latitudes of the stars having been altered following the alteration of the obliquity of the ecliptic', he stated, 'you write most learnedly. For there could be no other way of proceeding skilfully in confirmation of this matter than that by which you have proceeded. Because the places of the fixed stars are laid down wholly erroneously in the tables; and, I aver, were never observed by Ptolemy, but only copied from Hipparchus'.[171] Investigating the *Syntaxis'* stellar positions when still a student at Wittenberg, the *mathematicus* explained, he had noticed that the star Regulus was assigned a latitude to the north of the ecliptic. Yet that which could be derived from the given longitude and declination was actually $6^2/_3'$ to the south.[172] This was a difficult error to account for, and from Ptolemy's description of his observation of the star, Rothmann considered its declination, rather than its latitude, to be suspect.[173] Since the decrease in declination between Hipparchus and Ptolemy was in proportion to that from Timocharis to Hipparchus, he surmised that Ptolemy had not observed his own stellar declinations, but had merely extrapolated from the values his predecessors had recorded.[174] It was not at all evident on the basis of the text, he pointed out, how Ptolemy could have made such measurements. Of the instruments that he described, the parallactic rulers were adapted only for taking observations of the moon; and had he employed his zodiacal armillaries, this discrepancy in declination could not have resulted.[175]

For these reasons alone, then, Rothmann thought Tycho would be entirely justified in discounting the Ptolemaic values assigned to Sirius, and the consequent obstacle this star presented to the hypothesis of the latitude varying with the alteration over time of the ecliptic's obliquity.[176] Yet, in an ingenious variation on the notion that transcription errors

---

[170] *TBOO* VI, 99.34–100.3.   [171] *TBOO* VI, 115.7–12.   [172] *TBOO* VI, 115.15–19.

[173] *TBOO* VI, 115.20–23. For Ptolemy's observation of Regulus in ecliptic coordinates, see Toomer 1998, 328.

[174] *TBOO* VI, 115.23–33. Rothmann was not the only sixteenth-century astronomer to accuse Ptolemy of doctoring his data; on the position taken by Rheticus, see Grafton 1997a, 218. Similar criticisms of Ptolemy have been made in recent times; see Newton 1977, esp. 341–379.

[175] *TBOO* VI, 115.34–39.   [176] *TBOO* VI, 115.39–116.2.

were implicated, he also wondered whether a faulty text of Hipparchus had deceived Ptolemy himself. [177] According to the *Syntaxis*, Timocharis had found the declination of Sirius to be 16° 20', while Hipparchus had observed just 16°. But these values might be out of true by thirty minutes each. 'For the declination could have been handed down by Timocharis as ıs. ɣ. that is 16° 50', [and] by Hipparchus as ıs. ɣ, which is 16° 30'; and by the negligence of the copyists, who thought ɣ.ɣ doubled in error, an ɣ could have been omitted. On the basis of which erroneous declinations Ptolemy could then have drawn his up as well'.[178] To be sure, the scribal error could also have occurred subsequently, as Tycho had suggested; there were certainly many stars recorded with erroneous latitudes.[179] And Rothmann also agreed with the astronomer's conclusion; whatever explanation was adopted for the discrepancy of Sirius, 'we expend many words on discussion of this single star, when its contribution is easily outweighed by those of all the rest'.[180] Ultimately, Tycho and Rothmann were more concerned with the reformation of astronomy than the emendation of its texts. Conjectures such as these were just a means to this end.

On several occasions, Tycho also sought to improve his understanding of key astronomical works by reperforming the fundamental observations on which their data were based. Thus, as Rothmann learnt from Tycho's disciple Jacobi in 1586, the Danish astronomer had despatched one of his students to Frauenburg in Poland in order to check on the empirical grounding of the most important astronomical work of the century, Copernicus' *De revolutionibus*. Rothmann was keen as Tycho must have been to learn the results. 'I would very much like to know what was the accuracy of the instruments that Copernicus used in his observations', he wrote, 'and whether the elevation of the pole at Frauenburg was as Copernicus asserted in chapter II of book III'.[181] Tycho was underwhelmed by Copernicus' methods and instruments, and was able to report that Morsing, his student, had determined the elevation of Frauenburg's pole to be 54° 22' 15", almost three

---

[177] *TBOO* VI, 116.3–5.

[178] *TBOO* VI, 116.5–9. For the values recorded in the *Syntaxis*, see Toomer 1998, 331. Rothmann's reasoning is based on the fact that in the Greek alphabetic numeral system, fractions were represented as sums of unit fractions, and a fraction was distinguished from an integer only by a diacritical mark. Thus, the symbol ɣ appears to function here as both the unit six, and the fraction one-half; ɣ not as the unit three, but as the fraction one-third that, added to one-half, gives five-sixths of a degree. However, Dreyer notes, *TBOO* VI, 356, that different signs were used for six and one-half in the manuscript letter; and indeed, the symbol for one-half would not normally be a ɣ, but a beta, or a symbol such as C'. See Heath 1921, I, 31–32, 41–42. However, the printed text is consistent with Rothmann's original point, that the symbol representing one-half might erroneously have been omitted from both declinations in the copy of the records employed by Ptolemy.

[179] *TBOO* VI, 116.9–14.   [180] *TBOO* VI, 116.14–16.   [181] *TBOO* VI, 58.11–19.

minutes greater than the value the Polish canon had recorded.[182] Moreover, since Morsing had travelled to Königsberg as well, and had there identified a twenty-six-minute error in Erasmus Reinhold's determination of latitude, the *Tabulae Prutenicae* (1551) were also found to stand in need of some correction as a result of this process.[183] Tycho would have liked to extend his empirical evaluation of textual sources to the work of earlier astronomers. Thus, in later correspondence, he was to speak of sending someone to check on the Nuremberg observations of Regiomontanus and his disciple Bernhard Walther.[184] This should have been fairly straightforward to arrange, so it is somewhat surprising to realise that Tycho seems not have to managed it.[185] It is less remarkable, however, that he did not succeed in advancing a much more ambitious plan, one that would have seen an expedition despatched to Ptolemy's home, Egyptian Alexandria. Tycho first attempted to attract sponsorship for such a mission in December 1590, at which point he was looking to secure both Imperial funding and Venetian expertise in the eastern Mediterranean.[186] But he was still looking for the necessary support a decade later, in the year before his death.[187] Ultimately, therefore, while Tycho placed more faith in the capacity of carefully constructed instruments and well-trained assistants to uncover the truth about the heavens, he was forced to fall back upon his skill as a reader. It was not always possible to test the observations of other astronomers by empirical means, and even when such checks could be performed, explaining all but the most fundamental of systematic discrepancies still involved a conjectural process.

Although the Hven–Kassel correspondence is valuable in revealing the role that reading played in Tycho's astronomical work, exchanges with other scholars sometimes provide more evidence concerning particular interpretative strategies than the letters to Rothmann. Tycho's most determined incursion onto the contested ground of scriptural exegesis, for example, occurred not during his cosmological disputes with the Hessen *mathematicus*, but rather in the context of his somewhat more limited interactions with Peucer. The issue that concerned him on this occasion was not, however, the extent to which rival planetary hypotheses enjoyed, or required, the warrant of scripture. Instead Tycho found himself required to address the question of whether one ought to believe, as suggested by certain passages in Genesis and other biblical books, in the existence of supracelestial waters located

---

[182] *TBOO* VI, 103.21–26.   [183] *TBOO* VI, 103.34–38.
[184] *TBOO* VII, 286.1–9. On Walther's observations, see Beaver 1970.   [185] *TBOO* II, 39.33–40.
[186] *TBOO* VII, 286.27–288.5.   [187] *TBOO* VIII, 310.17–21.

beyond the sphere of fixed stars.[188] This was a topic of long-standing interest to practitioners of hexaemeral analysis,[189] and Tycho was able to support his polite but firm rebuttal of Peucer's arguments with reference to Protestant theologians and Hebraists such as Philipp Melanchthon, Jean Calvin, Sébastien Castellion (*c.*1510–1603), Immanuel Tremellius (1510–1580), and Franciscus Junius (1545–1602). Consequently, though far from uninteresting, Tycho's letter to Peucer is at first sight less enlightening than might have been hoped. In particular, it leaves unclear the extent to which Tycho's willingness to engage in such a detailed exegesis was because the practice was one he routinely engaged in and valued, or was due instead to the nature of the question at hand and the ease with which authorities could be found to endorse his conclusions.

Having said that, the somewhat anomalous character of Tycho's letter to Peucer may itself be significant. In rejecting the notion of waters existing beyond the visible heaven, Tycho employed both the criterion of cosmological proportionality found in his exchanges with Rothmann and the strong sense of the intelligibility of divine providence in the natural world that he and his Philippist interlocutors shared, and from which the notion of proportionality was clearly derived.[190] At the same time, however, he also had recourse to a version of the accommodationist principle that he had been disinclined to accept when employed by Rothmann to defend the Copernican planetary hypotheses. Thus Tycho declared that although Moses wrote of waters above the heavens, he employed the term 'heaven' not in the strict natural philosophical sense of the supralunar space inhabited by the celestial bodies, but rather in a way accommodated to the understanding of the common man, to mean the air over the heads of earth-dwellers, yet below the water-filled clouds to which he intended to refer.[191] Taken together, these features of the exchange suggest that, whatever role exegesis may have played in shaping Tycho's theological and natural philosophical ideas, it was on these, rather than exegesis *per se*, that his cosmology and astronomy were grounded. Certainly, Tycho seems not to have seen the interpretation of scripture as central to his public persona as a reforming astronomer. Though content to reject heliocentrism as contrary to the testimony of the Bible, his engagement with specific scriptural texts, not to mention his espousal of exegetical principles, was limited in scope, and can rarely be found in his extant writings except at another's instigation.

---

[188]   *TBOO* VII, 231.5–235.30.
[189]   Williams 1948, 184–185; Lemay 1977; Grant 1994, 103–104, 332–334.
[190]   *TBOO* VII, 232.30–235.30, esp. 232.30–233.18, 235.24–30.
[191]   *TBOO* VII, 231.10–232.29, esp. 231.23–30; Howell 2002, 105–106.

The recourse to exegesis was initiated by Peucer and Rothmann, not Tycho himself.[192]

The sort of critical reading that Tycho made central to his astronomical work, on the other hand, was analysis of the mathematically informed literature on recent celestial phenomena. In both the *De recentioribus phaenomenis*, which considered the comet of 1577, and in his re-examination of the 1572 nova in the *Progymnasmata*, Tycho devoted significant quantities of effort and paper to critiquing the accounts of these prodigies that other scholars had produced.[193] This strategy of incorporating discussion of rival works into his own treatment of these phenomena was not especially unusual. Some of the writings that Tycho studied in depth, such as the *Dialexis* of Hagecius, were themselves composite texts that incorporated and commented on a selection of the literature. Tycho's approach was distinctive, however, in respect of its attempt at comprehensiveness and its sustained focus on a single objective, the demonstration that both the comet and nova were celestial phenomena. In the long-term, this strategy clearly worked as well as Tycho had intended. The standard twentieth-century study of the 1577 comet, for example, was so strongly guided in its structure and agenda by Tycho's *De recentioribus phaenomenis*, that it concentrated on the issue of where various authors located the comet in the cosmos at the expense of the more prevalent sixteenth-century concern, its possible astrological and eschatological significance.[194] It was not that this was a question in which Tycho had no interest himself. On the contrary, it was a major focus of his earliest accounts of both phenomena, the *De nova stella* of 1573 and the German manuscript treatise of 1578.[195] But in the *De recentioribus phaenomenis*, Tycho deferred consideration of such matters to the final volume of the trilogy, never actually produced.[196]

That Tycho intended, with the help of his extensive critiques, for his works to be seen as definitive accounts of the new celestial phenomena, is indicated by his reaction to those contemporary readers of *De recentioribus phaenomenis* who remained unconvinced. Over the course of his epistolary dispute with John Craig, the Scots Aristotelian who rejected the claim that the 1577 comet was a supralunary and not a meteorological phenomenon, Tycho unilaterally came to the view that the tracts setting out their respective

---

[192] Thus Tycho's exegetical statements in his earlier letter to Peucer of 13 September 1588, evidently occurred in response to Peucer's discussion of scriptural warrant for the fluidity of the heavens. See *TBOO* VII, 133–134, esp. 133.22–26.

[193] *TBOO* III, 5–299; *TBOO* IV, 180–367.

[194] Hellman 1944. For a treatment that seeks to do justice to the range of sixteenth-century concerns, see van Nouhuys 1998.

[195] *TBOO* I, 30–44; *TBOO* IV, 389–396; Christianson 1979.   [196] *TBOO* IV, 377.7–14.

positions should be published as an appendix to the book prior to any commercial release.[197] In this way, Craig's critical reading of the text would be more than adequately answered within the volume itself, and editorial privilege would ensure that Tycho's final word in the debate appeared last overall. Rothmann, though his position was much closer to Tycho's than Craig, was another reader of *De recentioribus phaenomenis* who objected to parts of Tycho's cometary analysis. In his case, however, the situation with respect to a public response was a little more complex. The decision to publish the Hven–Kassel correspondence meant that a printed version of Rothmann's comments, along with Tycho's replies, was already on the cards. In theory, therefore, no additional provision had to be made to ensure that his objections would be properly answered in the eyes of the wider astronomical community. As it happened, however, some of the authorities and evidence that Rothmann adduced were not treated by Tycho until the *Progymnasmata*. In combination with the fact that cometary analysis was a topic of the Hven–Kassel correspondence prior to the completion of the *De recentioribus phaenomenis*, and that consideration of Rothmann's work on the comet of 1585 may have played a significant role in shaping the final form of this text, this means that three of the five books Tycho published form, in respect of their concern with the literature on celestial phenomena, a richly interwoven corpus of texts.

John Craig's challenge to Tycho's readings of other astronomers' works in *De recentioribus phaenomenis* occurred after the fact. Craig sought to turn Tycho's discussion of a wide range of cometary treatises against him by using the discrepancies in the data and conclusions they presented to contest his supposedly definitive demonstration that the comet of 1577 was a supralunary object.[198] Rothmann did not dispute this, the central thesis of *De recentioribus phaenomenis*. Prior to his receipt of the work, however, he did question one of the other key claims it contained, the assertion that *all* comets were celestial. Rothmann's objection to this claim, made after Tycho had advanced it in his letter to Wilhelm of January 1587, rested on the observations of Regiomontanus and the later Viennese astronomer Johannes Vögelin (*fl.* 1530).[199] A work attributed to Regiomontanus had recorded a parallax of 6° for the comet that appeared in 1472, while Vögelin had derived 35° for the one that was observed in 1532.[200] Both of these values seemed to indicate atmospheric phenomena. It was far from unreasonable for Rothmann to make use of these observations as evidence. Since he

[197] *TBOO* VII, 310.36–311.5.   [198] See Mosley 2002b.   [199] *TBOO* VI, 65.26–31, 119.22–26.
[200] Hellman 1944, 83, 97; Jervis 1980 and 1985, 117–120, 123.

admitted no physical distinction or barrier between sublunary and inter-planetary space, he had no *a priori* reason to suppose that comets might not inhabit each of these regions, or pass from one to another.[201] But for him to do so was hardly original. First published under Regiomontanus' name in Jacob Ziegler's *Commentarii* (1548), where the comet was misdated to 1475, and in Vögelin's *Significatio cometae qui anno 1532 apparuit* (1533), these two sets of observations had previously been yoked together in Hagecius' *Dialexis* for exactly the same purpose as Rothmann employed them, in order to show that certain classes of phenomena might be both atmospheric and celestial.[202] And the Imperial Physician had been followed by other writers on the comet of 1577, such as Cornelius Gemma and Michael Maestlin, the comets of 1475 and 1532 thereby becoming a standard point of reference for authors who wished to defend the possibility of aerial comets.[203] It is small wonder therefore that Tycho, who had read all of these authors, was fully aware that the parallax observations of ps.-Regiomontanus and Vögelin constituted the chief obstacle to his view that all comets were celestial, and had named them as such in his letter to Wilhelm.[204] He had already prepared his response.[205]

Tycho's answer to Rothmann appeared in his letter of 17 August 1588, following hard upon his despatch to Kassel of the *De recentioribus phaenome-nis*. Having rejected the observations of Regiomontanus and Vögelin three times in that volume (he was to do so twice more in the *Progymnasmata*),[206] he set out the bare bones of his critique in the following way:

The records of Regiomontanus are not self-consistent, and he could not in any way have observed the parallax of so swift a comet with respect to Spica in Virgo. In these, therefore, he was too attentive to the assertions of Aristotle, so that if through some carelessness he arrived at a conjecture about some parallax which rendered it sublunar, he held onto it tenaciously as the truer value. For he seems to have struggled a great deal in this case to elicit as large [a parallax] as possible. The observations of Vögelin were wholly incorrect, and do not agree with those which were published by Peter Apian and Fracastoro on the same comet, as I shall demonstrate more fully elsewhere.[207]

[201] *TBOO* VI, 119.29–32
[202] Hagecius 1574, 48: 'Ac quod in elementari regione flagrent seu fulgeant, probatione non eget. Iam enim id abunde ab Aristotele probatum, & pluribus conformatum est: sed omnium exactissimse demonstratum a Regiomontano in eo cometa qui luxit anno 1475: et a Ioanne Vogelino in alio cometa, qui luxit anno 1532 . . .'
[203] Gemma 1578, 39; Maestlin 1578, 17; *TBOO* III, 80.25–29; Hellman 1944, 182–183.
[204] *TBOO* VI, 65.31–36.
[205] Long before completion of the *De recentioribus phaenomenis*, Tycho took the matter up with Hagecius directly. See *TBOO* VII, 68.14–28.
[206] *TBOO* IV, 208, 290, 351; *TBOO* III, 26.4–14, 140.34–141.3.   [207] *TBOO* VI, 147.25–33.

Regiomontanus was widely seen as an authority on determinations of cometary parallax, thanks to his *Problemata* published at Nuremberg in 1531.[208] But Tycho argued in *De recentioribus phaenomenis* that the methods he and others advocated were unsuited to the study of objects whose parallax was slight: minor observational errors would tend to result in significant flaws in the final result, and in any case these methods were applied so as to assume that comets possessed no proper motion of their own.[209] Tycho had no qualms, therefore, about dismissing ps.-Regiomontanus' observations of the comet of 1475, especially since *De Cometa* did not even apply the methods that Regiomontanus had described, but instead derived the parallax from the apparent displacement of the head with respect to Spica, the nearby star.[210] And Vögelin's observations on the comet of 1532, Tycho thought, could be trumped by Apian's, as found in the *Astronomicum Caesareum*, and by Girolamo Fracastoro's, first published in his *Homocentrica* of 1538.[211] This was the way that texts concerning long-vanished phenomena had to be dealt with: they could be dismissed on the grounds of internal inconsistency and error, or rejected in favour of other accounts with which they failed to agree.

After he had read *De recentioribus phaenomenis*, Rothmann applied this strategy to another, similar, cometary problem. What Tycho's book lacked, he complained, was a thorough discussion of the material substance of comets, one that could provide a convincing explanation of their tails.[212] Rothmann had changed his mind about this topic since composing his treatise on the comet of 1585. As he chose to put it, he had elected to include a chapter on the subject in which he had recounted the opinions of Apian, Girolamo Cardano, and Julius Scaliger on the generation of cometary tails, and then, 'added myself at the end, so as to frankly reveal that in this opinion . . . I follow the judgement of others, rather than my own'.[213] Apian, in the *Astronomicum Caesareum*, Cardano in his *De subtilitate* (1551) and *De rerum varietate* (1557), though taking his cue from Fracastoro, and Scaliger in the *Exercitationes de subtilitate* (1557), a critique of Cardano, had all noted that comets' tails pointed away from the Sun, and took this to be evidence of an optical phenomenon: the body of the comet acted as a lens, focusing incident rays so as to produce a cone of light that extended

[208] Jervis 1985, 93–114, 170–193.   [209] *TBOO* IV, 82.14–83.40.

[210] *TBOO* IV, 133.41–134.5; Jervis 1985, 114–120. As discussed by Jervis, the identification of the star as Spica is an error.

[211] Apian 1540, Ov–O2r; Fracastoro 1538, 59r–60v; Hellman 1944, 87–88.

[212] *TBOO* VI, 155.9–14.   [213] *TBOO* VI, 155.16–19. See Snellius 1619, 130–139.

backwards from the head.[214] Rothmann now disagreed with this theory. He thought it, 'reasonable for the tail to be of the same substance as the body of the comet itself, yet rarer'.[215] For if solar light were refracted by the comet's head, there was nothing, he thought, onto which it could be projected, so as to render it visible.[216] The tail of the comet of 1585 had not been very prominent. Consequently, Rothmann explained, he had been reluctant to speak out against so many authorities, since he, 'had never observed tailed stars of this kind, and not yet carefully considered the shape and form of a tail in any instance'.[217] But he subsequently found, in a source that could not have been any closer to home, the observational record to compensate for his own lack of experience.

I learnt last winter that my judgement about the tail was, contrary to those great men, true. For when by a lucky chance I happened on that gilded copy of the Cyprian *Ephemerides*, which our Most Illustrious Prince was accustomed to carry with him always (along with the Holy Scriptures and a mechanical globe, and a small quadrant), I found written at the beginning of them the following: In the year one thousand five hundred and fifty eight after salvation had been offered to the whole sphere of the Earth, a comet appeared, which on the 20th of August was observed by the Most Illustrious Prince and Lord WILHELM, LANDGRAVE OF HESSE, with a torquetum, around the ninth hour, at 21 Degrees into Virgo, with a latitude 31 Degrees from the Ecliptic, the head extended towards the extreme tail of Ursa Major; while the thing itself was in the constellation Coma Berenices, where Venus and the Moon were.[218]

Rothmann took the note contained in Wilhelm's copy of Cyprian Leowitz' *Ephemerides* (1557) to be evidence that not every comet's tail was opposed to the Sun or another bright celestial object. Cometary plumes could not, therefore, be optical phenomena.[219] And he added, as an afterthought, that the alternative hypothesis of material tails could better account for some of the other comets previously observed. In particular, it might account for an immense comet described by Seneca in his *Quaestiones naturales*, and a more recent one of significant size mentioned by Giovanni Pontano (c. 1426–1503), the humanist poet and astrologer, in his commentary on the *Centiloquium* attributed to Ptolemy.[220] Did Tycho not think, Rothmann

---

[214] Apian 1540, N2v; Cardanus 1551, 97; Cardanus 1557, 3; Scaliger 1557, 122v; Jervis 1985, 121–123; Barker 1993.

[215] *TBOO* VI, 155.19–27.   [216] *TBOO* VI, 155.19–27, *cont.*   [217] *TBOO* VI, 155.19–31.

[218] *TBOO* VI, 155.32–156.2.   [219] *TBOO* VI, 156.8–10.

[220] See Corcoran 1971–1972, II, 258–261; Pontanus 1512, K6r–K6v. Pontano is discussed in Trinkaus 1985.

asked, that in respect of comets' tails, his theory of a material basis might be the true explanation?[221]

Tycho did not. He would go on in the *Progymnasmata* to point out that the records proffered by Seneca and Pontano could not be considered reliable.[222] And in the interim, he reminded Rothmann that he had demonstrated in the *De recentioribus phaenomenis* that the tail of the comet of 1577 pointed directly away from Venus. At that time he had openly wondered whether the appearances had in some way been distorted, for he did not wish to attribute the powerful illumination generating the tail to any celestial body other than the Sun. However, he was now prepared to accept that, as confirmed by his reading of the *Diversarum speculationum liber* (1585) of the Venetian-born mathematician Giovanni Battista Benedetti, Venus' rays were sufficiently strong to illuminate the side of the Moon turned away from the Sun.[223] Since he had also argued that the course of the comet was not very far from the circumsolar orbit of Venus, it now seemed to him that this tail could be optically explained without any difficulty, as could the many observed 'by Regiomontanus, Fracastoro, Apian, Gemma Frisius, and others'.[224] What Tycho could not see, however, was how the material hypothesis advanced by Rothmann could account either for the tails of all of these comets having being opposed to the Sun, or for the acutely observed fact that the closer comets approached to the solar course, the more their plumes grew.[225] Unfortunately for Rothmann, the first articulation of the 'dirty snowball' model of comets, which would both vindicate his view that the tails were similar in substance to comets themselves and explain the relationship between the variability of their appearance and the Sun's proximity, was still more than 350 years away.

None of these comments addressed the crucial 1558 comet witnessed by Wilhelm. But in respect of this object, Tycho had yet another textual ace up his sleeve. In the *De divinis naturae characterismis* (1575) of Cornelius Gemma, a work cited only briefly in the *De recentioribus phaenomenis*, there was a description of the same comet that diverged from Wilhelm's on crucial matters of detail, and that Tycho declared himself disposed to prefer to the Landgrave's. He gave as his reason the fact that in making his observations Gemma had employed an astronomer's staff, an instrument that, however inaccurate, was easier to handle and less liable to distortion under its own

[221] *TBOO* VI, 156.16–17.   [222] See *TBOO* II, 328–329, 435; *TBOO* III, 228.

[223] *TBOO* VI, 172.2–12, quoting Benedetti 1585, 257: 'Manifeste videtur, dum Luna reperitur secundum Longitudinem inter Solem & Venerem, quod pars Lunae lumine Solis destituta, a lumine Veneris aliquantulum illustratur, quod ego sæpe vidi, & multis ostendi.'

[224] *TBOO* VI, 172.20–24   [225] *TBOO* VI, 172.20–26.

weight than the torquetum Wilhelm had used.[226] In fact, though neither correspondent ever realised it, Tycho's retrospective analysis was marred in this instance by a typesetter's mistake: *20 die*, the twentieth day of August and the date on which the Landgrave had recorded the comet's location and appearance, was printed in Gemma's text in place of *eo die*, the seventeenth of the month.[227] Between the two observations, taken at different sites on different days, there was no real discrepancy that meant one or other of them had to be dispensed with.

Tycho considered more than twenty cometary tracts in his *De recentioribus phaenomenis*. Many of the authors of these works had also written something on the new star of 1572, and for this reason were also mentioned in the *Progymnasmata*. Several, as we have seen, featured in the Hven–Kassel correspondence, and a number were in epistolary contact with Tycho themselves, or were recipients of one or more of his books. A few of them, therefore, like Rothmann, not only drew on, discussed, and contributed to, the body of literature on celestial phenomena that Tycho strove to digest and master, but were themselves then in a position to comment on the outcome. In this way, the peculiar nature and chronology of Tycho's publishing programme produced circumstances conducive to a remarkable degree of intertextuality. More significant than the mere fact of this complex set of relationships between Tycho's printed works and his manuscript correspondence, however, is what it reveals about astronomical culture. It shows that even at the most empirical of sixteenth-century centres of astronomy, books continued to be indispensable to astronomical practice: as pursued at both Kassel and Hven, astronomy required not just well-made instruments and able observers, but also a library of astronomical treatises, works of geometry and optics, and natural philosophical texts. And it strongly indicates the role that books played in creating and sustaining an astronomical tradition and community. By means of correspondence, early modern astronomers were able to form a community of practitioners with common interests, a powerful strategy for overcoming the difficulties associated with the construction of a universally valid astronomy from any one centre. Through the study of texts, they were able to enlarge that community to include deceased individuals alongside living contemporaries. Hipparchus and Ptolemy, as well as Regiomontanus and Copernicus, were important members of the community to which not only Tycho and Rothmann, but also Hagecius and Maestlin, Wilhelm and Rantzau, Peucer and Scultetus, considered themselves to belong.

---

[226] *TBOO* VI, 173.26–174.20.   [227] Gemma 1575, II, 33–34; *TBOO* VI, 360–361, n. to 174.13.

The textual tradition provided the only way in which sixteenth-century astronomers could study astronomical parameters over the time-scale necessary to elucidate the slowest of long-term changes. And it was also the only means by which they could integrate accounts of past celestial phenomena into their studies of more recent events. Little could be done about predecessors' observational standards unless one was prepared, as Tycho was, to have fundamental measurements reperformed *in situ*. But often the limits of accuracy were amenable to estimation and, just as importantly, close study of a text, and comparison of one book with another, held out the prospect of distinguishing genuine data from unreliable reports and uncovering those mistakes that could be attributed to the carelessness of a scribe or printer. What emerges from consideration of the reading strategies evidenced in Tycho's correspondence and printed works is, therefore, a picture of an astronomer at work in his library – a picture that allows us to acknowledge the important role played by textual sources without subscribing to a simplistic model of the printing press as an unproblematic agent for the transmission and accumulation of knowledge. Texts were not uniformly reliable or valuable, and for this reason the great quantity of printed works available could prove to be as much a hindrance as a help. Certainly, the reproduction of ps.-Regiomontanus' and Vögelin's cometary observations made it more difficult, not easier, to persuade the astronomical community that comets were uniformly aethereal. For this reason, among others, Tycho expended nearly as much time and effort acquiring the *right* books as he did in collecting the most precise observations.

(ii) The astronomical library

For a long time I have searched far and wide for JOHANNES WERNER'S little work *De motu octavae sphaerae*, which was once printed, I believe, at Nuremberg. But I have not yet obtained it anywhere. Consequently, if you find this book there and make it available to me, then also add that manuscript *De directionum significationibus* by an unknown author, which I saw with you in Regensburg, having had it copied by someone, you will do me a great favour. I shall gladly remit all your expenses, and return the courtesy, whenever possible. (Tycho to Hagecius, 25 August 1585)[228]

Books required for the practice of astronomy had to be obtained before they could be read. This could be a problem for most early modern scholars, but the seclusion of Hven, conducive as it was to the pursuit of observational labours, must have rendered Tycho's acquisition of texts particularly

---

[228] *TBOO* VII, 95.10–18.

difficult. To some extent, the ways in which Tycho sought to overcome this obstacle to the practice of astronomy mirrored the strategy he adopted for the dissemination of his own books. For as we have seen, though intending to make the texts printed at Uraniborg commercially available, Tycho also chose to distribute them privately to scholars and patrons. Since this private distribution was managed alongside and through his correspondence, books and letters often travelled together on the journey from Hven. But large though this outward traffic was, the inward volume of letters accompanied by texts was probably greater. Uraniborg was a net importer of books, and Tycho's correspondents seem to have been particularly prominent among its suppliers.

In some cases, it is true, the association of books and letters in transit must have been purely accidental. It is difficult to believe that the merchants who travelled to and from the Frankfurt book-fairs, whom Tycho often used as couriers for his correspondence, played no role in securing the books that he wanted.[229] Nevertheless, scant evidence of their contribution exists. Tycho's orders can seldom have been placed with these commercial bookdealers through a third-party in writing, and he infrequently had cause to note his purchases in other correspondence. Some exceptional cases do exist. In May 1579, for example, Tycho sent his erstwhile tutor Anders Sørensen Vedel a copy of the recent Frankfurt book-fair catalogue, instructing him to pass it on to another individual, presumably a Copenhagen bookdealer, who was to obtain for him the texts, 'nine of them in Latin, three in German, and one in French', he had indicated by a mark in the margin.[230] And in 1598, he informed Magini that he had acquired the Bolognese mathematician's *Tabula Tetragonica* (1592) just three years previously, after one of his students had found it among the stock of a Copenhagen bookseller – a somewhat barbed remark given that, as dedicatee of this text, Tycho might reasonably have expected to receive a copy as a gift.[231] But given the rarity of such comments, there is little prospect of determining directly which books, of the many that the astronomer is known to have owned, were commercially sourced.[232]

Occasions when Tycho received a book as a gift can be identified more readily. This was not an infrequent occurrence given the economic

[229] One historian has suggested, however, that after an early unhappy experience with a dishonest courier, Tycho restricted himself to buying books through friends and associates. See Thoren 1990, 467.
[230] *TBOO* VII, 54.38–42.   [231] *TBOO* VIII, 120.27–121.13.
[232] For a partial reconstruction of Tycho's library, see Nørlind 1970, 336–366. The manner of acquisition of some extant copies is known from notes recorded within them. For example, in his copy of Jacob Christmann's *Observationum solarium libri tres* (1601), Tycho wrote, 'emi Pragae mense Maio'.

and strategic considerations that, the negligent Magini notwithstanding, favoured the practice of authorial presentation. Thus Rantzau sent Tycho copies of his *Catalogus* (1584), the treatment of illustrious practitioners of astrology which mentioned both Tycho and Wilhelm, and also his *Rantzovianum calendarium* (1590).[233] Other works that found their way into Tycho's library by a similar route were Giordano Bruno's *Camoeracensis acrotismus* (1588), Kepler's *Mysterium cosmographicum* (1597), and Joseph Scaliger's *De emendatione temporum* (1598), as well as the revised version of his edition of Manilius' *Astronomica* (1600).[234] Bartholomaeus Pitiscus (1561–1613) presented his *Trigonometria* (1595) via Tycho's former student Cunradus Aslachus (1564–1624);[235] while the Jesuit Jacob Monaw (1546–1603) only just stopped short of giving the astronomer a copy of his *Symbolum* (1595), an anthology of verses written by others that took his personal motto as their theme.[236] The majority of these authorial presentations seem, like Tycho's, to have been mediated by letters rather than personal contact. Given the predominantly epistolary nature of the evidence, it would perhaps be unwise to make too much of this fact. Nevertheless, it testifies to the role that the giving of gifts, in the form of books, played in developing and sustaining the notional community comprising the Republic of Letters.[237]

Members of the international community were also expected to display their friendship by performing and reciprocating favours and services, and it was primarily in this way that Tycho's correspondents were recruited to help him to acquire the books that he wanted. Very occasionally, this involved soliciting a book from an author rather than waiting for an unprompted presentation. Thus in early 1600, Tycho approached Magini for a copy of his augmented *Ephemerides*, albeit a little prematurely; the volume he wanted, it turned out, was still in production.[238] Sometimes Tycho would ask an individual for a work that, although he had good reason to think otherwise, they did not possess. In 1584, upon being told by Heinrich Brucaeus that he intended to produce a new commentary on the *Hypotyposes*

---

[233] Nørlind 1970, 359–360.   [234] Westman 1980a, 96; *TBOO* VIII, 14.27–33, 32.3–7, 264.10–11.

[235] *TBOO* VII, 377.5–6. On Pitiscus' work, see von Braunmühl 1900, 220–226. Aslachus is discussed by Christianson 2000, 252–253.

[236] Monaw stated that if Tycho desired the work, he could obtain a copy from Vincentius Mollerus. See *TBOO* VIII, 30.20–22. For the work, see Evans 1997, 148; for Mollerus, Jöcher 1750–1751, III, 575.

[237] On this subject, see also Goldgar 1995, 12–53.

[238] *TBOO* VIII, 230.30–36, 253.38–41. Magini's *Ephemerides*, covering 1582–1620 had first been printed in Venice. In his reply to Tycho, the mathematician noted that only one volume, covering 1598–1610, had been reprinted, and said this contained little new; the second volume, extending a further twenty years, was then in the press.

of Proclus, Tycho asked him to supply a Latin copy of the text.[239] On this occasion, however, Brucaeus could only provide his correspondent with some vague bibliographical help: 'The works of Giorgio Valla of Piacenza were published somewhere or other, to which were added, I believe, the *Hypotyposes*, translated by the same author'.[240] And it was not entirely unknown for Tycho to receive, unsolicited, a gift-copy of a book sent to him by someone who had not produced it. Hence at least three individuals sought to provide him with a copy of Ursus' *De astronomicis hypothesibus* (1597), a book whose importance for Tycho would have been evident even to a casual reader.[241] Typically, however, the books which the astronomer received from his correspondents were either texts that he had specifically requested, or were devoted to a topic, such as the nova of 1572, in which his perennial interest was known. And accommodating Tycho in this way seems to have represented a service to the astronomer rather than a gift. Or rather, he often expressed his willingness to recompense correspondents for the trouble and expense incurred in supplying these books. Whether he actually did so is harder to tell.

Frequently, the biannual Frankfurt book-fair and its catalogues played a key role in the identification of the works that Tycho received or requested.[242] In 1585, Brucaeus, who was writing from Rostock, informed the astronomer that no mathematical works had been published at the recent fairs, and none were being carried by any of the local dealers in books.[243] On a prior occasion however, towards the end of 1578, he had written that, having noticed Michael Maestlin's recent cometary treatise listed in the recent Frankfurt catalogue, he would send this work to Tycho if it was possible for him to obtain it.[244] In 1584, Johannes Maior, who had just despatched to Tycho's order a number of texts concerning the reform of the calendar, hesitantly noted that the Willer catalogues for Frankfurt contained a number of other works, on clocks and astrology, that he might

---

[239] *TBOO* VII, 78.12–14.
[240] *TBOO* VII, 85.37–39. The translation by Valla was first printed in Venice in 1501. See Manitius 1909, v.
[241] As Tycho related to Kepler in April 1598, the courier from Helmstedt who brought him the *Mysterium cosmographicum*, also brought him the *De astronomicis hypothesibus*. Georg Rollenhagen subsequently sent Tycho a copy, the second he had obtained, after (as he reported it) the loss of the first prevented him from sending a transcription more promptly. And Hagecius attempted to send a copy, which was apparently lost in transit. In replying to the Imperial Physician, however, Tycho revealed that it was a certain student in Helmstedt who had first sent him the work. See *TBOO* VIII, 46.37–40, 50.38–51.6, 56.5–13.
[242] See Blum 1958–1960; Fabian 1972–2001.   [243] *TBOO* VII, 92.30–31.
[244] *TBOO* VII, 48.31–49.1.

also wish to possess.[245] Aslachus sent the latest fair catalogues to Hven in October 1594, offering to obtain any texts that his former master wanted.[246] Herwart von Hohenburg, the Chancellor of Bavaria, did likewise for both the fairs of 1600; no doubt he did so on the basis of his experience of the previous year, when he had been asked to acquire six works advertised in the autumn catalogue.[247] And in March 1598, Tycho wrote to Christian Longomontanus, saying that although he had seen the *Ephemerides* (1597) of Martin Everart advertised in a catalogue, he could not find them for sale in Hamburg, the city nearest to his temporary residence.[248] This request for assistance would seem to have worked, because Tycho soon had a copy in his possession and was able to criticise it that May in a letter addressed to Holger Rosenkrantz, a Danish nobleman who advised and assisted Tycho with respect to his future patronage prospects.[249] Tycho was far from pleased that this work, dedicated to King Christian IV, might convey the impression that what he had spent so many years and so much money attempting to do, could have been carried out more swiftly and cheaply, and wanted Rosenkrantz to keep him informed about its reception.[250] Clearly, sixteenth-century scholars, having more than one reason to keep abreast of the contemporary literature, found that the fair-catalogues could be consulted with profit.

Tycho did not always limit his requests to just one of his contacts. Sometimes, therefore, he received multiple copies of a single publication. As already noted, Brucaeus promised to send Maestlin's cometary treatise in 1578, and he probably did so not long after it was published. But along with the treatments of Cornelius Gemma, Helisaeus Röslin, and other unnamed authors, and a manuscript letter from Maestlin to Wolf which contained the Kassel observations, this text was also forwarded to Tycho in 1579 by another correspondent, the Augsburg patrician Paul Hainzel.[251] In 1584, moreover, both Johannes Maior and Brucaeus sent the astronomer a copy of Joseph Scaliger's *De emendatione temporum* (1583), a work that, as we have seen, he later acquired from the author in its second edition.[252] But with so many books and letters going astray in transit, and with the lengthy delays that routinely attended a long-distance correspondence, duplication of items in the Uraniborg library was probably viewed as a hazard worth risking.

Tycho's great appetite for books concerning comets and the 1572 nova was a product of his desire to review these works comprehensively in his

---

[245] The suggestion was, however, declined. See *TBOO* VII, 83.34–38, 86.19.
[246] *TBOO* VII, 367.28–30.   [247] *TBOO* VIII, 294.4–9, 386.30–32, 202.17–21.
[248] *TBOO* VIII, 35.3–8.   [249] Christianson 2000, 344–346.   [250] *TBOO* VIII, 61.41–63.2.
[251] *TBOO* VII, 50.4–20.   [252] *TBOO* VII, 83.15–19, 86.2–3.

series on celestial phenomena. In 1581, having just received Hagecius' recent *Apodixis physica et mathematica de Cometis* from Bartholomaeus Scultetus, he asked to be sent Andreas Dudith's *De cometarum significatione* (1579), or similar works by any other author.[253] In 1588, when Brucaeus relayed reports of a book by Camerarius on the comet of 1577, Tycho instructed him to see that he procured this text quickly.[254] And in 1594, Aslachus knew enough to send the astronomer, without awaiting his approval, Christoph Clavius' commentary on the *De Sphaera* of Sacrobosco, that had, 'a certain little tract about the new star of 1572 inserted within it'.[255] But the correspondent badgered most persistently about this kind of material was Hagecius himself, whom Tycho began asking for such texts in 1585. He was still being asked to procure them as late as 1590.[256]

The difficulty of obtaining and using these texts was sometimes exacerbated by a problem of language. Thus Hagecius was asked not only to supply Tycho with Italian, Spanish, and French writings on the nova of 1572, but also, if possible, to have them translated into Latin.[257] In the case of at least one book that Tycho had specifically requested, the *Libro del nuevo cometa* (1572) of Jerónimo Muñoz, Valencian professor of mathematics and Hebrew, Hagecius was unable to accommodate his friend because he possessed the work but could not find a translator.[258] As a consequence, when Tycho turned his attention to Muñoz's study of the nova in his *Progymnasmata*, he was forced to base his analysis partly on a copy of a letter that Muñoz had written to Hagecius comparing their results, partly on the account to be found in Cornelius Gemma's *De divinis naturae characteristicis*.[259] Gemma's work gave him access to the Valencian's observational data and conclusions, the chief of which was that the nova was a celestial comet.

[253] *TBOO* VII, 64.14 18, 63.22 25.

[254] *TBOO* VII, 144.28–30, 164.13–15. Brucaeus was probably mistaken in referring to a work on the comet published by Camerarius at Frankfurt an der Oder. A text by Joachim Camerarius the Elder, *De eorum qui cometae appellantur, nominibus* was first published in 1558 at Leipzig, and reprinted several times, including in 1578. But the work was not produced at Frankfurt an der Oder, and was not specifically about the comet of 1577. However, a text on the nova of 1572 was published there by Elias Camerarius in 1573, and was discussed by Tycho in the *Progymnasmata*, following Hagecius' provision of a manuscript copy; it constitutes a likely source of confusion. See Hellman 1944, 343–344; Zinner 1964, nos. 2186, 2616, 2811; *TBOO* III, 205–212.

[255] *TBOO* VII, 367.12–14. Clavius' *In sphaeram Ioannis de Sacro Bosco commentarius* was first published at Rome in 1570, and material on the nova was introduced into the third edition of 1585. But Aslachus was probably sending Tycho a copy of the recent fourth edition of 1593. See Lattis 1994, 147–156.

[256] *TBOO* VII, 93.36–39, 227.29–30.   [257] *TBOO* VII, 120.36–40.

[258] *TBOO* VII, 165.38–41, 251.16–18.

[259] Gemma 1575, II, 267–274; *TBOO* III, 80–87; Brótons 1981, esp. 70–75; van Nouhuys 1998, 151; Gingerich 2002b, 326–327.

But it also required him to treat this material through the filter supplied by Gemma's own critical reading. Lack of a translator, therefore, meant that Tycho's assessment of Muñoz was, in large part, a critique of a critique.

The provision of translations was just one reason among several for the continued propagation and circulation of manuscript material.[260] Thus, sought-after texts continued to be transcribed from manuscripts that had never been published, and copies were made by hand of printed sources if these could no longer be readily purchased. Tycho may have proved unwilling to publish Hagecius' manuscript of George of Trebizond on Ptolemy's *Syntaxis*, but he did not, while casting around for a more amenable printer, neglect to have it transcribed. Ostensibly the copy was made to protect against loss of the original; but making it also, of course, enabled Tycho to add the unpublished text to his library.[261] Brucaeus offered to copy texts for Tycho in 1584 for essentially the same reason as Hagecius was asked to do so in 1590: in order to compensate for a shortfall in the number of printed copies to be found in circulation.[262] And some time previously, in the mid-1570s, the Rostock professor had returned to Hven a section of Johannes Homelius' unpublished *Gnomonica* that he was apparently borrowing piecemeal for the purpose of copying.[263] This manuscript does not even seem to have been Tycho's originally. In 1579, when he furnished the book-hunting Anders Sørensen Vedel with letters of introduction that might facilitate his 'scouring' of certain German libraries, Tycho charged his friend with returning the work to the Nuremberg physician Joachim Camerarius.[264] Alongside scribal reproduction, lending and borrowing played their part in making key works available.

Tycho's other instruction to Vedel was that if he found 'any books in Germany, whether about the comet [of 1577], apart from those I already have, or something noteworthy on astronomical matters', he should procure them and could expect to be repaid.[265] In this case it is somewhat unclear whether the astronomer envisaged Vedel stumbling across a mathematical rarity in the course of his archival researches, which were historical in nature, or simply keeping him in mind whenever he encountered a bookseller. On other occasions, however, Tycho left no room for doubt that he had a strong, and far from idle, curiosity about the holdings of existing libraries and archives. Thus in 1585, perhaps prompted by his receipt of

---

[260] Saunders 1951; Reeve 1983; D'Amico 1988a; Ezell 1999, 21–44.

[261] *TBOO* VII, 105.27–35.   [262] *TBOO* VII, 86.3–6, 227.29–30.

[263] *TBOO* VII, 33.8–10. On Homelius (1518–1562), see Zinner 1979, 388–389.

[264] *TBOO* VII, 56.17–19. The 'scouring' description comes from Tycho's contemporaneous letter to Hieronymus Wolf. See *TBOO* VII, 53.17–22.

[265] *TBOO* VII, 57.1–3.

Rantzau's *Catalogus*, Tycho wrote to the Holstein governor requesting a list of the astrological works he possessed, knowing that his library, which ran to more than 6,000 volumes at its fullest, was very extensive.[266] Whether this petition met with success is a little unclear; Tycho was to ask for the loan of specific astrological books, but he did so on the basis of their mention in a letter addressed to Rantzau by Georg Rollenhagen and forwarded to him, not with reference to a catalogue.[267] But the strategy of asking for a catalogue was one that he repeated on several later occasions. 'You write that in the Imperial Library, which you head in Vienna', he remarked in 1592 to Hugo Blotius, 'more than 90,000 volumes of books are preserved. If therefore it is possible, and at some point you have the time, I would like to have the names of the authors of those, together with the title of each book, particularly as pertains to Philosophical, Astronomical, Medical, Chemical, and similar works'.[268] Pressing his luck more than a little, one suspects, he also asked Blotius for guides to the libraries of Archduke Maximilian and other Austrian worthies.[269] Later still, his appetite for books sharpened, not dulled, by his logistically troublesome departure from Denmark, Tycho obtained from Herwart von Hohenburg a catalogue of the mathematical books in the Bavarian ducal library – from which, as he disarmingly put it following receipt, 'although very many of them are pleasing, I have selected only fifty'.[270] Tycho's list of *desiderata* has unfortunately not survived alongside his letter, but several of the volumes were named by Herwart von Hohenburg, with their catalogue numbers, as he despatched them to Tycho in a series of batches. The first of these parcels contained, 'the tables of Elisabeth, Queen of England, Gemma Frisius' *De radio astronomico*, and the Greco-Latin *Astrologia* of Joachim Camerarius No. 106; The perpetual tables of the longitudes and latitudes of the planets from Louvain, and the *Theoricae Planetarum* of Peurbach corrected by Oronce Finé No. 99'.[271] Subsequent loans included tables and texts by Johannes Virdung, Apollonius of Perga, Johannes Blanchinus, Nicolaus Pruckner, Reinhold, and Peurbach. Although the Bavarian chancellor was

---

[266] *TBOO* VII, 90.29–31. On Rantzau's library, see Gebele 1927, 550; Collijn 1933; Zeeberg 2004, 24–25.

[267] *TBOO* VII, 126.1–5; see also Oestmann 2004, 40, 53.

[268] *TBOO* VII, 329.9–14. Tycho and Blotius knew one another from time spent together as students in Basle, in 1568–1569; see Christianson 2000, 256–257.

[269] *TBOO* VII, 329.14–17.   [270] *TBOO* VIII, 174.42–175.5, 202.15–17.

[271] *TBOO* VIII, 293.9–14. The *Tabule regine Elisabeth* were printed at Venice in 1503, and had nothing to do with the yet-to-be-born Queen of England. See Gingerich 1987, 91. The other texts are identifiable as Frisius' *De radio astronomico et geometrico liber* (1545), H. Baers' *Tabulae perpetuae longitudinum ac latitudinum noviter copulatae ad meridiem Universitatis Lovaniensis* (1528), and one of the editions of Peurbach's *Theorica novae planetarum* produced by Oronce Finé. The work by Camerarius is most probably one of the several editions of his translation of, and/or commentary on, Ptolemy's *Tetrabiblos*.

not always able to find a courier who could carry such sizeable and valuable packages to Tycho along with a letter, he proved to be a very rich source of books.[272]

Tycho may have enjoyed owning items that had a strong association with illustrious astronomers. He especially cherished, for example, the 'Ptolemaic rulers' Copernicus had crafted and used, to the extent of composing a poem on their memorial value that was displayed at Uraniborg along with the instrument.[273] It seems likely, however, that Tycho's pursuit of specific predecessors' papers and books was not motivated by an interest in collecting for its own sake, but stemmed instead from the same conscientious anxiety about incorporating their work into his own that led him to propose an expedition to Alexandria in order to assess the legacy of Ptolemy. This was certainly the spirit in which he sought from Magini, in December 1590, any extant manuscripts or books of Domenico Maria Novara (1454–1504), the one-time teacher of Copernicus.[274] From Hagecius, having rather belatedly learnt of the physician's inheritance of the library of Rheticus, Copernicus' only disciple, Tycho sought to obtain the astronomical manuscripts of both him and his master in March 1592.[275] Joseph Scaliger, although unable to provide a better text of al-Battani's work when Tycho approached him in August 1598 and April 1599, did assist him by supplying the stellar observations of Ptolemy and Hipparchus in Greek, with his own annotations.[276] And through the increasingly indispensable Herwart von Hohenburg, the astronomer tried to gain access to the Fugger library in Augsburg only to acquire the manuscripts of the famed astrologer and astronomical tabulator Cyprian Leowitz (1524–1574), even as he recalled that it possessed many excellent works in a number of fields.[277] Similarly, though somewhat less specifically, when Tycho wrote in 1600 to Paul Melissus, the Heidelberg poet-librarian, he expressed an interest only in the *Bibliotheca Palatina*'s astronomical manuscripts, neglecting to enquire about its stock of printed astronomical material or other kinds of book.[278]

---

[272] *TBOO* VIII, 302.24–31, 366.10–13, 386.13–26, 393.18–22, 393.26–32.

[273] *TBOO* VI, 265–267; Turner 1995a, 142. Even possession of this object conferred a practical benefit, however, since knowledge of its flaws helped Tycho to explain some of the errors he detected in Copernicus' observational data. See *TBOO* II, 31–32.

[274] *TBOO* VII, 297.8–299.1, esp. 298.38–30.

[275] *TBOO* VII, 333.39–334.1. I say belatedly because Tycho had already benefited from this bequest by having obtained a copy of Copernicus' *Commentariolus* from Hagecius. See Rosen 1985b, 76.

[276] *TBOO* VIII, 102.12–22, 151.10–25, 152.33–38, 264.6–8.

[277] *TBOO* VIII, 202.23–203.11. On Leowitz, see Mayer 1903; Oestmann 2002.

[278] *TBOO* VIII, 250.3–6. Although there was a connection between Leowitz and Palatine Count Ottheinrich, Tycho wrote of the library's *veteres libros Astronomicos manuscriptos*, a term that he would probably not have applied to works by Leowitz.

On several occasions, Tycho sought through his correspondents to acquire whole libraries, or significant portions of them, following the death of their owner. In most of these cases he seems to have enjoyed very little success. An early attempt to obtain the books of one Dr Ambrose, a citizen of Augsburg, was frustrated in 1575, and Tycho was to learn of this collection's dispersal, by the heirs of its next owner, years too late to acquire, at a second attempt, what he had failed to before.[279] In 1576, with Joachim Camerarius as his intermediary, he tried to secure the mathematical books of a certain Brechtalius, despite knowing that the greater part of his library had already been sold.[280] And much later, in Prague, having being shown a catalogue of the library of Dudith, he sought to acquire selected works from the heirs of the bishop, or if that could not be straightforwardly achieved, the entirety of what was a substantial collection.[281] But the reply of Jacob Monaw, Tycho's chosen agent in this matter, could hardly have been more disheartening to any scholar or bibliophile. The library, he stated:

now hangs as if between heaven and earth, after it was not taken possession of by the heirs of Dudith, but rather was not so much sold as pawned to someone else, for a fixed sum of money. It now lies wretchedly nearby in a certain house, no differently (for I have seen it several times) than heaps of grain lie strewn in noble households. And neither buyer nor seller has full rights over it.[282]

It was, Monaw mournfully concluded, an intricate state of affairs; and one that he feared would result in some irreparable damage to the collection.[283] But the immediate result was that in this case, as in the others, Tycho's hopes were frustrated.

Dudith's was not the first Wroclavian library that Tycho had sought to obtain with Monaw's assistance. Indeed, rather more has been made of Tycho's interest in the books and manuscripts of another Wroclavian scholar, the mathematician Paul Wittich, than his ambitious pursuit of the whole of the substantial Dudith collection. It is, to be sure, striking how quickly Tycho moved to secure this part of Wittich's legacy, and how long he persisted in his attempts to obtain the items he wanted. Having learnt from Rothmann, in September 1587, of the mathematician's death the previous year, Tycho charged two noble visitors to Hven with the task of obtaining his library a few months later, at what may well have been his earliest real opportunity.[284] A little later, Tycho took up the matter with

[279] *TBOO* VII, 15.24–27, 84.1–2.   [280] *TBOO* VII, 42.14–16.

[281] *TBOO* VIII, 320.37–321.5. See Jankovics and Monok 1993.

[282] *TBOO* VIII, 325.39–326.4.   [283] *TBOO* VIII, 326.4–6.

[284] These men were not Danish but had visited Uraniborg whilst travelling through Denmark, and held the baronies of Herberstein in the Austrian province of Styria, and Liechtenstein. They were

Hagecius, who in turn recruited Monaw as the astronomer's agent. But a transaction that could, in theory, have taken weeks or months, was to drag on for years, with the result that the Jesuit was still to be negotiating on Tycho's behalf long after the astronomer had himself moved from Denmark to Prague.[285]

Informed that Wittich's library was in the possession of his sister, Tycho had made his wishes known to Hagecius in the following way:

I therefore ask you very kindly if you would do something through some reliable man living in Wroclaw, to negotiate with her in my name so that the entire stock of books and papers can be had at a fair price. I promise that I will send the money promptly, for I do not want to do anything that would cause those priceless books of that distinguished mathematician to be lost or come into the hands of people who do not value them. I especially want his Copernicus and what he wrote on trigonometry, and also other manuscripts if there are any. There should also be found among his other books a copy of the *Astronomicum* of Apian, of which mention was made before, which I bought for him for 20 florins and gave to him when he was here. I do not want it back unless it can be bought for a favourable price. But if his sister grants me that one thing along with the others, I shall give you the Apian *Astronomicum*, provided that it is still there, for I have another copy to hand . . .[286]

Tycho had presented Wittich with the *Astronomicum Caesareum* in October 1580, upon his departure from Hven. It was a gift, according to its inscription, 'to a friend and fellow lover of mathematics', and quite likely an attempt to ensure that Wittich would choose to return to Uraniborg, as he had promised.[287] Yet, despite its value and rarity, this was not the text Tycho sought to obtain from the Wroclavian's estate. Rather, as he reiterated in a subsequent letter to Monaw, it was Wittich's 'copies of Copernicus' *De revolutionibus coelestibus* that had certain annotations by him', and 'what he composed about triangles', that he was anxious to recover.[288] So keen was he, indeed, that he was prepared to purchase the whole library if he could not acquire these particular items, and expressed a wish to borrow them if Wittich's sister could not be persuaded to sell.[289]

---

asked to greet Hagecius on Tycho's behalf. See *TBOO* VII, 166.10–17; c.f. Gingerich and Westman 1988, 20.

[285] The campaign to acquire the books was not, however, sustained throughout this period. As Gingerich and Westman 1988, 21, indicate, Monaw's absence from Wroclaw at an early stage led to a pause in the proceedings of more than five years.

[286] *TBOO* VII, 214.36–215.7. Based, with slight modifications, on the translation by Gingerich and Westman 1988, 21.

[287] Gingerich and Westman 1988, 17.

[288] *TBOO* VIII, 237.9–14; Gingerich and Westman 1988, 21–22.

[289] *TBOO* VIII, 320.26–31.

As other historians have recognised, Tycho's pursuit of Wittich's papers and books was probably linked to his dispute with Ursus, the Dithmarschener autodidact with whom he quarrelled bitterly, if indirectly, in both manuscript and print.[290] One of the key issues in this conflict was what Tycho represented as Ursus' plagiarism of his geoheliocentric system of the world. Following the limited distribution of his *De recentioribus phaenomenis*, Tycho was initially surprised to hear that Georg Rollenhagen, one of those to whom Rantzau had sent its depiction of the Tychonic world-system, believed that he had previously encountered these planetary hypotheses.[291] He was almost equally astonished to learn that a geoheliocentric model of the planetary system had already been manufactured at Kassel.[292] Soon, however, Tycho had acquired a copy of the Dithmarschener's *Fundamentum astronomicum*, the work of 1588 that displayed the system his rival claimed to have invented.[293] This was identical to Tycho's except that the circumsolar orbit of Mars appeared to enclose, rather than intersect, the Sun's path around the Earth, and a diurnal rotation was assigned to the Earth instead of the sphere of fixed stars.[294] The Danish astronomer was quick to assert that Ursus was guilty of theft, claiming that during a visit to Uraniborg in 1584, he must have filched or caught sight of a diagram of the yet-to-be published Tychonic arrangement. He therefore began his epistolary campaign to blacken Ursus' name amongst the other members of the international astronomical community. However, with much of the *Fundamentum astronomicum* concerned with the resolution of trigonometrical problems, rather than the exposition of planetary hypotheses, Tycho also chose to attack Ursus for publishing, as his own, mathematical techniques that others had devised.[295] A particular allegation concerned the method of prosthaphaeresis, a technique for simplifying trigonometrical calculations that depended upon the substitution of addition and subtraction of functions for their multiplication or division. Although it was only of limited theoretical interest, its practical application by astronomers, who could employ it to speed up the reduction of observational data, gave it considerable value.[296] And since the first of two prosthaphaeretic identities

---

[290] Gingerich and Westman 1988. For the dispute more generally, see Rosen 1986; Jardine 1988a; Granada 1996; Launert 1999; Jardine and Launert 2005.

[291] *TBOO* VII, 135.38–42 and 408, n. to 135.42.

[292] *TBOO* VI, 156.40–157.16, 179.14–25.

[293] Tycho received a copy from Rantzau towards the end of 1588. But he reported that the text had previously been supplied to him by Rollenhagen. See *TBOO* VI, 387.16–19.

[294] Ursus 1588, 37r–40v; Schofield 1981, 109–112.

[295] E.g. *TBOO* VI, 180.5–8.

[296] The prosthaphaeretic method utilises identities for trigonometric products, for example the fact that (to use modern notation) $\sin A \sin B = \frac{1}{2}\cos(A-B) - \frac{1}{2}\cos(A+B)$, to simplify calculations.

that came to be used in this way had been disclosed to Tycho by Wittich, the Dane may well have thought that he had good grounds for denying Ursus any credit for the technique's invention, whilst attempting to secure some for Wittich and himself. Systematically, therefore, Tycho propagated accounts of how Ursus' theft of the prosthaphaeretic material and planetary hypotheses had supposedly occurred.[297] Ursus responded by doubting Tycho's mathematical competence, and by attempting to characterise his planetary scheme as a mere geometrical transformation, or inversion, of the Copernican system, unoriginal and of little real significance.[298] Critically, Wittich had been experimenting with such mathematical manipulations in the pages of his copies of *De revolutionibus*, and sight of these could conceivably have played some role in Tycho's genesis of his system.[299] There was, therefore, a 'Wittich connection' to each of the principal issues in the Tycho-Ursus dispute.

Given this context, a number of explanations for Tycho's pursuit of Wittich's papers and books would seem to be possible. Tycho might have wanted to obtain this material in order to suppress evidence, possibly or actually incriminating, of contributions by Wittich to his mathematics and cosmology. Thus, while the evidence would seem to suggest that Tycho did not seriously entertain the possibility of a geoheliocentric planetary scheme until some years after Wittich's departure from Hven, knowledge of the contents of Wittich's copies of *De revolutionibus* might have caused him disquiet. He may have feared, for example, that Wittich might himself have progressed from the geometrical manipulation of Copernican models to advocacy of a geoheliocentric system of the world. He may simply have felt that disclosure of the Wroclavian's work might undermine his case for the originality and value of his own planetary hypotheses, especially since these

---

Two such identities are to be found in Tycho's manuscript manual, the *Triangulorum planorum et sphaericorum praxis arithmetica*, the version of which known to us dates from no earlier than 1591. See *TBOO* I, 281–293; Dreyer 1916; Thoren 1988.

[297] E.g. *TBOO* VII, 281.24–38.

[298] Jardine 1988a, 29, 35–36. Jardine remarks on p. 32, n. 12, that Tycho's claims for originality in the field of prosthaphaeresis are hardly consistent with the request he made of Daniel Cramer in March 1600, that he prepare a response to these challenges. I am minded to agree. However, Tycho went on to explain in his letter to Cramer that he would consider it indecorous to publish a solution to this challenge under his own name, and had instead decided to take the matter to the courts. See *TBOO* VIII, 292.26–29. This explanation should not be too swiftly dismissed as no more than an excuse: partly because several of Tycho's correspondents stated that Ursus' *De astronomicis hypothesibus* was not worth dignifying with a response, partly because it can hardly be the case that Tycho asked Kepler to contribute an essay refuting Ursus' arguments on cosmology because he himself had done no original work in *that* field. I would suggest that here, as elsewhere, Tycho found a way to use the etiquette of the day to his advantage.

[299] See Gingerich and Westman 1988; Goulding 1995.

were already liable to be identified with the project of accommodating the Copernican models to the immobility of the Earth.[300] On the other hand, Tycho might have sought to obtain Wittich's papers in the hope that they would contain materials that he himself could divulge in order to discredit Ursus. He may have thought, for example, that by shedding light on the development and communication of prosthaphaeresis, Wittich's notes on triangles might provide him with some valuable ammunition against his antagonist.

Considerations other than those generated by the quarrel with Ursus could also have played their part in promoting Tycho's interest in Wittich's papers and books. Ever respectful of Wittich's mathematical ability, the astronomer would have had good reason to believe that he might benefit from any trigonometrical work Wittich had carried out after his departure from Hven.[301] And although seemingly a sparse annotator himself, Tycho was evidently amenable to following the insights of others when these were recorded in the margins of a book.[302] Wittich owned no fewer than four copies of Copernicus' work, and turned them over time into highly individualised volumes, crammed with cross-referenced corrections, diagrams and notes.[303] Of this manuscript material, moreover, while much was indeed Wittich's own, there was also a substantial component derived from the meticulously annotated *De revolutionibus* of the Wittenberg astronomer Erasmus Reinhold. Just as Wittich had felt it worth transferring Reinhold's marginalia into his copies of the work, other scholars chose to transcribe both sets of additions to these books into volumes of their own. John Craig incorporated annotations from two of Wittich's copies and the method of prosthaphaeresis into his *De revolutionibus* when the two men met at Frankfurt an der Oder in 1576. The Scottish mathematician Duncan Liddel (1561–1617) transcribed some of Wittich's marginalia in Wroclaw in 1582. And *c*.1604, Valentine von Sebisch, a city councillor of Liegnitz, copied from the book which remained in Wroclaw after Monaw, finally successful in meeting Tycho's request, had despatched the other three volumes to Prague.[304] So in prizing the manuscript content of Wittich's copies of *De revolutionibus*, Tycho was clearly far from alone. Given the effort he put into acquiring texts and collections that he considered to be useful on

---

[300] E.g. *TBOO* VI, 157.17–23.   [301] See Gingerich and Westman 1988, 8–10, n. 20.

[302] For the exceptional cases, see Gingerich 2002b; Jardine and Launert 2005, 97–106, 137–142; Sherman 1995, 79–100, provides a good introduction to sixteenth-century annotation practices.

[303] Gingerich and Westman 1988, 27–36.

[304] Gingerich and Westman 1988, 7, 11, 16, 27, 35. On Liddel, see Schofield 1981, 145–160; Molland 1995.

other occasions, it seems likely that he would have sought to obtain these items even if he had not been motivated by the need to build and defend his case against Ursus.

Tycho's pursuit of Wittich's books would seem, therefore, to have been overdetermined. Several plausible reasons can be adduced to explain his interest in the acquisition of these volumes, any one of which might have been sufficient by itself, and it is far from clear which would have mattered the most. Yet along with his interest in Wittich's 'writings on triangles', Tycho's avid pursuit of another printed book, the *De motu octavae sphaerae* of the Nuremberg astronomer Johannes Werner, suggests that he was particularly vulnerable when it came to challenging Ursus with respect to prosthaphaeresis. When analysed carefully, Tycho and Ursus' accounts of the evolution of this trigonometrical technique hardly differ at all. Both seem to have believed that Wittich revealed the first prosthaphaeretic identity to Tycho during his visit to Hven, and then to the Kassel astronomers, and that subsequently the second identity, and proofs for both, were found by Jost Bürgi, who divulged them to Ursus.[305] Tycho strove to portray this second communication of material in the worst possible light, but even he allowed that this interaction consisted of a fair exchange rather than a theft: in return for the material supplied to him, Ursus had translated *De revolutionibus* into German, so that the talented but poorly educated Bürgi might also study and make use of Copernicus.[306] Overall, the story raises the question of what basis Tycho could have had for demanding a share of the credit. Interestingly, in those accounts which he generated against his opponent, Tycho never actually advanced the claim that he had independently derived the second prosthaphaeretic identity, or geometric proofs of the method, although he may have underexplained his work in the field so as to convey this impression. But he was unambiguously to demand recognition as an 'originator' of the method, partly because of his supposed collaboration with the Wroclavian, but also because, as he would seem to have written to Hagecius in late 1580, Wittich acknowledged that his work on prosthaphaeresis was, 'prompted by those words which once he heard from me as I passed through Wittenberg, where he was a

---

[305] Ursus acknowledged his debt to Wittich and Bürgi in his *De astronomicis hypothesibus*; e.g. Ursus 1597, I3r: 'Prosthaphaereseos casus diversi duo in Fundamento Astronomico foliis 16. and 17. habentur: quorum quidem priorem, sed absque Demonstratione, Paulus Wittichius Silesius, in hac doctrina exercitatissimus, Cassellas circa annum Domini. 1584. attulit cuius casus Demonstrationem atque caussam Iustus Byrgi Helvetius invenit . . .' See also von Braunmühl 1900, 193–196; Jardine 1988a, 31–32; Thoren 1988.

[306] *TBOO* VII, 323.25–36. See Hamel 1998, 113–173.

student . . . although I cannot recall them'.[307] He was later to elaborate on the claim that he had been considering such matters prior to Wittich's arrival on Hven, and had even inspired his supposed collaborator's work in this field, by suggesting in a letter to Herwart von Hohenburg of November 1599 that he had followed up a hint of the prosthaphaeretic method to be found in the little treatise by Werner.[308]

Werner's *De motu octavae sphaerae* does indeed contain a reference to the application of prosthaphaeresis in simplifying the determination of a stellar coordinate, although procedural details are lacking.[309] Indeed, Werner is now generally considered to have been the true originator of the prosthaphaeretic technique. Wittich, though mathematically talented, is thought to have done no more than notice the trigonometric identity employed by Werner in manuscript materials in the possession of Johannes Praetorius.[310] But while Werner's tract might plausibly have inspired an attempt to derive his unpublished method from scratch, Tycho's correspondence reveals that he possessed no copy of the work until the late 1580s, despite having searched for it avidly.[311] Two of his contacts, Hagecius and Magini, were eventually able to procure him a copy, but not until November 1588 and September 1590.[312] And since Magini's letter announcing success in locating the book was reproduced in the *Mechanica*, by the time that Tycho composed his letter to Herwart von Hohenburg in 1599, it was a matter of record that he had not had ready access to Werner's text until long after he and Wittich had met.

Innocent explanations for this apparent discrepancy can be conjectured. Tycho could, for example, have consulted someone else's copy of Werner before 1580, although the remarks he made about its contents following receipt do not suggest any prior familiarity.[313] It is also the case, however, that Tycho's remarks about the importance of Werner's little book are rather ambiguous, perhaps deliberately so. He had taken up the work on resolving triangles prosthaphaeretically twenty years previously, he informed Herwart von Hohenburg, after reading a comment by Rheticus to the effect that one should look for calculational shortcuts to obtain the true places of the planets from their mean motions without the aid of the usual astronomical tables. He had therefore devised the alternative way of proceeding, to which Werner's book on the eighth sphere also provided an entry-point.

---

[307] *TBOO* VII, 58.36–59.1.   [308] *TBOO* VIII, 200.33–201.8.
[309] Werner 1522, k1v-k2r; Björnbo 1907, 155–156.   [310] Björnbo 1907, 167–171; Thoren 1988.
[311] *TBOO* VII, 83.31–34, 95.10–13, 104.1–3.
[312] *TBOO* VII, 147.23–26, 213.3–7, 295.17–18; *TBOO* V, 126.22–25.
[313] *TBOO* VII, 213.3–10.

Subsequently he collaborated with Wittich, who had been inspired by meeting him at Wittenberg to pursue similar work, in order to make the method clearer.[314] This phrasing is curious enough to suggest an attempt at misdirection. And the suspicion that something is amiss is strengthened by the realisation that while there are two extant copies of the 1580 letter to Hagecius amongst Tycho's papers, only one of these contains the reference to the words that the astronomer had supposedly once said to Wittich while passing through Wittenberg.

Both of the surviving copies of Tycho's letter to Hagecius are bound within a single codex, alongside other items of correspondence that Tycho is known to have prepared for publication. There are several ways to account for the omission in one copy of a passage highly significant to Tycho's dispute with Ursus, but it is tempting to suppose that they must have been inserted, or removed, as part of the editing process. And it is difficult to see why Tycho would have chosen to excise them.[315] Although, as we shall see, concerns about Tycho's editing proclivities have not always been warranted, this may represent an instance when Tycho opted to alter the historical record retrospectively, in order to strengthen his claim to have played a significant part in an important development. The possibility reveals, once again, what could be at stake in the timely acquisition of a text. For if Tycho did embellish his earlier letter to Hagecius, it was surely because he belatedly realised that, without prior access to Werner's treatise, his claim to have worked on prosthaphaeresis independently of Wittich might be difficult to credit without the support of some apparently contemporary evidence.

(iii) Readings hostile and authorial

I would have written more – especially about that dirty scoundrel Nicholas Reymers Ursus of Dithmarsch, who last winter worked, I believe, as a compositor at Your Excellency's; namely how here he did not cease to defame you with insults, and how I defended you – had lack of time not prevented me. (Christoph Rothmann to Tycho, 26 August 1586)[316]

'He took as his pretext for these insults the fact that there is to be found in the first volume of my *Epistolae astronomicae* a letter of Christoph Rothmann, *mathematicus* to the Landgrave, in which he calls Ursus 'that dirty scoundrel' . . . But it happened

---

[314] *TBOO* VIII, 200.33–201.8.

[315] It is possible that Hagecius would have chosen to excise such words from the original letter when he returned a copy to Tycho for potential publication. See *TBOO* VII, 255.16–19. If so, it would perhaps indicate that he did not believe in this supposed sequence of events.

[316] *TBOO* VI, 61.41–62.4.

that when I entrusted to certain of my students the task of correcting the printing of the book of letters, and handed the original manuscripts over to them just as they were, they inferred that nothing was to be changed and did not dare to disturb me, occupied as I was with more serious matters. And so they allowed everything to be printed just as in the originals. Thus it came about through a certain carelessness that his name, with the epithets applied to it, was retained. (Tycho to Kepler, 29 November 1599)[317]

In terms of its subject-matter, its capacity to convey information about the scope and the tools of the Tychonic astronomical project, and the identity of the individuals to whom it was presented as a gift, the *Epistolae astronomicae* was very much like Tycho's other publications. This is not quite enough, however, to establish the effect of the transformation of the Hven–Kassel correspondence from manuscript to print. As Tycho's willingness to purchase the entirety of Wittich's library in order to obtain just a handful of items serves to remind us, even in the sixteenth century, when books were more expensive than now, possession of a text did not always lead directly to use. The list of individuals to whom Tycho gave the *Epistolae astronomicae* may indicate how he himself viewed the work, but it tells us nothing about how it was received. To complete our understanding of the work, it is necessary to determine whether those who were given or who bought a copy of the *Epistolae astronomicae* actually studied the book, and the extent to which they acceded to, or resisted, Tycho's attempts to shape the range of interpretations that could be placed upon it. Just how successful was the book of astronomical letters as a Tychonic publication?

In seeking to find an answer to this question, one other lesson may be taken from the library of Wittich. Venturing beyond a text in search of its readers need not take us any further, his volumes of Copernicus suggest, than the pages and margins of particular copies. To date, however, in the absence of a complete census of the *Epistolae astronomicae* of the kind that has been produced for *De revolutionibus*,[318] only one volume with annotations attributable to a known individual has come to historians' attention. Purchased in 1610 by Johannes Broscius (1585–1652), later professor of astrology at Cracow,[319] its pages are marked according to whether the arguments to be found on them are *pro* or *contra Copernico*. The presence of these marginalia may, as one scholar has remarked, indicate the real value of the text in raising its readers' cosmological consciousness.[320] If so, then the book would seem to have worked at least partially as Tycho intended,

---

[317] *TBOO* VIII, 204.17–22, 204.38–205.2. Based on the translation in Jardine 1988a, 22–24.
[318] Gingerich 2002a.   [319] *DSB* II, 526–527.
[320] Westman 1980a, 98, including n. 42.

conveying the content of his dispute with Rothmann over the true system of the world to an audience outside his immediate circle, and continuing to do so after his death. Further evidence to support this view exists in at least some of the other extant copies of the *Epistolae astronomicae*, in the form of underlinings and annotations indicating an interest in the text's cosmological content.[321] In these cases, however, the identity of the reader or readers is difficult to establish.

While a systematic study of surviving copies may yet shed more light on the ownership and reception of the *Epistolae astronomicae*, substantial evidence of contemporary readings can also be obtained from a number of other texts, in both manuscript and print. Without doubt, the best-known reading of the *Epistolae astronomicae* to have survived from the period, that of Tycho's nemesis Ursus, was also the most antagonistic. Indeed, the Dithmarschener's response to the letters, contained in his *De astronomicis hypothesibus* of 1597, is so virulent, that it, and Tycho's reaction to it, have arguably exerted a disproportionate influence on historians' perceptions of the intended purpose of the book. Ursus clearly considered the *Epistolae astronomicae* a deliberate assault on his reputation and integrity. In the invective-filled pages of his counter-attack, he railed not only against the Danish astronomer, but also against Rothmann, for having, 'aroused, and stirred up, and incited Tycho against me'.[322] And he quoted, from one of Rothmann's earliest letters to Hven, a comment to which he took understandable exception: 'I would have written more – especially about that dirty scoundrel Nicholas Reymers Ursus . . . had lack of time not prevented me'.[323] Tycho seized upon this phrase as explanation for the tone of the *De astronomicis hypothesibus* – doing so, it would seem, precisely so as to forestall any thought that the *Epistolae astronomicae* had been intended to provoke. As he described it, the appearance in the letter-book of these remarks, and of Ursus' name in general, was due to an oversight on the part of the assistants to whom he had assigned the responsibility of preparing the letters for the press. Rightly suspicious of this claim, twentieth-century historians focused their attention on Rothmann's offensive words of August 1586, only to find that, in the extant draft of his letter at Kassel, this reference

---

[321] For example the copy of the 1610 edition, and the second copy of the 1601 edition in Det Kongelige Bibliotek, Copenhagen.

[322] Ursus 1597, E2v: 'ab eo [Rothmann] ipsum Tychonem in me esse commotum, instigatum ac incitatum.'

[323] Ursus 1597, E2v–E3r: 'Ita n. in me & de me ad Tychonem scripsit prius anno 1586, mense Augusto, Rotzmannus: Plura scriberem, præsertim de IMPURO ILLO NEBULONE Nicolao Raymaro Urso Dithmarso . . . nisi angustia temporis me impediret.'

to Ursus appears to be missing.[324] The fact of this discrepancy has led to the supposition that the crucial passage was supplied by Tycho with malicious intent. Although a plausible enough interpretation of events, more recent scrutiny of the evidence suggests that it is almost certainly false.[325] But in the absence of any general study of the *Epistolae astronomicae*, the long-standing belief that Tycho illegitimately tampered with the Hven–Kassel correspondence during publication has helped to foster the impression that the Ursus affair provides the key to understanding the letters' appearance in print.

Sufficient evidence has already been presented to demonstrate that Tycho had grounds for producing the letter-book other than his conflict with Ursus. And that this was not even the principal reason is suggested by the relative scarcity with which the Dithmarschener was mentioned in the Hven–Kassel correspondence. There were other, lengthier, discussions of Ursus that Tycho could have printed if publicising his grievances had been his priority. It would certainly be naïve, however, to suppose that the appearance of the *Epistolae astronomicae* was entirely unrelated to the Tycho-Ursus dispute. Deliberate or not, the failure to suppress the Dithmarschener's name left no room for doubt about the identity of the man Tycho accused, quite baldly, of having plagiarised his scheme of the cosmos.[326] Just as importantly, perhaps, Kassel and its denizens were thoroughly implicated in the history of the quarrel. It was at Kassel that Ursus had acquired the prosthaphaeretic formulae Tycho accused him of stealing, it was to Wilhelm that he had dedicated the diagram of his world-system in the *Fundamentum astronomicum*, and it was through Bürgi that his scheme had been realised in the form of an instrument.[327] Unobtrusively, Rothmann's letters to Tycho undermined the inflated claims Ursus had made for this planetarium in his *De astronomicis hypothesibus*, and challenged any suggestion that he had been a deserving recipient of the favour of the Landgrave. Moreover, by discussing the merits and empirical foundations of the Copernican and Tychonic world-systems, the correspondence indirectly provided a weight of evidence to support the view that Tycho, and not Ursus, had invented, studied, and developed the geoheliocentric hypothesis. Indeed, as we have seen, some of the content of the letters was probably shaped by Tycho's desire to construct and defend his claim to priority at the time they were written. So it seems rather unlikely that, during the lengthy interval between its conception and appearance, Tycho would

[324] *TBOO* VI, 351–352, n. to 61.41; Rosen 1986, 224–226.
[325] Mosley, Jardine and Tybjerg 2003.
[326] *TBOO* VI, 179.20–181.11.   [327] Ursus 1588, facing *4v, 37r.

have given no thought at all to the letter-book's role in his campaign against Ursus. It is a little difficult to believe that the inclusion in the printed text of the description 'dirty scoundrel' was neither deliberate nor malicious.

Among those scholars who have thought that Tycho inserted his own description of Ursus into Rothmann's letter, disapproval of his supposed editorial intervention has been motivated by considerations other than dislike of *any* attempt to alter the historical record. The inclusion of these words would seem to have had the effect of redistributing responsibility for the offensiveness of the *Epistolae astronomicae*, thereby making Rothmann just as much of a target of Ursus' anger as Tycho. There is also the fact that any alteration of the text of the correspondence would appear to sin against the editorial standards that Tycho had set himself in his dedicatory letter to Moritz, when he indicated that the printed letters might safely be compared with the manuscript originals. But these grounds for censure are weaker than they seem. Tycho's words to Moritz may have been very carefully chosen, for what he actually wrote was that to have interpolated anything *of his own* into the Kassel letters would have been criminal.[328] Such a remark does not actually rule out the omission or rearrangement of material, and it might also be taken to allow certain other changes conventional in the period, of a kind that Tycho showed himself willing to make on several occasions. Thus, when Hagecius was approached about the possibility of featuring in a volume of *Epistolae astronomicae*, he requested not only that grammatical errors in his letters were corrected, but also that certain passages, particularly those concerning a falling out between him and Scultetus, be omitted or altered.[329] Tycho reassured him that his edition of their exchanges would exclude anything that was of no interest to others, or might cause someone offence.[330] In line with these principles, *someone* at Uraniborg changed the text of one of Wilhelm's letters during its printing in order to conceal the nature of Rothmann's illness, substituting 'a serious and pernicious disease' for the more telling phrase in the original, the *morbus gallicus*, which identified syphilis.[331] And not only did Tycho admit that Ursus' name should not have been printed in the *Epistolae*

---

[328] *TBOO* VI, 14.24–29.

[329] Tycho asked Hagecius to forward copies of their letters several times, since he had not retained all of them himself. When he responded, Hagecius went so far as to cross out those lines that he thought should be 'omitted or altered' in one of Tycho's own letters. See *TBOO* VII, 218.37–38, 223.8–11, 225.36–40, 255.16–22.

[330] *TBOO* VII, 274.1–3.

[331] *TBOO* VI, 231.20–24, 233.4–6, 366 n. to 231.22. That this alteration occurred at Uraniborg is indicated by Tycho's claim to have been illuminated by this letter as to the nature of Rothmann's disease. See *TBOO* VI, 218.19–23.

astronomicae, it was indeed omitted from a letter by Jakob Kurtz reproduced in the *Mechanica*, where it was replaced by the conventional cipher *N.N.* meaning *non nominatus* or *nescio nomen*.[332] Clearly, therefore, some editing of the letters prior to publication was not considered improper and might even be expected.

By itself, this fact could not rescue Tycho from the charge of having behaved badly if he had indeed inserted his own description of Ursus into his correspondent's pen. The case against him in this respect might seem to be strengthened by the fact that in April 1600, Tycho was apparently receiving reports that Rothmann had seen the *Epistolae astronomicae* and was less than wholly satisfied with its fidelity. But Tycho's response to this complaint was strikingly indignant: 'If he considers that his letters were not honestly and frankly published by me', he wrote, 'undoubtedly he has copies to show in what place they were falsified; otherwise he should be ashamed to call my integrity into question without any occasion'.[333] Whatever the cause for Rothmann's concern, the evidence suggests that it was not due to a false representation of his opinion of Ursus. Notwithstanding their absence from the extant manuscript letter, we cannot be certain that Rothmann did *not* add the offensive remarks to the version that Tycho received at Uraniborg, since the surviving document is only an archival copy or draft. Another such manuscript, a draft of the letter written by Rothmann to Tycho on 22 August 1589, contains a passage about Ursus that is missing from the printed version of the text, in which he is again referred to as 'dirty' and incorrectly described as having worked at Tycho's press.[334] It has sometimes been suggested that Tycho transposed this material, moving it from the later letter to the earlier during the editing process.[335] Yet even this seems unlikely. Another letter preserved at Kassel, this time in the form of the document actually sent, was addressed by Tycho to Rothmann on 21 February 1589, and speaks of Ursus, 'on whom, in a certain letter sent to me, you once pinned a worthy label, calling him a dirty scoundrel'.[336] Unless Tycho was possessed of extraordinary powers of suggestion, sufficient to force Rothmann to adopt this language for speaking about Ursus and to overlook a false claim that he had previously done so, it would seem that at

---

[332] *TBOO* V, 121.34–41. As revealed, ironically enough, in Ursus' *De astronomicis hypothesibus, N.N.* replaced 'litteris Raimari Ursi'. See Ursus 1597, I3v. It is possible, if somewhat unlikely, that Tycho altered the *Mechanica* after seeing the *De astronomicis hypothesibus*, for he informed Kepler that he had received a copy of Ursus' work in a letter of 1 April 1598, and sent the first copies of his own to Holger Rosencrantz on 8 May. See *TBOO* VIII, 46.37–40, 64.23–35.

[333] *TBOO* VIII, 294.38–41.   [334] *TBOO* VI, 361–363, n. to 183.13.

[335] *TBOO* VI, 351–352, n. to 61.41.

[336] *TBOO* VI, 179.20–25. The translation is adapted from Rosen 1986, 225.

some point prior to this Rothmann had indeed expressed the unflattering opinion of the Dithmarschener that he evidently possessed.

Tycho was clearly not above shaping the *Epistolae astronomicae* for use as a weapon in a long-running dispute. Indeed, he clearly did so in his quarrel with John Craig, composing a lengthy letter to Rothmann in 1595, long after the process of producing the letter-book had got underway, in which he vented his frustration with the antagonist he identified, quite properly on this occasion, only as 'a certain Scotsman'.[337] The contents of this letter almost certainly had more to do with the realisation that the fuller repudiation, already prepared and designed to be appended to the *De recentioribus phaenomenis*, could not be distributed as swiftly as the letter-book, than with any desire to keep the *mathematicus* apprised of the debate.[338] But it was letters of a similarly self-conscious nature to this one that constituted the primary vehicle of Tycho's campaign against Ursus. Some of these items of correspondence, such as that addressed to Peucer on 13 September 1588, were written in a way that would have facilitated their eventual publication, while others were perhaps a little too forthright to be printed as written. For, while we cannot be entirely certain which letters Tycho would have included in his second and subsequent *libri epistolarum*,[339] there seems to have been this qualitative difference between manuscript and print, that Tycho was perfectly happy to disseminate, in writing, claims that he feared might incur the wrath of the Emperor if they were published.[340] It was really only after the escalation of the dispute effected by the *De astronomicis hypothesibus*, a work that would have seemed to Tycho to have been distributed widely, that Tycho laid plans for a publication of his own that

---

[337] *TBOO* VI, 317.13–336.14, esp. 317.39–42.

[338] Tycho possessed a sound pretext for discussing this material with Rothmann, however, since Craig had challenged Tycho's use of the Kassel observations and Rothmann's study of the comet of 1585 to support his claim that comets were supralunary phenomena. See *TBOO* VI, 331.40–332.24.

[339] Some evidence pertaining to this question does exist, however. In his *Astronomica Danica*, one of Tycho's former students printed, beside a discussion of the Ursus affair, in which he stated that the Dithmarschen had imprudently copied a diagram of Tycho's with the orbit of Mars enclosing that of the Sun, this reference: 'Vide Epistol. T. B. lib. 2 Epist. Astr. ad H. Ranzovium.' See Longomontanus 1622, 4. I take this to indicate that Tycho's letter to Rantzau of 21 December 1588, *TBOO* VII, 385–389, was slated to appear in the unfinished second volume. In addition, as mentioned above, n. 47, some of Tycho's editorial comments on the earlier letters he intended to publish have been preserved. And Zeeberg 1994, 998, has remarked on the fact that a letter written by Tycho's sister Sophie was to be published in this volume; see *TBOO* IX, 324–326.

[340] Writing to Hagecius in spring 1598, Tycho quoted from a letter of the student who had sent him Ursus' *De astronomicis hypothesibus* from Helmstedt. The author of this letter commented that it would be best if Tycho remained silent about the calumnies of this monster (*huius monstri calumnias*), since he was not worth provoking the anger of the Emperor. Tycho then asked Hagecius for advice on how he could pursue the matter without displeasing Rudolph. See *TBOO* VIII, 57.18–20, 58.16–27.

would directly engage his antagonist.[341] In the interim, however, letters served Tycho well enough that Ursus himself became aware of the charges against him. Indeed, one of the texts the Dithmarschener reproduced and responded to in 1597 was the very letter from Kurtz, in which the Imperial prochancellor named him as the plagiarist of Tycho, that subsequently appeared in expurgated form in the *Mechanica*.[342]

Given his knowledge of this epistolary campaign, it is hardly surprising that Ursus moved to attack and repudiate the *Epistolary astronomicae* as soon as it appeared. With the outcome that other historians have described, Ursus subjected the work to a ferocious reading of forensic intensity. On one level his *De astronomicis hypothesibus* made liberal use of sarcasm, ridicule, and imputations of scandal to belittle his opponents: Rothmann became Rotzmann, the 'Sniveller', whom no one in Kassel could bear to sit with at dinner;[343] Tycho was a 'mere mechanic' who 'made much of discerning double stars through the triple holes in his nose',[344] and in whose household visitors made free with his wife and his maidservant;[345] and Helisaeus Röslin, who had muddied the cosmological waters of 1597 by publishing his own geoheliocentric world-system, and by charging Ursus with having promulgated physical absurdities inconsistent with scripture, was called an 'absurd manikin' and a 'fibbing little man'.[346] Underneath its veneer of insults, however, the *De astronomicis hypothesibus* had a solid core of arguments intended not just to refute the criticisms and accusations contained

---

[341] As Tycho knew from Rollenhagen, Ursus had distributed copies in Prague. He subsequently learnt that it was known in Helmstedt, Wittenberg, Leipzig and Rostock and that when Duke Johann Adolf of Holstein-Gottorp had seen the book, he had ripped it up and commanded his Chancellor to burn it. See *TBOO* VIII, 50.32–37, 56.37–57.6, 59.9–15, 91.14–28. Tycho repeatedly denied intending to publish anything of his own against Ursus, and was not explicit to begin with about the purpose of the documents he solicited from various of his assistants; so the first clear indication that he intended to counter Ursus with a publication occurs in his letter to Daniel Cramer of March 1600; see *TBOO* VIII, 292.26–29.

[342] Ursus 1597, I3r-I3v; *TBOO* V, 119–122.

[343] Ursus 1597, E4r: 'in tota aula, inque toto atrio, ob eius arrogantiam, immodestiam, importunitatem, morositatem, oblatorumque ferculorum præposteram censuram, nullus eum in mensa suoque convictu ferre ac pati volebat . . .' Also Leopold 1986, 23 n. 87; Jardine 1988a, 36.

[344] Ursus 1597, B3v –B4r: 'Quod magnum vobis, est mihi vel minimum. / Cernere vel binas per terna foramina stella / . . . Mechanicus tantum est: tamen ipsa Mathematica nescit . . .'; Jardine 1988a, 35.

[345] Ursus 1597, F2r:'Plagium vero Hominum, ac proprie uxorum vel filiarum. Sed Tycho unquam uxorem duxit vel habuit: et filia ipsius, etiam inter omnes maxima natu, tunc temporis quo ego et adfui, nondum erat nubilis, ideoque mihi ad communem usum parum utilis, quomodo itaque quaeso plagium apud ipsum committere potuit. An vero mei festivi qui simul mecum adfuerant socii cum Concubina vel ancilla coquina Tychonis rem habuerunt, ego sane nescio . . .'; Jardine 1988a, 36. It is possible that Ursus' reference to Tycho's kitchen-maid was based on remembered Uraniborg gossip. See *TBOO* VII, 97.29–98.1.

[346] Ursus 1597, G4r-v 'mendaculus homulus . . . absurdus homuncio Roeslin'; Granada 1996, 157. On his world-system and involvement in the dispute, see also Schofield 1981, 136–144.

in the letter-book, and in Röslin's *De opere Dei creationis*, but also to thoroughly undermine the pretensions to eminence of each of their authors. Thus, Ursus claimed that, just as Copernicus' world-system was not his own invention, but that of Aristarchus of Samos, Tycho's hypothesis was previously known to Apollonius of Perga, as should have been evident to any competent reader of the *De revolutionibus*, and Röslin had merely stolen, ineptly, from the Danish astronomer.[347] So much for their claims of novelty and priority. But, in any case, planetary schemes were, he argued, merely mathematical devices; the presence in them of physical absurdities, or features contrary to scripture, was therefore irrelevant.[348] And if there was no truth in these hypotheses, necessary though they were to the practice of astronomy, then claims to have established or refuted them on the basis of observation were patently ridiculous. So much, then, for Tycho's emphasis on the construction and use of 'new' instrumentation.[349] Clever and sophisticated as these arguments were, Ursus' attacks were prompted, at least in part, by cues contained within the work of his opponent. For the claim that the Tychonic–Apollonian world-system could be found in *De revolutionibus* was undoubtedly inspired by Rothmann's notion, swiftly denied, that Tycho's was an 'inverted' Copernican cosmos.[350] And the sceptical account of astronomical hypotheses, which drew much from the writings of Petrus Ramus, may well have been suggested by the Danish astronomer's account, in one of the letters, of his meeting with the French philosopher in Augsburg.[351]

Much more could be said here, as it has been elsewhere, about the substance of Ursus' deadly little tract.[352] But what has received less consideration in the existing accounts, and is a point worth emphasising, is the extent to which Ursus himself constructed the *De astronomicis hypothesibus* as a close and critical reading. With the aid of two typographical devices, the Dithmarschener latched every barb of his own text deep into the two works he assaulted. Thus, when his own authorial voice was the dominant one, he provided marginal references to the passages that he was exploiting or answering. One can imagine, for example, the pleasure he took in marshalling the multiple references in the letter-book – 'pag. 198, 201, 203, 204, 207 &c'. – to the 'vile, disgusting, and unclean illness' of the Hessen

---

[347] See Jardine 1988a, 55–56.   [348] Ursus 1597, B4r–Cr; Jardine 1988a, 41–43.

[349] Ursus challenged Tycho's claims to have developed new instrumentation, mocking him, for example, for asserting that he invented both the form and name of the sextant. See Ursus 1597, F4r.

[350] *TBOO* VI, 157.42–158.1, 178.40–179.1.

[351] *TBOO* VI, 88.25–89.11. See, on Ramus' call for astronomy without hypotheses, Jardine and Segonds 2001.

[352] E.g. Rosen 1986, 152–157, 184–199; Jardine 1988a, 29–71; Launert 1999, 285–320.

mathematicus, 'which the wretch had already been infected and almost killed by'.[353] His other strategy was to reproduce portions of others' text along with his own *adversaria*; marginal notes that, as in the section of the letter in which Tycho denounced his plagiarism to Rothmann, consisted of a combination of sarcastic remarks, misquotations, and pointers to the fuller repudiation that followed.[354]

28. Let it be a theft but as described in no. 20.

29, 30, 31. See the following text.

32. I thought nothing could be done wrongly by Tycho.

33. But I do not know, if you do not.

34. Why incorrectly drawn? You did not know what was correct.

35. With all the Mars diagrams, you joker.

36. Certainly a most evident demonstration: of course!

37. See the following text for Nr. 31.

38. As Tycho himself did not before 1582, neither from Copernicus nor from Rheticus.

39. That does not follow for that reason.

40. But not so shamelessly as Tycho himself most impudently claims the Hypotheses of Apollonius as his own.

41. And I still do not blush. But you blush deservedly.

42. What things, I pray? Give instances.

43. See the following text.

44. Absolutely nothing at all, since what I sought was not found at your place.

45. From Bürgi in fact. Nothing from Rotzmann.

46. From whom, pray? Give instances.

47. But 'those German fellows are all half-cracked.' 'What, therefore from them?

48. What if I had done? For that was why we were wandering about.

But he betrays his theft[28] clearly enough, this[29] impostor, and at the same time[30] his ignorance of the merchandise he peddles. For in drawing the orb[31] of Mars completely enclosing the Sun's, he copied[32] a certain incorrect diagram which was drawn[33] in this way through I know not what[34] negligence, and for that reason was thrown away somewhere among my papers as inaccurate. Which diagram[35] I still possess, as it was tracked down in order[36] to make this matter absolutely clear. Since he could not correct[37] the error committed in the drawing of that diagram he makes it known to those with sufficient understanding that he does not comprehend[38] this invention, much less is its author,[39] although he shamelessly claims it as his own invention[40] and did not blush[41] to dedicate it to the Most Illustrious Prince Wilhelm as his own discovery. Moreover, he has many other things[42] in that[43] little book of his which although[44] he did not steal them all from here, however partly from you,[45] partly from other Mathematicians[46] in Germany,[47] either covertly or overtly,[48] so that if the material belonging to others is removed almost

[353] Ursus 1597, E4r: 'Ne aliquid addam de impuro, abominoso, & contagiosio ipsius morbo, quo iam infectus & pene confectus est miser: de quo passim sæpeque legenti occurret in Epistolis Astronomicis Tychonis . . .'

[354] For a comparable case of an invective treatise being built up from critical adversaria, see Grafton 1979.

49. *A little bit, however: for you say almost.*
50. *See the following text for No. 22.*
51. *Of course: we are dispensing fruit.*

52. *But not his nose ruptured, like Tycho.*
53. 54. *See the following text.*
55. *And slanderously, and unjustly and abusively enough.*[355]

nothing[49] remains which is his. But he persists in his plagiarism,[50] while to you and other scholars[51] with whom he once tarried, who he is, and how his forehead has been rubbed[52] clean of shame,[53] and what an arrogant[54] trifler he is, is quite clear enough.[55] [356]

It is enough, indeed. Enough to show that, in order to undermine the claims Tycho made in the *Epistolae astronomicae*, Ursus was compelled to pay the text the compliment of reading it most carefully. Such was his thoroughness, in fact, that he opposed the *De astronomicis hypothesibus* to the letter-book not just as a record of discovery and invention, nor even as a form of astronomical manifesto, but also as a gift and an honour to the Landgraves of Hesse. Like Tycho's text, his was dedicated to Moritz; and in extending to the prince an invitation to choose between the two works, he did not neglect to mention that, 'in his letters, *passim*', the Danish astronomer, had 'detracted from the authority and observations' of

[355] Ursus 1597, Fr–Fv: '28. Sit furtum: sed tale, quale id Nro 20 / 29. 30. 31. Vide poste Textum. / 32. Putabam nil falsi a Tychone fieri. / 33. At ego scio, quia ipse tunc nescivisti / 34. Cur non recte depinxisti? Recte nesci / visti. / 35. Zu allen marzeichen: o nugator / 36. Certissima sane Demonstratio: scilicet. / 37. Vide post Textum ad Num. 31. / 38. Ut neque ipse Tycho ante annum 1582. / neque ex Copernico, neque ex Rhetico. / 39. Id propterea non sequitur. / 40. Sed non tam impudenter, quam/ impudentiss. ipse Tycho Hypotheses / Apollonii sibi arrogat./ 41. Et adhuc non erubesco: sed tu me / rite erubesces. / 42. Quaenam quaeso? da demonstrandi / 43. Vide post Textum. / 44. Sane omnino nihil, quæ n. ego quæ/ ro, apud te non reperiuntur / 45. A Byrgi quidem: nihil a Rotzmanno / 46. A quibus quæso? da demonstrandi / 47. At. Den Tyske Karle er allsammel / all gall. quid ergo ab illis / 48. Quid si fecissem? ideo n. peregri / namur. / 49. Aliquantulum tamen: fere n. ais. / 50. Vide post Textum ad Num. 22. / 51. Scilicet: Nos poma natamus. / 52. Sed non perfracti nasi, ut Tycho. / 53. 54. Vide post Textum. / 55. Et satis criminose, iniuriose, cont / tumelioseque.' The remark, 'those German fellows are all half-cracked' was, according to Ursus, a comment that Tycho had made to Erik Lange during his visit to Hven. See Ursus 1597, A4v; Dreyer 1890, 274.

[356] Ursus 1597, Fr–Fv: 'Sed satis prodit[28] Furtum suum iste[29] impostor, / unaque[30]. ignorantiam rei, quam venditat. Dum / enim orbem[31]. Martis totaliter Solarem ambi / entem delineat, imitatur[32]. falsam quandam de / signationem,[33] quæ nescio, qua[34] incuria hic ita / depicta erat, atque ob id tanquam inepta alicubi / inter chartas meas proiecta, quam[35]. etiamnum ad / manus habeo, eodem nomine requisitam, ut / [36]. certo idipsum pateret. Cumque vitium istud in / Schemate depingendo commissum emendare[37]. ne / quiverit, sufficienter intelligentibus notum red / dit, se eam inventionem non[38] intelligere, multo / minus eius[39]. Autorem esse, utut hic conquisi / tam sibi[40]. impudenter arrogare & Illustrißi / mo Principi WILHELMO, tanquam / propriam excogitationem dedicare non[41]. eru / buerit.[42]. Multa, insuper alia habet in isto[43]. li/ bello, quæ licet[44]. non omnia hinc abstulerit, ta / men partim a[45]. Vobis, partim ab[46]. aliis Ma / thematicis in[47]. Germania, sive clam sive pa / lam[48]. corrasit, adeo ut si aliena demantur, nihil / [49]. fere restet, quod suum erit. Sed facessat iste / cum suo[50]. Plagio, tibi et aliis[51]. Eruditis, apud / quos aliquando fuit, quis sit, & quam perfrictæ[52]. per / frontis,[53]. arrogansque[54]. nugator, satis per/ spectus[55].'

the Landgrave's father Wilhelm.[357] Tycho may not have liked the analysis offered by his opponent, but he could hardly fault its attentiveness.

Carried out by the man who preceded Tycho as Imperial Mathematician, this powerful and subversive reading of the book was to help structure that of his successor in the post, the great Johannes Kepler. In fact, it is evident that Kepler consulted the letter-book on more than one occasion. His optical treatise of 1604, for example, the *Ad Vitellionem Paralipomena*, contained several references to the work, including an entire section, in a chapter on the measurement of atmospheric refraction, entitled, 'On the wrangling between Tycho and Rothmann on the matter of refractions'.[358] It was, he said, not intending to be complimentary, 'a debate so intricate that I could scarcely disentangle myself'.[359] His earliest consideration of the work, however, was carried out very much in the context of the Ursus dispute, and with him occupying the partisan role of Tycho's assistant and advocate. The way in which Kepler became involved in the quarrel has again been very well documented. By citing the author of the *Mysterium cosmographicum* (1597) in the *De astronomicis hypothesibus*, Ursus completed his reference to that triumvirate of cosmologists – Tycho, Röslin, Kepler – whose works had coincidentally appeared together at the recent spring fair in Frankfurt.[360] But since he had not seen Kepler's text, and therefore could not criticise it, Ursus chose instead to do him the disservice of praising its underlying concept, 'for I know it is truer than the hypotheses of Röslin'.[361] More damagingly still, he reproduced a sycophantic letter that the young mathematician had sent to him two years before, in which Kepler had immoderately praised Ursus as a great teacher, from whose *Fundamentum astronomicum* he had learnt much of value. Kepler, who had since come to realise that Ursus was, 'not in the least a serious writer', was thus captured for the party of Tycho by his considerable embarrassment.[362] Whatever hope he may have had of letting the matter drop quietly diminished with

---

[357] Ursus 1597, A4v: 'V. Cels. laudatissimi Principis parentis . . . autoritati & observationibus ipse passim in suis Epistolis detrahit . . .'

[358] *KGW* II, 78.29–30; Donahue 2000, 93. Kepler also mentioned the letter-book in his posthumously published *Somnium*, with reference to Tycho's exchanges with Wilhelm about elk. See Rosen 1967, xx, 12, 44.

[359] *KGW* II, 80.17–18: 'disputatio adeo involuta est, ut me vix expediam'; c.f. Donahue 2000, 95.

[360] Fabian 1972–2001, V, 370, 372, 373. Kepler's name, he later complained, was incorrectly rendered Repler and this was the name that Ursus gave in some places of the *De astronomicis hypothesibus*. See Ursus 1597, Dr; *KGW* VIII, 20.15–20.

[361] *KGW* XIII, 124.24–125.32; Ursus 1597, Dr: 'veriorem nam scio Hypothesibus Roeslini . . .'; Jardine 1988a, 53.

[362] *KGW* XIII, 143.126–128; Rosen 1986, 90–91.

the growing realisation that, as the Counter-Reformation gathered pace in Styria, threatening his employment as a Protestant schoolteacher, the Danish astronomer represented his best hope of finding another, safer, position.[363] 'But if he [Ursus] assigns to me the part of a judge', Kepler wrote in 1599, 'and he could not have done so more openly than by publicising my letter in this way, he should know that a disposition other than that of a pupil suits me now'.[364] Tycho was to hold him to his word, making cooperation in the campaign against Ursus a condition of Kepler's employment as one of his assistants in Prague.[365]

The famous outcome of that stipulation, a work enjoyed by historians of astronomy and philosophy, if not readily available to Kepler's contemporaries, was the text since dubbed the *Apologia pro Tychone contra Ursum*: a treatise, rightly prized for the epistemological and historiographical sophistication with which it answered the sceptical challenge of the *De astronomicis hypothesibus*, that was composed between October 1600 and April 1601.[366] Partly because he was never one to stay his pen, however, and partly because he was asked to do so by Herwart von Hohenburg, Kepler had previously assumed the role of Ursus' judge, as Tycho was aware.[367] Having received a copy of the Dithmarschener's work from the Bavarian Chancellor, Kepler had produced a detailed analysis of the Tychonic and Ursine hypotheses, and their relationship to earlier cosmological schemes, in May 1599.[368] Yet the *Apologia* made only slight reference to the *Epistolae astronomicae*, and the appraisal sent to Herwart did not cite it at all.[369] Kepler did not see the letter-book until sometime after September 1599;[370] and then, reluctant to validate Tycho's geoheliocentric hypothesis (or, for that matter, his account of Ursus' plagiarism) he constructed his defence in the form of a treatise, 'hardly mathematical, but rather philological'.[371]

---

[363] Caspar 1993, 77–85, 96–115; Pörtner 2001.   [364] *TBOO* VIII, 142.23–25; Rosen 1986, 140.

[365] *TBOO* VIII, 343.22–344.16; Jardine 1988a, 26.   [366] Jardine 1988a, 9.

[367] During his visit to Prague in early 1600, Kepler composed a brief tract on the dispute, which referred to the document he prepared for Herwart von Hohenburg. Tycho wrote asking the Chancellor for a copy of this document. See *KGW* XX.1, 65–69; *KGW* XIII, 341.102–348.358; *TBOO* VIII, 352.18–28; Jardine 1988a, 21–27, 58–65, 67–71.

[368] This account was composed with great rapidity: Herwart von Hohenburg sent the text with his letter of 16 May 1599, and Kepler received it on the 25th, and despatched his analysis on the 30th. See *KGW* XIII, 332.28–32, 339.2–4.

[369] Jardine 1988a, 44, 146, 193, 206.

[370] The *Epistolae astronomicae* was despatched by Herwart von Hohenburg on 29 August 1599, but Kepler subsequently complained that he had not received it. Tycho sent the letter-book and the *Mechanica* to Kepler in December 1599, and Kepler confirmed that he had received them the following year. See *KGW* XIV, 61.80–81, 75.518–522, 93.170–172, 130.100–101.

[371] *KGW* XIV, 165.165–166; Jardine 1988a, 27.

At some point, however, as the extant manuscripts make clear, he sat down with both the *De astronomicis hypothesibus* and the letter-book, and made his own assessment of the Dithmarschener's reading of Tycho's published correspondence.[372]

A. III. Rothmann writes to Tycho that Ursus sold *Tychonias* as his own at Kassel: from this Ursus takes it that in this letter his hypothesis was reported to Tycho by Rothmann.

Also he deprives Rothmann of what Tycho concedes to him. See fol. 90, 127, 131. of the *Epistolae*, and what Tycho replies. Tycho initiates a case against no one, he arrives at the same thing in a different way.

It is not difficult, he judges, *to understand any hypotheses, hypotheses are nothing but imaginary and fictitious things.*

Mathematical study is not academic. He does not understand Rothmann, thinking that he has hawked different hypotheses: however on fol. 131 he indicates otherwise.

The hypotheses of Copernicus are understood by artisans, [are] well known and very familiar: because he disdains Rothmann, who complained about this difficulty.

He expounds very badly the words of Tycho about Rothmann from the *Epistolae* fol. 206, 209, as about the hypotheses themselves, whereas there Tycho speaks about the new inequality of the planets against maintaining the scheme of Copernicus.[373]

Thus Kepler began: on sig. A3r of the *De astronomicis hypothesibus*, where Ursus first jabbed his sharpened pen at Rothmann and commenced citing the letter-book – '*Epist. pag. 90 & 127. item 131*'; '*Epist. pag. 186 & 187*'; '*Epist. pag. 90*'; '*Epist. pag. 130*'; '*Epist. pag. 206 et 209*' – pages to which his hired critic dutifully turned, noting the wilful misreadings which enabled Ursus to deny that Rothmann had himself adapted the theories of Copernicus to the immobility of the Sun, and marking the way he ridiculed the *mathematicus* for asserting, reasonably enough, that astronomy was poorly taught in the schools, that heliocentrism was barely understood anywhere, that Copernicus could and should be defended on physical grounds. And he continued in like manner through to sig. D4; finding, however, progressively less to comment upon with respect to Tycho and Rothmann, and more concerning Ursus' use, or ignorance, of authors that were rather more ancient.[374] Kepler's reading of the *Epistolae astronomicae* was not so much an assessment of the letter-book itself as an attempt to measure the distortions in the cracked lens through which others might view it.

---

[372] For a fuller exposition of Kepler's reading, see now Jardine 2006.
[373] *KGW* XX.1, 77.1–15, reading *venditasse* for *venditasque*, and *salvanda* for *salvandam*.
[374] *KGW* XX.1, 77.16–82.32.

That few of the known readers of the *Epistolae astronomicae* did see the text in this way has several explanations. Some encountered the work before the controversy had fully erupted, in the months between its publication and the appearance of Ursus' response. Others made use of the book at a sufficient interval, geographical or temporal, that they either remained unaware of the *De astronomicis hypothesibus*, or found its contents and its tone easy to ignore. The distance required for this to be the case was greatly foreshortened by Tycho's success in getting the undistributed copies of his opponent's work confiscated and incinerated as an illegitimate text.[375] And a number of individuals were sufficiently close to Tycho, even more so than Kepler, to be highly partisan readers. But almost all who read the text, and who left traces of doing so, displayed a sympathy for the *Epistolae astronomicae* as Tycho had framed it.

Of the reader's remarks preserved in correspondence, not every set is expansive, but even some of the briefer ones repay close attention. The comment by Helisaeus Röslin, for example, that Tycho was possessed of, 'a natural, easy style, and an exceptional, undoubtedly divine grace; so that anyone, provided that they love the truth, must give way to him', so vaguely suggests the cosmological debates with Rothmann, that it is difficult to be certain that this was intended to be a reference to the *Epistolae astronomicae*.[376] Röslin had, after all, previously read the *De recentioribus phaenomenis*, having borrowed the copy presented to Maestlin.[377] Yet Röslin had recently received the letter-book from Jacques Bongars, and it was he who, having overseen the production of Röslin's *De opere Dei creationis* at Frankfurt, forwarded Tycho a copy of this text at about the same time that he was sending Tycho's to Röslin.[378] Röslin made his remarks, moreover, in a letter to Herwart von Hohenburg of July 1597 whilst giving his verdict on Kepler and Kepler's *Mysterium cosmographicum*, a work that purported to provide an *a priori* demonstration of the truth of the Copernican universe.[379] Tycho, Röslin was suggesting to Herwart, would swiftly put Kepler right regarding the secrets of the cosmos, just as he had had already refuted earlier Copernican arguments; for this reason, Herwart might wish to have

---

[375] Tycho wrote to Kepler on 28 August 1600, and to Rollenhagen, on 26 September 1600, that the Archbishop of Prague, whose responsibility it was to see to the licensing and censorship of books, had been charged with seeking out copies of the work and having them burnt. See *TBOO* VIII, 343.40–344.5, 371.39–372.7; *KGW* XIV,149.143–151; Rosen 1986, 301, 307. Rosen also notes, on p. 310, that a decree of 20 October 1600 required that Ursus' widow be paid 300 florins in compensation for the books seized in compliance with this order.

[376] *KGW* XIII, 129.40–44.   [377] Granada 2002, 284–288.

[378] *TBOO* VII, 383.3–9; Rosen 1986, 59; Granada 1996, 125, 127 n. 27.

[379] See Duncan 1981, 92–101; Martens 2000, 39–56; Barker and Goldstein 2001, 88–113.

a copy of the *Epistolae astronomicae* brought to him from Frankfurt.[380] As well as affording a number of ironies, therefore, given the way that Ursus would treat this particular trio of authors, Röslin's brief remarks about the *Epistolae astronomicae* show us just how tightly interwoven the astronomical community mediated by manuscript correspondence and printed texts could actually be.

Of two readings of the letter-book we know very little. In the case of Bongars himself, although he confessed to having read the *Epistolae astronomicae*, we are left, as Tycho was, to infer from his actions alone that he had appreciated its cosmological content. And the remarks of Jakob Monaw are even less helpful, except to the extent that they suggest he (or someone close to him) had scanned the work thoroughly. He wrote only to say that rumours of his death, as relayed on one page of the text, were greatly exaggerated.[381] Other correspondents, however, were much more forthcoming. Thus, for Herwart von Hohenburg, the letter-book evidently worked as an advertisement of the Tychonic astronomical project. He had learnt from its prefaces, he informed the Dane, 'that you, from a new restoration of the courses of the Sun and the Moon, perceive the lunar Eclipses of this age of ours very differently, and much more precisely, than any tables have previously shown'.[382] And the pastor and sometime astronomer David Fabricius (1514–1617), found a use for the letter-book that, on Tycho's recommendation alone, would have been difficult to credit.[383] As he put it, writing to Tycho in November 1599:

in the case of certain stars, the intervals emerge a little differently from calculation than the intervals recorded in the book of letters. For the interval between Aldebaran and the right shoulder of Orion is given in the book of letters as $21°$ $23'$, but [through calculation from the given latitudes and longitudes] comes out as $21°$ $26'$. And the interval between both shoulders is reported there as $7°$ 30 $1/2'$, but comes out through calculation to be almost $7°$ $32'$. And so on for certain others; which, whether they were made so by a mistake of the calculators, or through a typographic error, or because those intervals were afterwards found to be otherwise, escapes me . . . I examined these very few intervals for this reason, that I had decided to adjust and test my sextant according to those values and others. Since, however,

---

[380] *KGW* XIII, 128–129, esp. 128.14–17.

[381] Monaw suggested that he and his brother had been confused. See *TBOO* VI, 317.24–36; *TBOO* VIII, 28.22–26.

[382] *TBOO* VIII, 158.4–7.

[383] On Fabricius, see Rosen 1967, 226–232; Thoren 1990, 431 n. 61, Christianson 2000, 273–276. According to his letter to Kepler of 14 January 1605, Fabricius was sent a copy of the letter-book by Tycho before his departure from Denmark, and this was done for the noteworthy reason that he might thereby be set right by the text on the matter of atmospheric refraction. See *KGW* XV, 121.230–242.

typographic errors are easily committed with respect to numbers (as in the book of letters, when on page 52 the meridianal altitude of the left shoulder of Orion is given as 40° 59', when it ought to be 39° – and that was committed through a typographic error) I wanted first to see whether they agreed with the distances by calculation. Because if that were the case then there would have been no doubt that they were free from typographic error.[384]

Clearly a man who attended to details, Fabricius was both unabashed to apply to the text the same methods of data verification that Tycho had advocated within it, and willing to follow through on the suggestion that it be employed in the calibration of his instruments. Unlike Ursus, therefore, he represents an example of a critical reader who was both responsive and respectful.

One reader who expressed an opinion of the *Epistolae astronomicae* addressed the letter containing his remarks not to Tycho, but to Rantzau. However, the Holstein nobleman promptly forwarded a copy to his Danish correspondent.[385] It is evident, moreover, that while the writer received the letter-book through Rantzau, it was sent to him at Tycho's behest for a particular reason. Ernst, the Archbishop and Elector of Cologne, Duke of Bavaria, and Bishop of Liège, Freiburg, Hildesheim, and Münster, had been asked to exert his influence in obtaining for Tycho a printing privilege valid in the southern part of the Netherlands controlled by the Spanish. And indeed, although he apologetically indicated that, given the recent history of the southern provinces, the Spanish authorities were unwilling to grant a privilege for unseen works, wishing to be 'assured, that in the guise of mathematics, nothing should be treated therein that would be against religion in the slightest', Ernst seems to have been able to procure some such document for Tycho, perhaps one that applied to Iberia rather than, or as well as, Spanish territories further north.[386] In February 1597, the astronomer expressed his gratitude to Rantzau, and through him the Archbishop, for help in obtaining this *Hispanicum privilegium*.[387]

Notwithstanding their confessional differences, the Archbishop was enthused by Tycho's book of letters. He thanked Tycho and Rantzau both for it, and for an otherwise unknown document referred to as, 'the

---

[384] *TBOO* VIII, 191.23–39.   [385] *TBOO* VII, 379.29–33.

[386] Ernst's letter has been published in at least two forms: as transcribed by Dreyer, from the copy in Munich, and as transcribed by the editors of Christoph Clavius' correspondence, from the copy in Rome. See *TBOO* XIV, 310.6–14; *CCC* IV.1, 103–105. I have followed the Munich version, as the one most likely to be a draft of the original, or based upon it, except in the rare case where it appears as if the Rome copy supplies text missing from the other. In such cases, the composite text is given below, with material supplied from the Clavius in angled brackets.

[387] *TBOO* VII, 381.6–7, 382.36–42.

resolution by Tycho Brahe of our questions put to him'.[388] Writing from Arnsberg, he expressed the intention of sending to the astronomer, as soon as he returned to Liège, 'one of our servants, whom we employ in mathematics', who would 'communicate all of our queries to him personally', and, 'if it should not be repugnant to the man, look at his mathematical devices and instruments'.[389] In fact, the interest which Ernst displayed in the practical aspects of the Tychonic enterprise revealed by the *Epistolae astronomicae* extended beyond what was purely mechanical. 'We have at our leisure', he wrote:

diligently read over his *Epistolae Astronomicae*. This is such that the expectation we always had of Tycho is not only entirely met, but is almost surpassed . . . But we have also quite happily understood from his letters, that he amuses himself daily not only in mathematics and astronomy, but also in the spagyric [arts], a study worthy of such a man; and we would not keep from your Grace that over almost 22 years we have also laboured [at this study], and not so much through others as by our own hands, and we have seen many strange things, both in transmutations and in medicines.

If we had also known that this was pleasing to Tycho Brahe, we would have exchanged confidential correspondence with him on it, and would have sent him a catalogue of some spagyric experiments.[390]

It is difficult, when interpreting this passage, to know what significance to attach to the word 'confidential' (*vertreuliche*). Although Ernst may have intended nothing more abstruse than an exchange of *epistolae familiares*, he could in fact have wanted to indicate that the contents of any alchemical correspondence should remain private. This was a policy that Tycho certainly endorsed in his reply via Rantzau, suggesting that 'certain arcane circumlocutions', or 'some private alphabet, known only to us', be employed to prevent alchemical secrets falling into the hands of those who might use them

---

[388] *TBOO* XIV, 309.43–310.4. It is possible, therefore, that it was Ernst who initiated contact with Tycho, approaching him through Rantzau as Wilhelm had done previously.

[389] *TBOO* XIV, 310.18–23.

[390] *TBOO* XIV, 310.24–28; <*CCC* IV.1, 104>: 'Sonstenn haben Wir seine *Epistolas Astronomicas* mit fließ *per otium* uberlesen. Befindenn dieselbe dermassen beschaffen, das sy der *expectation*, so Wier alzeit von ihme *Tychone* gehabt, nicht allein genug thun, sondern auch schie ubertreffen . . . So habenn Wier auch ganz gerne auß seinen *Epistolis* verstandenn, das er sich nicht allein in *mathematicis & astronomicis*, sondern auch in *spagyricis* täglichenn *delictirt, studium tali viro dignum* und könnenn euch gnediger meinung nicht vohaltenn, das auch fast in die 22 jar nit so viel <*per alios* alss *per proprias manus* laborirt haben und viel> seltzame und wunderbarliche sachenn so wol in *transmutationibus* alß *in medicis* gesehen. / Wan Wir auch wistenn, das dem *Tychoni de Brahe* annehmlich wehre, woltten Wir auch in diesem mit ihme vertreuliche correspondenz haltenn, auch einenn *catalogum* etlicher spagirischenn *experientz* zuschichenn.'

improperly.[391] It also possible, however, that having noticed Tycho's claim that he would continue to publish letters dealing mostly with astronomy, but 'with other philosophical questions interspersed sometimes, pertaining to physics and, particularly, to pyronomic study',[392] Ernst was responding to the invitation to help, 'by providing the material for this volume to be enlarged by many books'.[393]

Ernst displayed his appreciation of the letter-book, and of the whole Tychonic project, in another fashion as well. He was, he informed Rantzau, sending him some books written by the Jesuit Father Christoph Clavius:

who among those Fathers who writes on mathematics is unique. He is held among us as the Father of mathematics, and as such among the Spanish, French and Italians, and also is adored among very many of the Germans. He is a man seventy years old and more; and it causes us great wonder that we hear no mention of him in the *Epistolae astronomicae* of Brahe.[394]

Supposing that, because of his location in Rome and his use of Roman presses, the two men might not know of the 'first mover and promoter of the Gregorian calendar', the archbishop offered to initiate a correspondence between him and Tycho; a correspondence that he must also have imagined being published, since he expressed the hope that it would be of service not only to the participants, but also to the 'whole study of mathematics'.[395] Tycho thought Ernst's comments were strange. He knew Clavius and his works of course, but as he pointed out to Rantzau there had simply been no occasion to mention him in the letters, particularly as he, 'did not treat astronomical matters according to the heavens themselves, as was discussed for the most part therein'.[396] Nevertheless, the Ernst's remarks constituted a highly favourable response, and one upon which Tycho later sought to capitalise. In the midst of his patronage crisis, Ernst was one of the princes to whom he turned for support; and from him Tycho received not only testimonials to present in Prague, but also assurances that if, for some inexplicable reason, the Emperor did not offer to retain the astronomer, he

---

[391] *TBOO* VII, 380.4–13.   [392] *TBOO* VI, 22.23–27.

[393] *TBOO* VI, 22.15–22. Ernst's next statement would seem to suggest this, since he wrote, *TBOO* XIV, 310.38–40, that: 'If now some things about this, that were pleasing to him, should be sent by us, then we would also provide them for his correspondence.' Ernst, like Rantzau and Herwart von Hohenburg, is one of those noble figures who seems to have patronised, and brokered relations between, a number of astronomers and natural philosophers of the period. His interest in Paracelsianism is also well attested. See van Helden 1989b, 104; Halleux and Bernès, 1995; Granada 1996, 113 n. 5; Trevor-Roper 1998, 120–121; van Cleempoel 2002, 65–69.

[394] *TBOO* XIV, 310.41–46.

[395] *TBOO* XIV, 310.49–311.2, 311.11–12. For Clavius' role in the Gregorian calendar reform see Baldini 1983.

[396] *TBOO* VII, 382.15–23. Equally, Clavius was aware of Tycho; see Lattis 1994, 163.

would find a place with the Archbishop himself.[397] But Tycho's later attempt to use Ernst's letter to stimulate an exchange with Clavius was somewhat less successful. He sent the Jesuit a copy, but the good impression it might have created was perhaps undermined by Tycho's insistence on telling him that he had contributed, albeit inadvertently, to Ursus' wronging of himself and Wittich in the matter of who deserved credit for the invention of prosthaphaeresis.[398] As far as we know, Clavius deigned not to reply.

Not all who showed their approval of the letter-book did so through correspondence. The first of several Uraniborg alumni to cite the Hven–Kassel exchanges, Cunradus Aslachus, did so in his encyclopaedic *De natura caeli libri triplicis libelli tres* (1597), a work that he dedicated to Tycho.[399] Aslachus quoted his master's letters to Rothmann twice: once on the difficulty of ascertaining the nature of celestial matter,[400] and once on Ramus' call for an astronomy without hypotheses.[401] But while he conceivably consulted the *Epistolae astronomicae* to do so, the dating of his dedication to 16 November 1596, suggests that he, like Magini and Hagecius, had prior access to the letters, perhaps in the form of partial proofs, but most likely, since he also cited items from Tycho's unpublished correspondence, in the original manuscripts.[402] If Aslachus wrote too promptly to quote from the letter-book, however, other students of Tycho cited it explicitly. Johannes Isaacsen Pontanus (1571–1639) followed him in reprinting Tycho's remarks

---

[397] *TBOO* VIII, 80–83, esp. 81.24–42; XIV, 141.

[398] Tycho wrote to Clavius on 5 January 1600, enclosing a copy of Ernst's letter to Rantzau. As the opening to this letter made clear, Tycho had previously attempted to establish relations with the Jesuit through Tengnagel. See *CCC* IV.1, 96–103; Nørlind 1954 and 1970, 376–381. Clavius had published on prosthaphaeresis in his *Astrolabium* (1593) and Ursus, in his *De astronomicis hypothesibus*, quoted from a letter of 1593 in which his correspondent stated that he had shown the Jesuit the Dithmarschener's *Fundamentum Astronomicum* and that Clavius intended to incorporate this material into his work; hence Tycho made a point of informing Clavius of the injury done to him and to Wittich. See Clavius 1593, 1/8–180; Ursus 1597, F2v–F21; von Braunmühl 1900, 193; Jardine 1988a, 64 n. 13.

[399] The three heavens of the title are the airy, the sidereal, and the eternal. For discussion of the work, see Moesgaard 1972b, 122; Donahue 1981, 77–79; Shackelford 1991, 238–245.

[400] Aslachus 1597, 40–41: 'Et Nobilissimus ac praestantissimus hoc aevo Uraniae Mystes, Tycho Brahe, in epistola quadam ad Rothmannum de Caelo hoc fatetur, inquiens: *Nemo mortalium de ipsissima Caeli materia (si modo e materia aliqua proprie constare dici potest) certi quid in medium proferre potest. Est enim Caelum abstractum quid & immateriali simile, nostrum captum fugiens; quod sane potius, quid non sit, quam sit percipitur.*' Cf. *TBOO* VI, 136.9–21.

[401] Aslachus 1597, 160–161: 'De qua re gravissimam Nobilissimi viri Tychonis Brahe censuram adscribere non piget, qui ait: Quod celeberrimus ille nostri aevi Philosophus Petrus Ramus existimarit sine hypothesibus per Logicas rationes Astronomiam constitui posse, caret fundamento. Proposuit quidem ille mihi ante annos elapsos 16, cum Augustae Vindelicorum una essemus, hanc opinionem, & hortatur simul erat, ut postquam per hypotheses Siderum cursum in exactum ordinem redegissem, idem sine his tentare affectarem . . .' Cf. *TBOO* VI, 88.25–40.

[402] See Aslachus 1597, 122, 137, 149, where he references letters from Tycho to Craig, Tycho to Peucer, and Peucer to Tycho.

on Ramus in 1617, noting that they came from the sixtieth page of the 'book of astronomical letters'.[403] Christian Longomontanus followed him also in his *Astronomica Danica* (1622), when he invoked Tycho's views on the ether. Rejecting the 'opinion of Johannes Pena, Christoph Rothmann, and others' that the universe was filled throughout with air, he printed in the margin, 'As may be seen in Bk. I. of the Astronomical Letters of Tycho Brahe, where he replied to Rothmann on this question with very many arguments'.[404] And Longomontanus also referenced the debate on atmospheric refraction, taking an *en passant* swipe at Kepler's treatment of the subject as he did so.[405] He even recalled Tycho's comments on the implausible immensity of the Copernican cosmos, thereby making use of the lengthy disquisition his master had inserted expressly in order to provide arguments against heliocentric planetary hypotheses.[406] A fourth assistant, the cartographer Willem Janszoon Blaeu, appears to have combined parts of the work and of the *Mechanica* into a description of Hven and its instruments published in his and his son's *Atlas maior* (1662), although the evidence of this has become somewhat confused.[407] Wilhelm's son, Joan, reported that his father had served as Tycho's amanuensis in 1591, so it would seem to be possible that he was the assistant who compiled the *Synopsis* that was sent to Wilhelm and featured in the *Epistolae astronomicae*.[408] However, other sources, such as Tycho's weather diary, offer no evidence of time at Hven prior to the six months he spent at Uraniborg in 1596 – just when the letter-book was being

---

[403] Pontanus' comment was among the additions he made to the later additions of Robert Hues' *Tractatus de globis et eorum usu*, later translated into English by John Chilmead. See Hues 1639, 2; Markham 1889, xl–xli; van der Krogt 1993, 625–626; Christianson 2000, 337–339.

[404] Longomontanus 1622, I, 45: 'Videatur li. I Epist. Astro. Tych. Brahae, ubi etiam pluribus rationibus Rothmanno in hac controversia responsum est.' On this work and its author, see Moesgaard 1972b, 126–134, and 1975; Christianson 2000, 313–319.

[405] Longomontanus 1622, I, 142: 'Doctrina Refractionis Siderum, etsi in Opticis Vitellionis, & Alhazeni quodammodo sit indicata: nostra tamen aetate primum per Tychonem Brahe nostrum ab experientia coelesti est reperta, & causae eiusdem luculenter cum Rothmanno disceptatae. Has postea Iohannes Kepplerus numeris demonstrationi alligatis explicare conatus est, sed non satis generali modo . . .', with printed marginal notes 'Tomo I Epi. Astron.' and 'Io. Kepplerus in Opticis.'

[406] Longomontanus 1622, II, 19: 'Quoniam itaque tam ex immensa stellarum fixarum a tellure remotione, atque inter ultimum planetarum & earundem fixarum orbem intercapedine, quam incredibili, quae hinc sequitur, ad orbem annuum terrae, item Solem, ac multo magis ad terram, fixae primi (ut supposuimus) honoris, magnitudine, omnis bene constituta mundanarum partium symmetria facile tollitur; praeter absurdum, quod supra ex conditione rerum primario creaturum pro argumentis ex sacris adduximus, & quod postea de motu trepidationis terrae, ex latitudine potius quam declinatione alterabili stellarum fixarum, destruendo, adducturi sumus: idcirco hypothesin Copernicaeam de annua praesertim telluris motione, deque eius super polis suis libratione iure eximendum puto, & cum feliciore Tychonis Brahae inventione permutandam . . .' with a printed marginal note 'Vide l. prim. Epist. Astronom. T. B. pag. 192'. Cf. *TBOO* VI, 222.12–223.3.

[407] Blaeu 1662, I, 53–98.

[408] Koeman 1970, 57; Kejlbo 1995, 38–39.

produced.[409] It seems likely, therefore, that Joan, recognising the *Synopsis* as one of the sources of the material on Uraniborg contained in the *Atlas*, wrongly inferred that it was the work of his father and supposed that had visited Tycho at an earlier date than was actually the case. This would make Blaeu a reader, and not the author, of the letter-book's account of Tycho's observatory.

Coming from Tycho's closest associates, such testimonials and appropriations might be thought unremarkable. But if so, the cases of two readers further afield, with no connections to the astronomer, are a little more striking. The first of these, the English gentleman-soldier Sir Christopher Heydon, displayed an early familiarity with Tycho's published work in his *A Defence of Iudiciall Astrologie* (1603), a lengthy answer to an earlier printed attack on divination by means of the stars that constructed its case as a point-by-point refutation of his adversary's views, and primarily by introducing and interpreting a number of authorities, both ancient and modern.[410] Tycho featured on several occasions: in reference to the supralunarity of comets, the phenomenon of atmospheric refraction, the faults contained within both the Alfonsine and the Prutenic catalogues of the stars, the precision of timekeeping devices, the precession of the equinoxes, the non-reality of the celestial orbs, and the status and variety of astronomical hypotheses.[411] And while his *Progymnasmata* was explicitly mentioned as well as the letter-book, the fact that Heydon frequently paired him with Landgrave Wilhelm or Rothmann strongly suggests that the *Epistolae astronomicae* was Heydon's principal source for his understanding of Tycho's astronomy.

In the *De Situ et Quiete Terrae contra Copernici Systema Disputatio*, a short text produced by the Italian cleric Francesco Ingoli (1578–1649) in 1615 or 1616, the *Epistolae astronomicae* played a more central role still.[412] Occasioned by a debate with Galileo, and circulated in manuscript in Rome, the *Disputatio* marshalled its objections to heliocentrism under the three headings of mathematics, physics, and theology, considering first in each category the position of the Earth, and then its putative motion.[413] Many of its arguments were taken directly from Tycho. Thus, in the fifth chapter, 'Mathematical arguments against the Copernican motion of the Earth', Ingoli noted that:

[409] *TBOO* IX, 140; Christianson 2000, 254–256.
[410] See Chamber 1601; Heydon 1603; Allen 1966, 126–135.
[411] Heydon 1603, 124–125, 137, 143, 267, 344, 370–371, 375.
[412] See Drake 1978, 292–295, 453; Blackwell 1991, 62–63; Biagioli 1993, 283–285; Sharrat 1996, 145–148.
[413] *OdG* V, 403–412.

Many objections against the diurnal motion of the Earth can be raised, some of which Tycho relates in two letters in the book of Astronomical Letters, page 167 and 188, against Rothmann, a defender of the Copernican hypothesis. Namely from the perpendicular fall of a globe of lead from the highest tower, unimpeded by the concomitant forward motion of the air; when it ought to be the case that, on the contrary, it is moved with the diurnal motion of the Earth one hundred and fifty paces more in a second, even at the more northern altitude of Germany. Also from a cannon discharged from the east to the west, and from the north to the south, especially from one discharged near the pole, where the motion with the Earth is slowest; for, from the given diurnal motion of the Earth, a most evident difference would be noted, however none is observed.[414]

By citing page 167, Ingoli was making use of Tycho's letter of the 24 November 1589; page 188, however, was not as he stated a letter, but once again part of Tycho's account of Rothmann's capitulation in their face-to-face debate over Copernican cosmology.[415] Ingoli employed this material in other forms too, invoking, for example, in his chapter of 'Mathematical arguments against the Copernican position of the Earth', Tycho's *reductio ad absurdum* regarding the lack of observable parallax: 'Either the fixed stars could not operate on this inferior [world] on account of their excessive distance, or the fixed stars are of such a great magnitude, that they equal or exceed the size of the deferent circle of the Earth'.[416] Thanks to Ingoli, therefore, we can be certain that Tycho's former students were not alone in treating the letter-book as a detailed cosmological source, and that Broscius was not the only one to notice its arguments *pro et contra Copernico*. And thanks to Ingoli, too, we can be certain that Galileo, who may have been the chance recipient of a gift-copy of the letter-book, read it with care.[417] In his long-deferred reply to the *Disputatio* of 1624, also circulated in manuscript, the Tuscan astronomer countered the Tychonic arguments as thoroughly as he did the other non-theological arguments used by his opponent, devising responses that he would later work into his *Dialogo* on the two chief world-systems.[418] Galileo's Salviati, in other words, took up the debate where Rothmann had left it.

So far, then, seventeen readers or putative readers of the *Epistolae astronomicae* have been identified from the late sixteenth and early seventeenth

---

[414] *OdG* V, 408.   [415] *TBOO* VI, 197.6–198.9, 218.1–219.4.   [416] *OdG* V, 406.

[417] See the Appendix. There is uncorroborated evidence in the correspondence to suggest that Galileo had read the work before the dispute with Ingoli. See *TBOO* VIII, 311.20–312.1.

[418] *OdG* VI, 509–561; Drake 1953, 125–127, 141–142, 154–155 and 358–361. Only in the last of these sections, where Galileo treats the apparent size of the stars, does he make it clear that Tycho is one of his opponents, but he still does not mention that the Dane's arguments may be found in the *Epistolae astronomicae*.

centuries: Aslachus, Blaeu, Bongars, Broscius, Ernst of Cologne, Fabricius, Galileo, Herwart von Hohenburg, Heydon, Ingoli, Kepler, Longomontanus, Monaw, Pontanus, Röslin, Rothmann, and Ursus. This, while hardly constituting a triumphant reception for an edition of 1500, is not so bad a tally considering that it derives from only the most accessible and obvious sources. Moreover, while no list of this kind can ever, by its nature, hope to be definitive, this one already includes the two most famous astronomers of the next generation. Taken together, these disparate readers of the *Epistolae astronomicae* collectively recognised the text's multivalency. They found within it, among other things, a personal and an epistemological challenge, needing to be answered; an invitation to correspond with its author, or to recruit others for this purpose; a tool for calibration; a treatise in physics; and a treatment of optics. Most of these readings were legitimate ones from Tycho's point of view, interpretations that resonated not only with the original content of the letters, but also with the emphasis that he had given to the work during the editing process. And thus, although we are often told that the authorial intention underlying a text does not in itself define its meaning, the case of the letter-book illustrates how the constraints of a shared vocabulary, and common literary conventions, largely circumscribed the range of interpretations that readers attempted.[419] Even Ursus, after all, who sought to undermine the *Epistolae astronomicae* in any way that proved possible, did so by treating the work as an astronomical text. This convergence of readings hostile and authorial, suggests that perhaps it is historians, who have been more willing to consider the book of letters a source on, than a part of, Tycho's astronomical project, who have actually given the text the most discordant reception.

III A MELANCHOLY CONCLUSION

It is much controverted betwixt Tycho Brahe, and Christopher Rotman, the Lantsgrave of Hassia's Mathematitian, in their Astronomicall Epistles, whether it be the same Diaphanum, cleerenesse, matter of aire and heavens, or two distinct Essences? . . . Tycho will have two distinct matters of Heaven and Ayre; but to say truth, with some small qualification, they have one and the selfe same opinion, about the Essence and matter of Heavens, that it is not hard and impenetrable as *Peripateticks* hold, transparent, of a *quinta essentia but that it is penetrable and soft as the ayre it self is, and that the Planets move in it, as Birds in the ayre, Fishes in the sea.* (Robert Burton, *The Anatomy of Melancholy* (1621))[420]

---

[419] See Chartier 1989, esp. 154–158. [420] Blair, Faulkner and Kiessling 1989–1994, II, 48.

Tycho Brahe's *Epistolae astronomicae* was just one of a handful of publications produced during the course of a very long career; just one, in fact, of a much larger number of texts that he would have published, if time and circumstances had allowed him to complete his astronomical work as he had intended. It was an unusual book in several respects, not least because, unlike the *De recentioribus phaenomenis*, *Mechanica* and *Progymnasmata*, it was commercially sold while Tycho still lived. But it was less peculiar for being an edition of letters than we, as modern readers, might tend to suppose. And like Tycho's other publications, and those of many other contemporary astronomers, it was a work that was intentionally multivalent and complex. Positioned as a memorial and a gift for an audience of cultured nobility, it also functioned as a serious scholarly text, and as a rhetorically charged vehicle for propagating the Tychonic vision of astronomy. It fitted well into the astronomer's publishing programme, and a number of readers, all of whom knew the conventions that Tycho, as its author and editor, had adhered to and employed, received it accordingly.

Reading and writing books, like the reading and writing of letters, was an important part of the practice of astronomy in the early modern period. In the course of this chapter we have seen some of the ways in which Tycho and other astronomers functioned within the production-consumption cycle of books, the so-called 'communications circuit', as authors and readers, producers and users. And the behaviour we have observed suggests that, to the extent that they apprehended its role in relation to scholarship, the existence of printing technology caused them a degree of anxiety. Certainly, individuals such as Tycho strove in a number of ways to control and to mitigate the effects of the medium. They read widely but critically, and were frequently at pains to sift out errors of transmission and reproduction from errors of various kinds to be found in the earliest manuscripts. They struggled to obtain and use the texts they wanted and needed from an ever-increasing supply of material, produced at a widening range of locations and in an enlarged number of languages. And they attempted to exert an influence not only over who made use of their works, and how, but also to shape the ways in which the texts of others were read and perceived. Tycho's magisterial surveys of the current literature in his *De recentioribus phaenomenis* and his *Progymnasmata* were, I am suggesting, of a piece with his attempts to secure the manuscripts of his predecessors, his interest in contemporary readings preserved in annotated texts, his considered approach to the writings of ancient astronomers, his careful editing and publication of his own correspondence, and his determined assaults on the books and standing of an individual like Ursus. That is to say, all were born, to some extent, out of a deep-rooted concern about the nature and attribution of textual authority, its relation to astronomical scholarship and truth, and the role it would play in determining his reputation among his contemporaries and later generations.

The existence of such an anxiety should not really surprise us. In the sixteenth century, as now, the obvious way to go about reconstituting a discipline was to revise or replace its canonical texts. Inevitably, therefore, reforming astronomers perceived the importance of printing technology. But Tycho was far from the first astronomical scholar to see that command of the medium might assist him in achieving control of the message. He was not even the first reader of astronomical literature to realise that astronomy also needed to be reformed by means of systematic observation. Both of these insights were shared by Regiomontanus, who had acquired his expertise through the study of manuscripts.[421] And while much astronomical material was to be found in print at the end of sixteenth century, and was made very good use of, many of the techniques for managing it, including the critical readings conducted with a pen in one's hand, the writing of letters that dictated the movement of books, the copying of rarities and unpublished materials, and the drafting of new contributions to scholarship, continued to be primarily scribal. Considered with regard to the scholarly practices they deployed as texts' producers and consumers, Tycho and his contemporaries stood in the midst of a well-established tradition that would continue to persist for many years, not the beginning of a new one.

The fecundity of the printing presses contributed to concern about textual authority as much as it assuaged it. It is worth reflecting on the form of apprehension displayed by another reader of Tycho's *Epistolae astronomicae*: Robert Burton, the cheerfully saturnine author of *The Anatomy of Melancholy*, who knew all about early modern anxiety, and a great deal about how it related to books. In Burton's work, in 'The Digression on Ayre', the letter-book was mentioned, but with some of the finer distinctions between its interlocutors strangely elided; a fact that itself suggests some of the problems of being just one author among a crowded ensemble.[422] It would not have pleased Tycho, for example, to have the ether, the carrying medium for the planets' diurnal rotation in his alchemically informed conception of the cosmos, conflated with Rothmann's rarefied air. And if it would

[421] Zinner 1968, 163–165; Pedersen 1978b; Swerdlow 1990; Wingen-Trennhaus 1991.

[422] Whose copy of the *Epistolae astronomicae* Burton read is not known. Possibly he owned the letter-book himself, but no such copy is listed in Kiessling 1988. Burton's citation of Tycho was previously noted in Hellman 1963, 310–311.

perhaps have mollified him that Burton referred to the letter-book along-side such authoritative works as his own *Progymnasmata*, Kepler's *Epitome*, and Clavius' *Commentarius* on Sacrobosco, and did not fail to notice some of his reasons for rejecting Copernicus,[423] we can presume that he would have been aghast at how the Oxford scholar concluded his account of the various opinions about the arrangement of the planets and the means of their conveyance. For not himself writing as a student of astronomy, Burton declared that the authors of texts on this subject:

disagree amongst themselves, old and new, irreconcileable in their opinions; thus *Aristarchus*, thus *Hipparchus*, thus *Ptolomeus*, thus *Albategnius*, thus *Alfraganus*, thus *Ticho*, thus *Ramerus*, thus *Roeslinus*, thus *Fracastorius*, thus *Copernicus* and his adherents, thus *Clavius* and *Maginus, &c.* with their followers, vary and determine of these celestiall orbes and bodies; and so whilest these men contend about the Sunne and Moone, like the Phylosophers in *Lucian*, it is to be feared, the Sunne and Moone will hide themselves, and be as much offended, as shee was with those, and send another message to *Jupiter*, by some new-fangled *Icaromenippus*, to make an end of all those curious Controversies, and scatter them abroad.[424]

It is a striking rendering of the mood that I have described, and not least in the extent of its equivocation. For in the very act of bemoaning the vast and contradictory output of the printing press, Burton displayed his familiarity with, and mastery over, a great number of its products. And these remarks, so ambivalent about the ready availability of texts and the many debates they contained, were not only published themselves, but repeatedly reprinted.

[423] Blair, Faulkner and Kiessling 1989–1994, II, 53: '*Tycho* in his *Astronomicall Epistles*, out of a consideration of their vastity & greatness, breake out into some such like speeches, that he will never beleeve those great and huge Bodies were made to no other use, than this that we perceive, to illuminate the Earth, a point insensible, in respect of the whole.'

[424] Blair, Faulkner and Kiessling 1989–1994, II, 55 and 300. The clause from 'disagree' to 'bodies' appeared only in the edition of 1638. C.f. Heydon 1603, 370–371.

CHAPTER 4

Instruments

Largely as a consequence of the use made of the Uraniborg astronomical data by Johannes Kepler, and Tycho's concomitant reputation as the pre-eminent observational astronomer of the sixteenth century, the instrumental technology developed and used on Hven has long been considered of interest to historians of astronomy. Arguably, indeed, Tycho's instruments have received more attention within the history of the exact sciences than the apparatus of any other individual of the pre-telescopic age. The nature of the dominant historiographical tradition is such that the non-survival of Tycho's instruments has imposed remarkably few constraints on the form or the outcome of scholars' study of these objects. Preoccupied, to a great extent, with the issue of observational accuracy, historians have concentrated on those parts of Tycho's documentary legacy that most readily promise to shed light on the evolution of his observing instruments, and that confirm and account for their degree of precision. Having considered the errors in Tycho's published star catalogue, and having matched the descriptions of Uraniborg's equipment in the *Mechanica* with the entries in the observation logs, they have been content to emerge with an analysis of the accuracies, both aggregate and individual, of Tycho's observational instruments.[1]

Given Tycho's own concern with such matters, this interest in the development and performance of his observational technology cannot be said to be entirely misplaced. Even within their own limited terms of reference, however, historians have done themselves a disservice by paying slight attention to such sources as Tycho's correspondence. While the *Mechanica* is quite vague about the total number of observational instruments installed on Hven at any particular moment, a series of letters written during the years of the observatory's foundation and enlargement provide a

---

[1] Tupman 1900; Dreyer 1917; Wesley 1978; Thoren 1979, 1989, esp. 9–12, 1990; Chapman 1983b, 1989.

cumulative tally of Tycho's equipment.[2] And the most detailed of the epistolary accounts, the document Tycho had drawn up for Wilhelm in 1591,
describes the instruments of Uraniborg according to their location within
the observatory, thereby providing information that cannot be found anywhere else. One of Tycho's most recent biographers asserted that, having
found the traditional zodiacal configuration for the observational armillary
sphere unreliable, and having designed an equatorial form to replace it,
the astronomer remounted his zodiacal sphere in or before 1581, in order to
produce a second equatorial instrument. In his opinion, therefore, the zodiacal armillary and the second equatorial armillary depicted in the *Mechanica* (Fig. 4.1) were actually one and the same.[3] But in 1591, Wilhelm was
unequivocally informed that Tycho still possessed two zodiacal armillary
spheres, one of which was no longer used for observing.[4] If one such instrument was reconfigured in the early 1580s, then this document tells us that
at least three were constructed and tested on Hven during the long process
of developing Uraniborg's observational equipment. This is two more than
was previously thought.

The fact that Tycho retained possession of his zodiacal armillaries even
after they had ceased to be used for observations is just one of several
indications that the yet-to-be-written comprehensive account of Tycho's
instrumentation will have to be much broader in scope than the studies
now available. Astronomical performance was not the only basis on which
Tycho and contemporaries evaluated his instruments, but their symbolic
attributes have received little attention, despite being given considerable
emphasis in the *Mechanica*'s pages. In any case, Tycho's involvement with
instruments was not restricted to the apparatus that he constructed for use
at Uraniborg. He also participated in the exchange of instruments as gifts,
receiving several from friends in Augsburg and elsewhere, and presenting
others to nobles such as Rantzau. A study that examines the full range of
uses to which Tycho put scientific instruments, and explains their multiple
values within the courtly and scholarly cultures of the period, seems long
overdue.

While this chapter does not aspire to be entirely comprehensive, it does,
in the course of pursuing the topic of communication, rectify some of
the omissions and errors in earlier accounts of Tycho's instrumentation.
Communication *about* instruments was discussed in Chapters 2 and 3.

---

[2] *TBOO* VI, 36.8–37.25, 37.41–38.10; *TBOO* VII, 62.1–23, 72.32–40, 132.9–14.
[3] Thoren 1990, 165 including n. 45, 173. On this, see also *TBOO* VII, 62.13–18.
[4] These instruments were numbered VII and XXII in the *Synopsis*. See *TBOO* VI, 263.36–264.14,
281.35–283.8.

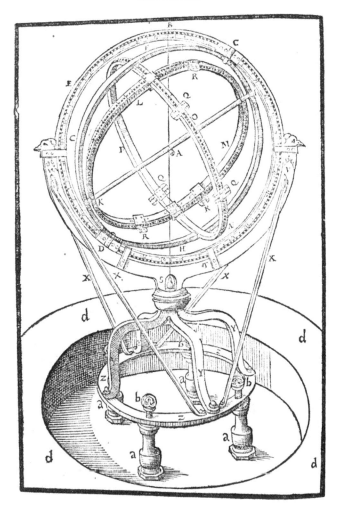

4.1 (a) The zodiacal armillary and (b) the second equatorial armillary depicted in Tycho's *Astronomiae instauratae mechanica* (Wandsbek, 1598), reproduced here from the later 1602 'edition' of the work sold under the Nuremberg imprint of Levinus Hulsius. Tycho found the zodiacal form of the instrument unreliable, and developed the equatorial configuration; hence it has been argued that these two pictures are 'before' and 'after' illustrations of the same instrument, reconfigured *c.*1581. But the Hven–Kassel correspondence indicates that Tycho had more than one zodiacal armillary sphere as late as 1591. Images courtesy of the Whipple Library, University of Cambridge.

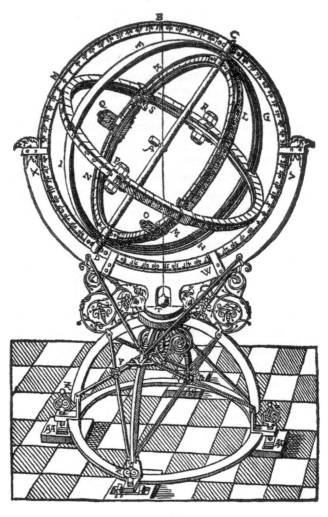

4.1 (b)

There we saw that the design and use of observatory equipment was a major topic of the Hven–Kassel correspondence, as well as the subject of what Tycho considered to be the illicit and inept disclosures to Wilhelm, and his technicians, of the mathematician Paul Wittich. Other instances of information concerning instruments being transmitted through letters, books and people, will be mentioned here in passing. The real subject of this chapter, however, is not that of communication *about* instruments, but rather the communication *through* them of knowledge, data and theories. Like almost any kind of object, an astronomical instrument can mediate the transfer of knowledge and ideas in one of two ways. By acting as a template or prototype, an instrument may convey information about itself, or about the larger class of objects of which it is a single example, simply because it embodies its own principles of design. These principles may be productively recovered from it, though not always infallibly, by individuals into whose possession it enters, or who have some other opportunity to study it closely. But an instrument may also convey information by serving some representational purpose, either one that is intrinsic to its instrumental function, or one that follows from additional meanings attached to the object by its makers and owners.

Perhaps the most interesting cases of instruments mediating the communication of design-principles are those which can genuinely be said to be instances of the transfer of technology. As has now been recognised by a number of historians and sociologists of science, novel objects and instruments of some complexity are often difficult to replicate without direct access to a working example, or the assistance of one of the object's original constructors.[5] In fact, it is clear that the amanuensis who prepared the 1591 *Synopsis* of Tycho's instruments for Wilhelm already knew of this problem. He wrote that:

> to describe all these with the necessary accuracy, and to explain all their parts so that their construction and use could be understood in every respect, would be a more troublesome effort than this brief synopsis can admit. Since not even very many words would suffice for this, but rather manual handling and visual inspection is required; and that by those who understand these things shrewdly and have learnt to use them correctly.[6]

In this respect it is interesting to note that, whilst studying medicine in Italy, Gellius Sascerides, formerly an assistant at Uraniborg, helped the Bolognese mathematician Magini to construct a sextant according to the Tychonic design; something that would likely have proved very much more difficult without his advice and assistance.[7] The movement of instruments and instrument-makers alike, therefore, had the potential to contribute in important ways to the distribution of new technologies. But the capacity of an instrument or set of instruments to guide the production of further objects of the same sort is probably more commonly evidenced in respect of details that are, strictly speaking, extraneous to function.

---

[5] See, e.g., Collins 1985; Shapin and Schaffer 1989, 225–282.
[6] *TBOO* VI, 288.1–8. [7] *TBOO* V, 125.8–10, 128.23–37; Dreyer 1890, 213–214.

The examples best known to historians of scientific instruments concern the planispheric astrolabe, a highly sophisticated device whose origins, though obscure, can be traced back to antiquity.[8] Without necessarily varying in their instrumental performance, astrolabes may exhibit very different stylistic characteristics, a fact used by historians to help them identify their era and area of origin. Sometimes it has been possible to identify features characteristic of a particular workshop or craftsman.[9] But several cases of cross-cultural fertilisation and craft-diffusion have also been established. During the sixteenth century, for example, a tulip-shape within the rete or star net of an astrolabe (Fig. 4.2) was characteristic of instruments produced by Flemish craftsmen such as Gerard Mercator (1512–1594) and Walter Arsenius (*d.* 1580).[10] But with the increasing unrest in the Low Countries that attended the Spanish policy of religious repression,[11] skilled men of all descriptions moved to other parts of Europe. Mercator relocated his Louvain workshop to Duisburg in 1551, while the migration of Thomas Gemini (*c.*1510–1562) to London in the previous decade is considered by some historians to have been the act that first established a scientific instrument-making trade in England.[12] Naturally, these makers took their designs, including the tulip-shaped rete, with them when they travelled.

The movement of astrolabes themselves, rather than their makers, also fostered the distribution of styles previously specific to a particular region or person. One reason for instrument-makers coming into contact with astrolabes others had made is that the standard instrument was equipped with a number of plates that allowed it to function properly at various latitudes. Owners of instruments acquired as gifts, or obtained for some other reason from a distant location, would often have cause, therefore, to commission additional plates as part of the process of personalisation.[13] Such modifications of instruments, capable in principle of propagating features of design, sometimes occurred across significant temporal as well as geographical distances.[14]

Designs were not, however, only susceptible to transmission via instruments and makers. With the basic technology well established, such information could also be conveyed by means of illustrations and texts. The astrolabe-maker Georg Hartmann (1498–1564) is known to have adopted

---

[8] Neugebauer 1949; Schechner Genuth 1998.
[9] E.g. Dekker and Turner 1993; Turner 1994, 1995b.
[10] Dekker and Turner 1993, 432–433; van Cleempoel 2002, 35, 46.   [11] Israel 1995, 74–230.
[12] Taylor 1954, 165–166; Brown 1979, 1–2; Turner 1994, 347–348, 1996, esp. 241–242, 2000; van Cleempoel 2002, 14.
[13] Dekker and Turner 1993, 415–420; Turner 1994, 342, 349–350; Turner 1995, 166 and plate 20.
[14] For an example, see Turner 1988.

4.2 A 'Louvain-style' planispheric and universal astrolabe, *c.*1570, displaying the characteristic tulip-shape within the ecliptic-ring of the rete. As yet, this instrument has not been definitively associated with either the Louvain school proper or the instrument-making tradition transplanted to London by Thomas Gemini. Image courtesy of the Whipple Museum of the History of Science, University of Cambridge [Wh 1467].

not only features of the instruments produced by Regiomontanus, but also elements from the *Elucidatio fabricae ususque astrolabii* (1524) of the mathematician Johannes Stöffler (1452–1531).[15] In fact, the distinction between printed or manuscript books and instruments can be drawn too strongly. Astrolabes are just one example of an instrument the illustration of which

---

[15] King 1993, 51; King and Turner 1994; Lamprey 1997.

on paper could, if of high enough quality, be transformed into a working model with very little effort. Books *about* instruments could, therefore, become books *of* instruments as well.[16] And works such as the
Astronomicum Caesareum (1540) of Peter Apian, which contained paper
volvelles, were clearly intended to be hybrids of book and instrument from
the outset.[17]

The capacity of an illustration of an instrument to duplicate its function is something that must also be borne in mind when considering the
ability of instruments to communicate through their powers of representation. Of the first class of interest, instruments with a representational
ability intrinsic to their function, planispheric astrolabes are again a good
example. An astrolabe is a model of the celestial sphere, albeit one which
mathematically transforms the three-dimensional sphere so as to render it
in only two dimensions. As such, it may convey the geocentric principles
of celestial motion: the diurnal rotation of the sphere of the fixed stars,
the annual course of the Sun through the ecliptic, and the zodiacal movement of the planets. It carries information about the position of certain
stars as well, often giving their names and their magnitudes; indeed, along
with other astronomical texts, astrolabes, and treatises about them, may
be held responsible for the number of Arabic-derived names that are still
to be found in modern star catalogues. Astrolabes also bear calendrical
information; typically they show the correspondence between the civil calendar of months and the zodiacal calendar, and occasionally they give feast
days of saints, golden numbers and dominical letters, and tables of epacts.[18]
In fact, the ability of an astrolabe to carry useful information relating to
its function was limited mainly by the ingenuity of its maker or owner, in
exploiting the otherwise empty spaces on the instrument. But not every
mark on the surface of an astrolabe necessarily contributed to its instrumental function. Others, including signatures, dedications, and ownership
inscriptions, constituted the instrument as a representation in the second
sense of interest, rendering it a symbol of the maker's accomplishments,
or the status of the owner. As was the case with early modern books, the
materials and decoration of an instrument were of great importance in
granting it a value according to the aesthetic sensibilities of the period. And
in the context of a culture of display that favoured silver and gold over
cheaper but more practical materials, excellence and variety of function

---

[16] Turner 1986, 42–43; Schechner Genuth 1998, 8.
[17] Apian 1540; Gingerich 1971a. For other examples, see Lindberg 1979; Gingerich 1993; Bennett and
Bertoloni Meli 1994, *passim.*
[18] Dekker 1993.

might themselves be appreciated more for their symbolic associations, than out of any desire to actually make use of the instrument.[19]

Whatever communicative role individual instruments fulfilled, they did so either as a consequence of the mobility of individuals, or as a result of being transported between two separate locations. And like early modern books, once again, the movement of instruments was mediated either by commercial transactions and mercantile channels, or through the courteous exchanges of friendship and patronage. But personal and commercial interests led instrument-makers, as it led authors and publishers, to seek means of retaining control over materials that were made available to the public. One important strategy for doing so, the privilege, is discussed in this chapter with particular reference to a type of astronomical instrument that, in one of its forms at least, proceeded from both the printing press and the artisan's workshop. It is with this class of object, the celestial globe, that we shall begin our consideration of communication through instruments as it is exemplified by Tycho and the international community to which he belonged.

I GLOBI TYCHONICI

In the study (where the students of TYCHO pursue their studies by day and perform astronomical calculations; where also his library is found, and which is on the middle floor of a round south-facing tower) there stands a great GLOBE, which is covered all over with choice and closely compacted sheets of brass, and so assiduously that it is as if it were cast from solid metal; moreover, it is of excellent sphericity all over . . . In addition other globes great and small, six or eight in number, some celestial and some terrestrial, are found in this study, and all sorts of smaller astronomical instruments and automata as well. (Tycho Brahe's amanuensis, *Synopsis of the Astronomical Instruments which the Dane Tycho Brahe has had placed here and there on the Island of Hven* (1591); printed in the *Epistolae astronomicae* (1596))[20]

Of all astronomical instruments to which a communicative function might be attributed, the celestial globe, as an object whose capacity to impart information is intrinsic to its nature, is perhaps the first to spring to mind. Serving as a demonstrational model displaying the divisions of the celestial sphere and the constellations, a celestial globe is, at least in principle, a visual embodiment of empirical data. Familiarity with its geographical

---

[19] Olmi 1985; Scheicher 1985; Thornton 1997, 99–125; Daston and Park 1998, 255–302; Watanabe-O'Kelly 2002, 71–99.

[20] *TBOO* VI, 268.16–22, 269.1–3.

counterpart, the terrestrial globe, is widespread even at the beginning of the twenty-first century, and may help to make evident the pedagogic value of such a representation, even to those with little knowledge of positional astronomy. This familiarity should also assist the realisation that, as is the case with a planar chart, the information conveyed by a three-dimensional model of spatial relations is both qualitative (names and shapes) and quantitative (distances and angles). But contemplation of the modern functions of globes is likely to prompt a third insight as well: that instruments of this sort, particularly when found outside the classroom, often carry connotations of ostentation and wealth. It remains within the bounds of modern experience, therefore, that globes can serve different representational functions, and may vary in size, materials, and extent of decoration accordingly, even if what is depicted on their surfaces remains substantially the same.

This feature of modern globes is very much a legacy of the early modern period. In the sixteenth and seventeenth centuries, celestial and terrestrial globes had great importance both as demonstration devices, and as signifiers of status.[21] Their production and distribution, in a range of forms, served the different needs and interests of a variety of individuals: astronomers and pedagogues, craftsmen and businessmen, clients and patrons. So when Tycho's amanuensis wrote of the 'globes great and small, six or eight in number, some celestial and some terrestrial' that the astronomer kept in his library, he could have been referring to a group of objects quite diverse in their natures and origins. Lacking further information about these instruments, it is unlikely that we shall ever know much about them, although we could speculate if we chose. Fortunately, however, there is sufficient evidence concerning other *globi Tychonici*, instruments either owned by Tycho or constructed to his design, to allow them to be used to illustrate the range of characteristics of early modern globes and their communicative role.

The most famous of Tycho's globes, 'covered all over with choice and closely compacted sheets of brass', was described and depicted (Fig. 4.3) in the *Mechanica* under the title of the Great Brass Globe (*Globus Magnus Orichalcicus*). According to this account, the instrument had a diameter of nearly six (Tychonic) feet, or 1.462 metres, and it was marked with the positions of the stars as Tycho and his assistants had observed them on Hven, uniformly precessed to the year 1600.[22] Its production had been a long and painstaking process. The wooden shell was commissioned in 1570 from an Augsburg instrument-maker described by Tycho as, 'a clever

[21] Dekker 1999, 2002.
[22] *TBOO* V, 103–105; Raeder, Strömgren and Strömgren 1946, 9; Kejlbo 1995, 32–33.

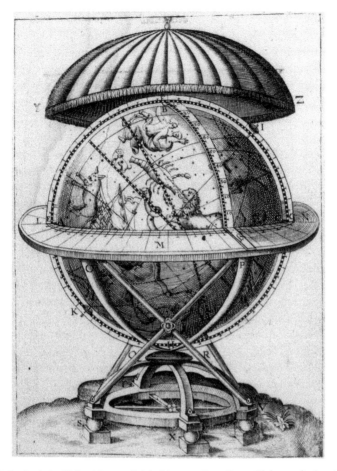

4.3 Tycho Brahe's *Globus Magnus Orichalcicus*, or Great Brass Globe, as depicted in the *Astronomiae instauratae mechanica* (Wandsbek, 1598), reproduced here from the later 1602 'edition' of the work sold under the Nuremberg imprint of Levinus Hulsius. A precision instrument, used for calculating and recording, the globe was provided with a cover, YZ, to protect it from dust and dirt. Image courtesy of the Whipple Library, University of Cambridge.

craftsman, a person I had spent a long time searching for in vain in other places'.[23] But when the globe was finally received in Denmark in 1576, it was found to lack the required degree of sphericity, and was marred by several cracks. These imperfections were carefully corrected, and the globe was then

---

[23] *TBOO* V, 103.7–10; Raeder, Strömgren and Strömgren 1946, 103. This instrument-maker is normally identified as Christoph Schissler, but see Chapter 1, n. 36.

observed over a two-year period to ensure that it would retain its shape, and survive the temperature variations of the changeable Danish climate. Subsequently it was covered with brass, marked with the ecliptic and the equator, divided to a minute of arc, and mounted within a meridian and horizon ring, likewise divided. Only then were any star positions marked on its surface.[24] An inscription placed on the horizon of the globe dated its completion to 1584.[25]

Tycho provided little information about the function of the globe in the *Mechanica*, stating only that its use was 'similar to that of other celestial globes', and that he intended to write a special book on the subject.[26] Like other projected works, however, this treatise was never produced. It is likely, however, that Tycho had in mind something akin to manuals such as Johannes Schöner's *Globi Stelliferi, sive sphaerae stellarum fixarum, usus et explicationes* (1533) or Robert Hues' *Tractatus de globis et eorum usu* (1594), works that were written to accompany a specific globe (or, since terrestrial and celestial globes were frequently produced and sold together, a specific pair of globes) but whose level of generality is attested to by their subsequent reissue on numerous occasions.[27] Since they contained explanations of the fundamentals of positional astronomy, and also, when appropriate, outlined the basics of terrestrial geography, these texts constituted elementary introductions to cosmographic study. Despite this, they gave less emphasis to the utility of globes as demonstrational models for teachers of cosmography, than to their versatility as practical problem-solving devices.[28] Hues' text, listing sixteen functions of the globe, stressed its calculational uses by including such sections as 'How to finde the Latitude of any place, by observing the Meridian Altitude of the Sunne, or other Starre' and 'How to make a Sunne Diall by the Globe, for any Latitude'.[29] And Schöner's more comprehensive treatise not only contained instructions on the same procedures, but also described how the globe could be used for astrological purposes, for example in the calculation of the boundaries of the horoscopic houses.[30]

---

[24] *TBOO* V, 103–104.   [25] *TBOO* V, 104.26–33; *TBOO* VI, 268.34–42.

[26] *TBOO* V, 105.3–4. Cf. Raeder, Strömgren and Strömgren 1946, 105, where the use of the globe is described as being 'similar to that of other instruments'. Their translation is unlikely to be correct in the context of the *Mechanica*, which *was* Tycho's long-intended description of the construction and use of his observing instruments. Like me, Kepler also construed this statement to mean that Tycho intended to publish a globe-manual; see *KGW* XX.1, 91.21–22; Grafton 1997a, 195.

[27] On Hues' work, see Markham 1889, xxiv–xlixi, although his claim that the fifth section of the work was contributed by Thomas Harriot (1560–1621) seems a misreading of Hues, 1594, 111. On Schöner's globes and globe-manuals, see van der Krogt 1993, esp. 30–33.

[28] See also Dekker 2002.

[29] Hues 1594, 67, 86; Hues 1639, 185, 251.   [30] Schöner 1561, CIIIr *sqq.*, chs. LX–LXIII.

Neither Hues nor Schöner neglected more elementary uses of the instrument. Their works indicated how the location of any point on the surface of a celestial globe could be found with respect to either of the universal coordinate systems, the ecliptic and the equatorial; or indeed, if the instrument were appropriately set in its mount, how the same determination could with be made with respect to the horizontal system of altitude and azimuth. Almost by inspection, therefore, a globe could be used to convert between these systems without recourse to the mathematical procedures for resolving spherical triangles. As Hues acknowledged, the owner of a globe might thereby sacrifice precision for the sake of convenience. That everything which could be done with the instrument could 'be performed farre more accurately, by the helpe of numbers, and the doctrine of Triangles, Plaines and Sphæricall bodies' was, he wrote, 'a thing very well known to those that are acquainted with the Mathematickes'. However:

> this way of proceeding, besides that it be tedious and prolixe, so likewise doth it require great practise in the Mathematickes. But the same things may be found out readily and easily by the helpe of the Globe, with little or no knowledge of the Mathematickes at all.[31]

By constructing a very large sphere with painstaking care, Tycho hoped to produce an instrument that could eliminate the calculational labour of his observational project, but without incurring the loss of accuracy experienced by users of the commercially available devices.

Tycho said as much himself in the *Mechanica*. 'On account of its great size', he wrote:

> this globe has the advantage over others that everything can be carried out with the greatest accuracy, indeed even to a minute of arc. Thus it is possible to determine mechanically, with very little trouble and without difficult calculations, everything concerning the doctrine of the *primum mobile* and the observations of the heavenly bodies relative to the ecliptic and the equator, or any other of the celestial circles.[32]

This assessment of the accuracy with which calculations could be performed using the globe is of considerable interest. On a great circle of the globe-sphere, which had a circumference of nearly 6π feet, or 4.69 metres, an angular distance of one minute of arc equated to just over a fifth of a millimetre. So slight a distance makes it intuitively difficult to credit the claim that computations to this limit of precision were possible, and it

---

[31] See Hues 1639, 164–165, translating Hues 1594, 60–61.
[32] *TBOO* V, 105.4–11; adapted from Raeder, Strömgren and Strömgren 1946, 105. See also Tycho's comments to Wilhelm, *TBOO* VI, 37.32–41.

might therefore seem as if Tycho was failing to distinguish between the functioning accuracy of the instrument and the limit of division of its scale. It is noteworthy, therefore, that Tycho had himself criticised Wilhelm IV's use of a globe to avoid trigonometrical calculations, stating that it was better to trust to the 'scrupulosity of Geometry and Arithmetic', and declaring that he did not trust his own instrument to provide an accuracy of one or one-and-a-half minutes of arc in all places.[33] Could it be that this slightly less-generous estimate represents Tycho's determination of the globe's limit of accuracy proper?[34]

Studies of the graduations of the arc scales of extant astronomical instruments have shown that early modern instrument-makers were capable of engraving divisions with remarkable precision.[35] Using the transversal scale he also employed on his observing instruments, Tycho's instrument-maker had to directly divide the circular arc only to sixths of a degree, equivalent to a spacing of about 1.3 millimetres. While still small, this linear value is comparable to that to which contemporary astrolabes were divided: one produced *c*.1585 in the workshop of Erasmus Habermel (*d.* 1606), for example, and now in the Museum of the History of Science in Oxford, was divided into degrees with a linear spacing of approximately 1.5 mm, and a mean error in the placement of each mark of less than 0.06 mm.[36] A similar level of competency in construction of the globe-scales would have resulted in a mean error in the marking of each sixth of a degree of about a fifth of a minute. Assuming, for the sake of argument, that no significant errors were introduced into the scale via the transversal divisions, this would have produced a mean greatest uncertainty in the length of arc between any two graduation marks of less than thirty seconds.

The validity of such a comparison may be called into question on a number of grounds. In particular, since the globe-scales were much larger than those of the Habermel astrolabe, and were more finely divided, it might be supposed that the process of graduation was necessarily different. In fact, we know little about the practical techniques used for marking instrument arcs prior to the invention of the circular dividing-engine in the eighteenth century, and even this was not used to graduate the scales of the finest observatory instruments.[37] However, the information we do possess suggests that the basic principles and dividing instruments would have

[33] *TBOO* VI, 68.8–26. On the Kassel globe and its use on the occasion in question, see Leopold 1986, 27, 56–59.
[34] Cf. Tycho's remarks to Scultetus in 1581, *TBOO* VII, 62.10–13.
[35] Chapman 1983a, 1983b, 1983c.
[36] Chapman 1983a, 476, Table 1. On Habermel, see Eckhardt 1976, 1977.
[37] Chapman 1993, 423–424.

been the same in both cases. What is possibly the earliest representation of the division of an astronomical circle portrays the graduation of a large globe, and is contained in the *Liber Organicus* (*c*.1674) of the Jesuit missionary Ferdinand Verbiest (1623–1688), a man whose astronomical work in China drew much inspiration from Tycho's observational project.[38] It shows the use of beam-compasses and a rule or set-square: tools also used in the construction of Persian astrolabes at exactly this period, and probably common to both Latin and Islamic workshops in previous centuries.[39] As an order of magnitude estimate, therefore, it seems reasonable to suppose that the scales of the *Globus Magnus Orichalcicus* could, in principle, have functioned to the limit of accuracy to which they were divided.

Inspection of the globe-scales themselves could perhaps settle the issue, but these are no longer available. After leaving Hven, Tycho transported the *Globus Magnus Orichalcicus* to Prague with his other instruments. Following his death it found its way to Poland; and in 1632 was returned from there to Denmark. But in 1728, it was destroyed by a fire in the Rundetårn, the site of the University of Copenhagen's observatory.[40] In any case, in practice, the error resulting from inaccuracies in the division of the instrument's scales is likely to have been much smaller than that contributed by other imperfections in the globe's construction, or by the fallibility of those who were using it. As Tycho seems to have recognised, anything less than perfect sphericity would have diminished the accuracy of the instrument; and it would clearly have been difficult to mark the globe and read off distances with a precision equal to that of its divisions.

The uniqueness and importance of the *Globus Magnus Orichalcicus* among Tycho's globes is attested to not only by its description in the *Mechanica*, but also by its appearance on the plan of Uraniborg (Fig. 4.4) included both in that work and in the *Epistolae astronomicae*.[41] A student of Tycho's, Johannes Isaacsen Pontanus, wrote that it was 'a vast and magnificent peece of work: insomuch that many strangers came out of diverse parts into *Denmarke*, while it was there, only to see this *Globe*'.[42] Since this comment appeared amongst interpolations in later editions of Hues'

[38] Chapman 1984; Halsberghie 1994; Iannaccone 1994; Libbrecht 1994.

[39] Chapman 1983a, 475–476, cites the *Voyages* (1686) of John Chardin, who visited workshops of Persian astrolabists in 1674, and was informed by his European escort that the same practices and tools were used in the West. This probably refers to division in general, rather than the inscribing of astrolabes, the production of which was in decline in Europe at this time.

[40] Dreyer 1890, 366; Kejlbo 1969–1971, 63–64; Kejlbo 1995, 33–36.

[41] The plan appeared in the *Epistolae astronomicae* in virtue of having been sent to Wilhelm IV in 1591 with the account of Tycho's instruments. See *TBOO* VI, 290.

[42] Hues 1639,) (ʳ.

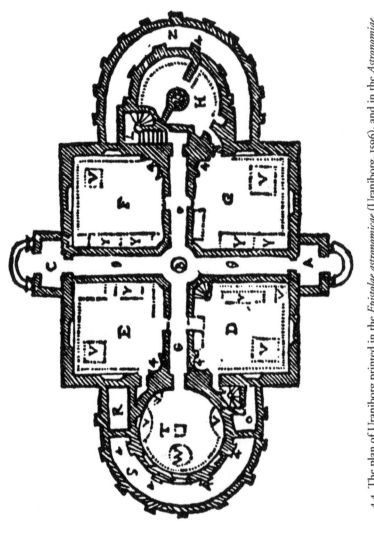

4.4 The plan of Uraniborg printed in the *Epistolae astronomicae* (Uraniborg, 1596), and in the *Astronomiae instauratae mechanica* (Wandsbek, 1598), reproduced here from the later 1602 'edition' of the *Mechanica* sold under the Nuremberg imprint of Levinus Hulsius. The circle marked W shows the *Globus Magnus Orichalcicus* standing in Tycho's library. Image courtesy of the Whipple Library, University of Cambridge.

Tractatus de globis et eorum usu, the implication that there was less interest in Tycho's other observatory equipment need not be taken seriously. But it seems clear that no other globe on Hven matched the Great Brass Globe in either size or accuracy. And since even it could not entirely satisfy Tycho's desire for a precise working instrument, no other globe could have made a significant contribution to the observational and calculational work undertaken on his island. What then was the role played by the other celestial globes to be found in Uraniborg's library?

The earliest extant reference to a Tychonic celestial globe concerns an instrument the astronomer gave to the University of Rostock in the late 1560s, perhaps as an inducement to overlook the duel that had taken place between him and Manderup Parsberg, another Danish student.[43] Nothing more is known of it, although it is possible that it was the globe 'only the size of a fist' that Tycho mentioned in the brief autobiography to be found in the *Mechanica*. This globe, he noted, had been employed by him when he was learning to distinguish the constellations, and when he first began comparing the motions of the planets with published ephemerides.[44] Even a small globe had its uses, therefore; indeed, Tycho clearly found the size of the instrument an advantage, because it made it easier not only to take it outside for comparison with the night-sky, but also to conceal the fact that he was doing so from his tutor. Although such secrecy would not have been needed at Uraniborg, some of the smaller globes in the library may have been used in a similar fashion by Tycho's assistants. Not all of them arrived on Hven with much experience of gazing at the heavens.

Yet another *globus Tychonicus* is discussed in letters dating from 1573. These make it clear that, at the same time as he was supervising the printing of Tycho's *De nova stella*, the Copenhagen academic Johannes Pratensis was also mediating between the astronomer and craftsmen he had commissioned to produce a number of instruments. Pratensis wrote that:

All the figures on your globe are finished. The meridian and the horizon are divided, the base is prepared, but the difficulty in this matter is not small with respect to the gilding and the colouring or painting that you specified. First, for example, the goldsmith was not confident that he could represent golden stars with the required form of the figures on top of the silver covering of the whole globe . . . As pertains to the painter, he will never portray those circles and divisions as precisely as they are now engraved, but will in fact cover the clean and smooth surface with an overlay of colours that will obscure those things which the goldsmith has accurately delineated. I shall say nothing . . . of how quickly part of the painted covering might

---

[43] *TBOO* XIV, 4.6–11. See also Dreyer 1890, 26–27; Thoren 1990, 22–24, 29–30.
[44] *TBOO* V, 106.33–107.6, 107.8–16.

be removed through some injury or just a casual touch. But listen to my advice, because I am also making known to you the opinion of the goldsmith. Gilding is necessary for a noble object, or indeed a royal one. And since the horizon and the meridian and the feet ought to be gilded, well then, we could gild these rather than the whole surface of the globe; although in this plan, the fact that those figures would, together with the stars, appear less correct, *makes a difference*, as they say. But the intermediate parts or the very surfaces of the globe could be polished, and brought to a gleaming and reddish lustre; this shine could also be applied to the stars themselves (if we wanted), so that that golden radiance springs forth as if sparkling from the glimmering light of the figure. This way it will also show the constellations precisely engraved, the acuteness of the points of which it is difficult to represent with paints. For they will besmear those slender points and hollows, which add grace to the whole figure, and fill up the crafted beauty of those angles. This way the globe will in the end be golden, resembling the substance and appearance of heaven simply, so that the feet will match the head; far removed from those parti-colourings and effeminate visual delights that amuse some . . .[45]

The details of Tycho's original commission, which called for the globe to be painted, may well have been shaped by his reading of the instructions given in Ptolemy's *Syntaxis*. 'We make the colour of the globe in question somewhat deep', the Alexandrian astronomer wrote, 'so as to resemble the night sky'.[46] But Pratensis and the goldsmith were clearly of the opinion that not only was this inappropriate for the type of globe being produced, and the status of its owner, it presented considerable practical difficulties. It also appears that the initial choice of a sphere constructed from metal, rather than the wood that would readily support a delicately painted surface, was dictated not by aesthetic grounds, but by the fact that the instrument was to have a mechanical function. 'I shall press STEPHANUS as much as I can, and encourage him to apply the little wheels today or tomorrow, and test the progression', Pratensis said in his letter, clearly referring to one of the Copenhagen craftsmen involved in the production of the instrument.[47] While the exact function of these 'little wheels' is not self-evident, Tycho's letter of reply, printed in *De nova stella*, asked that the professor 'stir up the craftsmen, so that our bronze globe is gilded well and, as I enjoined them, exquisitely represents the contrary course of the *primum mobile* and the

---

[45] *TBOO* VII, 10.29–11.24.

[46] Toomer 1998, 404; although Ptolemy also described, see p. 405, depicting the constellations in painted lines of a similar colour to that of the background, so as 'not to lose the advantages of this kind of pictorial description, and not to destroy the resemblance of the image to the original by applying a variety of colours'.

[47] *TBOO* VII, 11.27–29. Stephanus has been identified with the Copenhagen clock-maker Steffan Brenner. See Maurice and Mayr 1980, 215–216; Thoren 1990, 123; Christianson 2000, 265.

luminaries'.[48] It therefore would seem that the globe incorporated some clockwork mechanism, designed to allow it to represent the course of the Moon across the sky, the annual movement of the Sun, and the diurnal rotation of the heavens.

Whether such a mechanism would have made the instrument a better demonstrational model is debatable, but it would certainly have increased its worth as an *objet d'art* and showpiece. The description in Tycho's letter strongly favours the presumption that this globe was the same as the one represented in the famous mural of Uraniborg's great quadrant, and shown in the woodcut of that instrument that appeared in the *Mechanica* (Fig. 4.5).[49] 'Above my head at X', wrote Tycho:

a certain gilded brass globe is depicted, which is freely revolved on skilfully fitted internal wheels, and imitates the diurnal motion, and at the same time shows the contrary motion of the Sun and the Moon about the poles of the ecliptic: to the extent that the Moon even changes itself with respect to its form and illumination, by waxing and waning. But the Sun, besides showing its own motion, revolving around the equatorial axes in 24 hours in its diurnal course, shows each hour of the day as well, and also its times of rising and setting, and its crossing of the meridian.[50]

The fact this 'automaton', as Tycho called it, functioned as a celestial globe, a model capable of displaying and demonstrating heavenly phenomena, and a form of decorative clock, even indicating the place of the Moon in its cycle, identifies it with a class of objects increasingly common to the courts of early modern Europe.[51] In Hesse-Kassel, a major centre for the production of such instruments, Wilhelm's instrument-maker of the 1560s, Eberhard Baldewein, oversaw the construction of several mechanical globes, some free-standing and some crowning astronomical clocks, while his successor, Jost Bürgi, was to produce at least seven free-standing mechanical globes between 1579 and 1594.[52] Other self-moving globes were produced in cities such as Nuremberg, Strasbourg, and Vienna. The best of these beautiful instruments incorporated mathematical models of the

---

[48] *TBOO* I, 15.25–28.

[49] This identification has also been made by Kejlbo 1995, 75. However, he also suggests that the object referred to is the gilt armillary sphere now in the National Museum, Copenhagen, made by Josiah Habrecht (1552–1575) at Strasbourg in 1572. See King and Millburn 1978, 77, fig. 15.4; Maurice and Mayr 1980, 294–295; Kejlbo 1995, 72–75. This instrument, despite the tradition associating it with Tycho, does not seem to fit with either the description in Pratensis' letter of a globe marked with the constellations, nor the depiction in the *Mechanica*.

[50] *TBOO* V, 30.18–26, reading *facies* for *faces*.

[51] *TBOO* V, 30.26–32; King and Millburn 1978, 62–89; Leopold 1997.

[52] Leopold 1986; Leopold 1997, 7–9.

4.5 Tycho's mural quadrant, as depicted in the *Astronomiae instauratae mechanica*
(Wandsbek, 1598), reproduced here from the later 1602 'edition' of the *Mechanica* sold
under the Nuremberg imprint of Levinus Hulsius. At F the observer, perhaps Tycho, is
shown sighting on a celestial object through the aperture in the wall opposite. Two
assistants are present, one acting as recorder, one taking readings from the clocks at the
bottom right. The portrait of Tycho, and the section through Uraniborg above the
quadrant arc, were painted onto the wall on which the instrument was mounted. Note the
mechanical celestial globe at X, between the portraits Y and Z of King Frederick II of
Denmark and his consort Queen Sophie. Image courtesy of the Whipple Library,
University of Cambridge.

celestial motions with extraordinary fidelity, accommodating the non-uniform annual solar motion, for example, with gears in an epicyclic arrangement. Indeed, from two of Bürgi's instruments, it has been possible to retrieve the otherwise lost value for the length of the tropical year employed by the Kassel observers, and with a precision which shows that it had been determined to within ten seconds of the value now accepted.[53] Had such precision been replicable at the scale of the *Globus Magnus Orichalcicus*, then the resulting instruments would have made outstanding models of the heavenly motions, usable for the purpose of retrieving the positional data produced by the mathematical models they embodied. They could have functioned as calculating instruments of extraordinary accuracy. But with both technical and economic constraints on their size, the mechanical precision of these globes greatly exceeded their capacity to act as visual analogues of the heavens from which astronomical data might be recovered.[54]

Although these instruments were never intended to be of direct assistance to the practising astronomer, the same expertise required to construct them could also be put to use in furnishing an observatory. It is worth remembering, however, that pursuit of an observational reform of astronomy was the active concern of only a small proportion of astronomical practitioners, and the standards established at Hven and Kassel were more exacting than those achieved, or sought for, at other locations. Mechanical globes were not necessarily worse at communicating star positions, the form of constellations, and stellar magnitudes, than the commercially produced globes available to the majority. But the chief motivation for their construction was their value as exquisite showpieces and gifts. In the *Mechanica*, Tycho revealed that he had presented his mechanical globe to Christian IV during the monarch's visit to Hven in 1592; in return, he received a gold necklace adorned with the portrait of his king.[55] Wilhelm too, gave away many of the globes and clocks produced at his court to other ruling princes, and even the Emperor.[56] In the gift-economy of the early modern courts, the capacity of noble astronomers to conceive and commission such devices was a powerful asset.

---

[53] Leopold 1986, 143–173, esp. 153–156 and 164–170; Leopold 1997, 7.

[54] Although there is evidence that Bürgi used one of his mechanical globes, unfinished as a result of an irreparable error in the engraving of the horizon, in order to record his own stellar observations. See Leopold 1986, 42, 135–145; Leopold 1997, 7–8.

[55] *TBOO* V, 30.26–34; Raeder, Strömgren and Strömgren 1946, 30. The date given in the text is 1590; that it should be 1592 is made clear by Dreyer 1890, 214–216.

[56] Leopold 1986, esp. 36.

The instrumental functionality of the celestial globe was valued all the more in the case of mechanical examples, not only because they displayed the ingenious movements of their clockwork, but also because they were constructed from precious materials. Yet globes were also appreciated as symbols of earthly dominion and, like other objects associated with a particular scholarly discipline, as signs of erudition.[57] Even commercial printed globes, therefore, which indeed were not lacking in a certain beauty of their own, could be used as princely gifts, particularly if their presentation was reinforced by means of dedication.[58] In remarking upon Tycho's presentation of his mechanical globe to Christian IV, several commentators have made a connection to the king's receipt of a pair of printed globes in 1589, the supposition being that these objects were commissioned by the astronomer as a gift for his monarch. Although, as we shall see, the evidence does not readily support this inference, there is nevertheless a clear connection between the globes received by Christian, and the first *globi Tychonici* to be commercially produced. Both were manufactured by the van Langrens, a globe-making family based in Amsterdam. And as Tycho indicated in the *Epistolae astronomicae*, the van Langren Tychonic globes, first issued in 1594, were soon followed by others; in particular, by those manufactured in 1598 by Willem Janszoon Blaeu. 'Even now there are globes fashioned in the Low Countries according to the aforesaid reformation of the stars', he stated, referring to his own stellar observations, 'and more are being prepared'.[59] Since Blaeu was actually on Hven when this comment was made,[60] and the production of the earlier globes had been preceded by visits to the island from two of the van Langrens, Tycho had enjoyed the opportunity to contribute much more to the preparation of these instruments than just the basic stellar data that his statement might be taken to imply. But how did manufacture of these instruments come about? And how exactly was Tycho involved in their production?

The instruments themselves convey some of the facts. In the case of the van Langren globes, only a single example, now in the collection of the Historisches Museum in Frankfurt am Main, is known to be extant (Fig. 4.6). It consists of a papier-mâché sphere 0.33 metres in diameter, covered with hand-coloured printed paper (twelve gores and two polar caps), and mounted in a wooden stand with an inset compass in the base. A cartouche on the instrument states that it is:

A CELESTIAL GLOBE indicating the positions of the fixed stars, a true representation of the heavens for the year 1600 according to the exact observations made

---

[57] Lippincott 1999, 76–78.   [58] See van der Krogt 1993; Brotton 1997, *passim*.
[59] *TBOO* VI, 16.17–26.   [60] *TBOO* IX, 140; van der Krogt 1993, 142.

4.6 The single extant celestial globe produced by the van Langren firm of Amsterdam in 1594. Photograph by Ursula Seitz-Gray, reproduced by permission of the Historisches Museum Frankfurt [HMF X 14609].

recently in Denmark by the labour and astronomical instruments of the nobleman TYCHO BRAHE.[61]

Elsewhere on the globe, one can read that:

JACOB FLORIS, citizen of Amsterdam, had this Celestial Globe made, having sent his son HENDRIK to Denmark to carefully map the stars according to the confirmed observations of the Nobleman TYCHO BRAHE. With a privilege, granted to the same Lord TYCHO, and kindly shared by him. 1594.[62]

Prepared by Jacob Floris; his son Arnold engraved the notes, and his son Hendrik the figures.[63]

---

[61] Holbrook 1983, 69; van der Krogt 1993, 425; Glasemann 1999, 15–20.
[62] Holbrook 1983, 69, van der Krogt 1993, 425–426.
[63] Holbrook 1983, 70; van der Krogt 1993, 424 fig. 2.1.7, 425.

Jacob Floris van Langren (*d.* 1610) was a cartographer and globe-maker of the southern provinces of the Netherlands, who moved to Amsterdam, in or before 1585, in order to escape persecution for his Anabaptist beliefs.[64] His first known globes were produced in 1586. In 1589, working with his son Arnold (*c.*1571–*c.*1644), he issued a celestial and terrestrial globe-pair;[65] it was examples of these instruments that were sent to Christian IV in that very year, and that some historians have taken as evidence of a previous relationship between the van Langrens and Tycho. On receipt of these objects, the king, then aged twelve, did indeed write to inform the astronomer of their arrival and invite him to inspect them, and shortly afterwards Tycho was ordered to recompense Jacob van Langren with the sum of 100 thalers.[66] But since the van Langrens had themselves dedicated this edition of globes to Christian, the payment need not indicate, as some historians have suggested, that Tycho was involved in procuring these instruments for the king, or that his recommendation persuaded the monarch to acquire them.[67] Nor is it likely that it was prompted by the astronomer's acquisition of the objects: Christian's injunction, that if the globes were sent to Tycho to examine, he should return them promptly after one or two days, would rather seem to indicate that the monarch was set upon keeping them.[68] The most probable explanation for Tycho's payment to the van Langrens, therefore, is that Christian called upon him, as the *de facto* cosmographical expert of the Danish court, to comment on the value of his (unsolicited) gifts, and then commanded him, as was the royal prerogative, to render their manufacturers appropriate remuneration on behalf of his king.

It is likely then, that it was the van Langrens' presentation of a globe-pair to Christian of Denmark which first brought about the contact with Tycho that eventually led to the production of the Tychonic globes of 1594. Arnold van Langren spent six weeks on Hven in 1590, less than a year after this presentation, and the visit mentioned on the Frankfurt globe, that of his younger brother Hendrik (*c.*1574–1644), occurred in 1593.[69] It seems plausible, therefore, that the manufacture of a Tychonic globe was either the deferred objective of Arnold's trip to Uraniborg, or a project conceived in the course of his stay. Yet the necessity of Hendrik's subsequent visit has not been understood by historians, largely because of the confusions, omissions and errors of the earliest accounts of the van Langren–Tycho relationship.[70]

[64] Keuning 1956; van der Krogt 1993, 135.   [65] van der Krogt 1993, 89–96.
[66] *TBOO* XIV, 48.24–29; van der Krogt 1993, 96; Kejlbo 1995, 48.
[67] Dreyer 1890, 216 n. 1; Holbrook 1983, 72; Kejlbo 1995, 48.   [68] *TBOO* XIV, 48.24–29.
[69] Arnold stayed with Tycho between 6 April and 23 May; Hendrik from 13 April to 25 May. See *TBOO* IX, 83, 86, 118, 119.
[70] E.g. Christianson 2000, 142–143, 168, 230, 308–311.

To set the record straight, therefore, and since the interaction between the noble astronomer and these representatives of commercial enterprise is highly pertinent to the topic of the distribution of astronomical knowledge, it is worth taking the time to unravel the complex chronology relating the globes, the journeys of the van Langrens, and Tycho's statements about them. Only then does it become possible to correctly understand the events leading to the manufacture of these instruments and their significance to the Dane's astronomical programme.

Tycho's nineteenth-century biographer John Dreyer wrote that Jacob Floris van Langren sent his son (singular) to Uraniborg to obtain the correct places of the stars for a new instrument, and that while 'Tycho declined to give [them] in writing . . . he allowed him to examine the great globe in the library'.[71] In support of this claim, he cited a passage from Tycho's *Progymnasmata*. But when we turn to the relevant page of that text, we read only that:

JACOB FLORIS, citizen of Amsterdam, a great expert in the making of celestial and terrestrial Globes, having sent here his son, skilled in the construction of the same, decided to prepare a certain celestial globe following this rectification of ours of the places of the Fixed Stars, which in its certainty and skilful elaboration is about to greatly surpass those hitherto employed. Which I mention because if anyone is delighted by such things, they should seek out this completed mechanical work above all others, by which their desires will be more readily satisfied.[72]

This comment does not license the assertion that the van Langren representatives were only granted access to the *Globus Magnus Orichalcicus*, but it does explain why reference is only made to one of them visiting Hven: Tycho's biographer was following what Tycho had written himself. But how should this fact be interpreted? It is certainly not the case that Tycho was unaware of the relationship between Arnold and Hendrik, since his diary noted the latter's arrival on the island as that of the 'brother of Arnold'.[73] Recalling that much of the *Progymnasmata* was written between 1590 and 1593, we might suppose that at the time this particular passage was composed, Hendrik had yet to visit Hven, and therefore that Arnold was the van Langren son it mentioned. Certainly, it is the long interval between composition and publication which best accounts for Tycho's failure to mention here, as he did in the *Epistolae astronomicae*, the Tychonic celestial globes that Blaeu would be producing. But if this section of the *Progymnasmata* was written after May 1593, it seems likely that Tycho intended to refer to Hendrik van Langren, and had good reason to omit mention of Arnold.

---

[71] Dreyer 1890, 261 n. 1.　[72] *TBOO* II, 282.4–12.　[73] *TBOO* IX, 118.37–39.

One explanation could be that the elder son's visit was not motivated in any way by the collection of data for a new celestial globe. But the other possibility is that what Arnold set out to accomplish in 1590 could not be achieved until another three years had passed.

This idea finds some support in what is also the earliest traceable source of the idea that Tycho allowed Arnold, or Hendrik, to copy the stellar positions marked on his Great Brass Globe, but would not grant them access to the actual catalogue: the account of the astronomer's seventeenth-century biographer Pierre Gassendi (1592–1655).[74] Gassendi wrote that:

Jacob Floris an esteemed Amsterdam maker of celestial and terrestrial globes, because the rumour of the places of the fixed stars corrected by Tycho had spread abroad, sent his son into Denmark and begged Tycho to allow the catalogue of all the restored fixed stars to be copied out, which would be marked on new globes, and promised that he would inscribe his name on them as well; and they would then be called the Tychonic and Brahean celestial globes. And indeed this pleased Tycho. But since, although the calculation of eight hundred was almost finished, he had not yet begun to print them, and had it in mind that he would trust this [the catalogue] to no one unless it was printed (doubtless as, having been made the common property of all, no one would appropriate it for himself alone), he did not let it be copied out, but nevertheless expressed the hope of sending a copy as soon as it was printed; although he proposed that the father should not produce anything until the soon-to-be-undertaken calculation of the other two hundred was completed, lest those should either be missing from the globe, or be placed incorrectly amongst the others. But he allowed him [the son] to examine the great globe in the library, and inspect the stars on that, some of which were already engraved, some of which were being engraved from day to day, as each of their true places was determined by calculation.[75]

That nearly 800 of the stars on the van Langren instrument were placed according to Tycho's observations, with the remainder derived from the catalogue of the 'ancients', was information reported on the surface of the

---

[74] On Gassendi, see *DSB* V, 284–290; Jones 1981.

[75] Gassendi 1654, 139–140: 'Iacobus Florentius Artifex Globorum caelestium, terrestriumque Amstero-dami eximius, sparso rumore de correctis a Tychone Fixarum locis, destinavit in Daniam filium, ac Tychonem precatus est, ut pateretur ab eo exscribi restitutarum omnium Affixarum Catalogum, quas novis Globis inscriberet, pollicitus nomen ipsius illis quoque inscriptum iri; ac fore, ut deinceps caelestes Globi Tychonici, Braheanique appellarentur. Et gratum quidem id habuit Tycho; verum, quia licet pene absolutus octingentarum calculus foret, nondum tamen commiti praelo occoeperat, ipseque nisi typis expressum nemini illum credere in animum induxerat (nempe, ut plurium com-munem factum, sibi privatim nemo arrogaret) idcirco non dedit quidem exscribendum, sed spem tamen fecit transmittendi exempli statim, ac foret editus; tametsi author fuit, ne Pater quidquam praecipitaret, quousque aliarum praeterea ducentarum calculum mox suscipiendum absolvisset; ne aut illae globis deessent, aut perperam, caeteras inter bene adnotatas, collocarentur. Permisit vero, ut magnum Globum intra Musaeum inspectaret, inque eo Stellas consideraret, quae partim insculptae iam erant, partim in dies insculpebantur, prout germanus cuiusque locus erat calculo designatus.'

globes.[76] Gassendi might also have surmised this fact from the *Progymnasmata*, which contained the coordinates of just 777 stars, or even by reading a letter of 1599, in which Kepler described a copy of the instrument that he had seen in Graz.[77] But the claim that the globe data were not taken directly from Tycho's star catalogue finds no obvious basis in any of the contemporary documents of which I am aware.

Gassendi's motivation for composing his biography of Tycho is itself somewhat uncertain.[78] As an observational astronomer himself, he may well have been interested in promoting and shaping the discipline by means of an edifying history, and in this context it is worth noting that the preface to the work contained an account of astronomy prior to Tycho, and that summaries of the lives of Peurbach, Regiomontanus, and Copernicus were also appended.[79] But as a Catholic priest and crypto-Copernican, Gassendi's natural sympathies could not have lain entirely with his own choice of subject. He was nevertheless, diligent in seeking out information about Tycho, not only hunting out all his published works and enquiring of Ludwig Kepler, Johannes' son, concerning his manuscripts, but also seeking the help of Ole Worm, Denmark's famous antiquarian. And Worm, who was the brother-in-law of Christian Longomontanus, Tycho's former disciple, was able to provide Gassendi with some material he could not otherwise have obtained.[80] It is possible, therefore, that the Frenchman had access to documentary evidence no longer available, including some that pertained to the construction of the van Langren *globi Tychonici*.

It is also possible, however, that Gassendi's claim was based on no more evidence than is currently available. One pertinent source is a letter written by Tycho in May of 1600, in which, having been asked about globes, he stated that:

Two Dutch makers of such contrivances, one of whom is named JACOB FLORIS of Amsterdam, the other WILHELM JANSZOON of Alkmaar, have carefully constructed globes following our restoration of the fixed stars. The latter spent half a year with me on our Uraniborgian Island, and there marked on his own sphere, which he brought with him, the approximately 1000 stars I allowed him.[81]

---

[76] van der Krogt 1993, 426.   [77] On this letter, see below.

[78] It is known that Gassendi was asked to produce such a study for his friend Henri-Louis Habert de Montmor; but it is not self-evident why he would have acceded to such a request. Joy has discussed the importance of history to Gassendi's natural philosophy; but of the *Vita Tychonis* she picks out only his account of the meeting between Ramus and Tycho at Augsburg. See Joy 1989, 216–217. Thill 2002, 321–322, suggests that the *Vitae* were written as part of the quarrel between Gassendi and Jean-Baptiste Morin (1583–1656).

[79] For disciplinary histories see, besides Chapter 1, Graham, Lepenies and Weingart 1983.

[80] Gassendi 1658, 518–521, 526–527; Christianson 1970.   [81] *TBOO* VIII, 322.32–38.

Although there is no indication here of the means by which the stellar data were communicated to the van Langrens, the information that Blaeu recorded the stars' positions on a globe of his own might be taken to imply that he was not permitted to leave Hven with an actual copy of Tycho's manuscript catalogue. And if Tycho was determined to prevent any version of this document from escaping Uraniborg, then he might perhaps have refused to let Blaeu consult it; requiring instead that he obtain the information by study of the *Globus Magnus Orichalcicus*. Since it is unlikely that Tycho was any more generous with the van Langrens than with his later visitor, this letter might be taken to lend a measure of credibility to the account presented in Gassendi's *Vita Tychonis*.

On the other hand, there are clearly some problems with this line of reasoning. Let us suppose that Hendrik van Langren, like Blaeu, copied the Tychonic star positions onto a small globe of his own which he brought to Uraniborg. Neither he nor Blaeu would then have been able to recover from this record the precision of its source, whether that was the *Globus Magnus Orichalcicus* or a manuscript catalogue: a portable globe, likely to have been even smaller than the globes these cartographers subsequently published, could not have been marked with the accuracy sufficient to accomplish such a reconstruction. All Tycho had to ensure, in this case, was that neither man had the chance to copy his star catalogue, or abscond with it, and his observational data remained as secure as it had ever been. Unless there was a problem supervising their work on Hven, forcing the van Langrens or Blaeu to derive positions from the Great Globe offered no particular advantage if they were transferring the data to a portable sphere.[82] At the same time, however, this method could have jeopardised the precision claimed for the resultant globes by the van Langrens, by Blaeu, and even by Tycho. If, on the other hand, as Gassendi's *Vita Tychonis* suggests, Hendrik was allowed to compile a list of star positions on the basis of those recorded on the Great Globe, then he would have departed Hven with a reasonable approximation, not quite 'to one or one-and-a-half minutes of arc in all places', to Tycho's stellar catalogue. Barring a significant error during the retrieval process, or the engraving of the globe gores,

---

[82] The idea suggested by Warner 1979, 28, that the use of a portable recording-globe by Blaeu may indicate that the constellation figures on his commercial globes, which were *stylistically* innovative, were derived from those on the Great Brass Globe has been comprehensively rejected by van der Krogt 1993, 160. But questions of style aside, it could have been that a recording globe was used for indicating which stars Brahe placed within the constellations, and which were *informata*, that is, outside the constellation-figures, but associated with them. A recording-globe could have supplemented a transcription of the catalogue in this fashion. See the discussion by Leopold 1986, 39–43, of this topic with respect to the Kassel globes.

any differences between the two lists would have indistinguishable at the scale to which the printed instruments were produced, thoroughly justifying the assertion that they were based on Tycho's data. But Tycho would then have been forced to permit a version of his unpublished stellar catalogue to leave his domain, creating the possibility that others might obtain it before it had been printed.

There is, it is true, a certain psychological plausibility to the claim that Tycho would have been jealous of his stellar data prior to the publication of his stellar catalogue, and would not have allowed this information to leave Hven in any form whatsoever. Sufficient evidence exists, in the opinion of some historians, to establish that Tycho was by nature intrinsically secretive.[83] In the *Mechanica*, for example, Tycho said of his alchemical work that he would discuss it, 'frankly with princes and illustrious men, and other distinguished and learned people, who are interested in this and know something about it', provided that, 'I feel sure of their good intentions, and that they will keep it secret'.[84] And in Prague, Tycho made absolute secrecy regarding his observations and inventions a condition of Kepler's employment as his assistant.[85] Even so, Kepler was to complain that his access to the observational data was very much restricted.[86]

Extrapolation from evidence such as this, however, does not necessarily justify claims about Tycho's secretiveness in general. Tycho gave as his reason for not disclosing alchemical secrets, that, 'it serves no useful purpose, and is unreasonable, to make such things generally known. For although many people pretend to understand them, it is not given to everybody to treat these mysteries properly according to the demands of nature, and in an honest and beneficial manner.'[87] Within the context of the alchemical tradition, such a position was hardly unusual. It is clear, moreover, that Tycho was willing to share his knowledge with other initiates, provided that a secure method of communication could be found.[88] And with respect to Kepler's complaints, it must be taken into account that, increasingly preoccupied by his dispute with Ursus, Tycho's fear of an alliance between his would-be-assistant and the Dithmarschener mathematician could have induced him to act with more than his usual caution during their collaboration.[89] Moreover, even if the attitude he displayed

---

[83] E.g. Dear 1995, 50–51.
[84] *TBOO* V, 118.7–11. Translation adapted from Raeder, Strömgren and Strömgren 1946, 118.
[85] *KGW* XIX, 40–42.   [86] E.g. *KGW* XIV, 161.17.
[87] *TBOO* V, 118.11–14; as translated in Raeder, Strömgren and Strömgren 1946, 118.
[88] See Chapter 3.
[89] This point has been made by Thoren 1990, 441. For evidence that such caution was typical, however, see Christianson 2000, 151–153.

here is taken to be typical, it is plausible to suppose that, although the stellar catalogue at issue in the 1590s was also part of the Tychonic restoration of astronomy, the astronomer considered his planetary observations to be that much more precious, in virtue of their uniqueness, during the period in Prague when he was attempting to complete his analysis of all the planetary motions. Cautious Tycho certainly was; but that he was unusually so, or habitually secretive, is harder to demonstrate on the basis of this evidence.

The Ursus affair, and the other instances of alleged plagiarism discussed in the previous chapters, have also played a role in fostering the impression that Tycho was unduly possessive. But in disputing Ursus' claim to priority in the invention of the geoheliocentric world-system, and to credit for the discovery of prosthaphaeresis, the impression of secretiveness is largely a product of the assiduousness with which he pursued the quarrel, once it had started. Historians quite rightly view with suspicion the detailed accounts given by Tycho of the circumstances of Ursus' visit to Hven and supposed theft; in all fairness, then, they should not convict Tycho of paranoid secretiveness on the basis of the claims, contained therein, that having surmised that the Dithmarschener was a man of poor character, he refused to discuss planetary hypotheses whilst Ursus was in earshot.[90] The case of Wittich, it is true, prompted Tycho to write that, 'he has made me more careful, and ensured that I shall not in future entrust my things to anyone'.[91] But the fact that he apparently did so in 1581, long before he possessed any evidence that Wittich had transmitted details of his instruments to Kassel, might suggest that it should be interpreted as an instance of hyperbole, coloured by his evident sense of betrayal at Wittich's departure from Hven.[92] A more literal interpretation is hardly consistent with the continued reception on Hven of numerous visitors, nor indeed with Tycho's use of his own assistants and instrument-makers as long-distance couriers, any one of whom might, in theory, have absconded and divulged Tychonic astronomical 'secrets'.[93] And as we have already seen, at issue was not whether communication of these matters was improper *per se*, but whether Wittich and Ursus had any entitlement to reveal them to others and, in particular, to claim any credit for them. Thus, when discussing prosthaphaeresis, Tycho never attacked Wittich's communication of the technique to the Kassel astronomers, nor

---

[90] E.g. *TBOO* VII, 387.39–388.9.   [91] *TBOO* VII, 63.8–9.

[92] Although see also the discussion in Chapter 2 of the remark of Wittich reportedly overheard by another of Tycho's assistants.

[93] E.g. *TBOO* VII, 106.11–14.

even Bürgi's disclosure of it, following its substantial development at his hands, to Ursus. He criticised only Ursus' publication of it, as if he were its inventor.[94]

When the issue of credit is brought to the fore, it becomes clear that Tycho's stellar data were much less vulnerable to theft than his world-system, or even the designs of his instruments. Who, having appropriated this catalogue, could have justified publication of it with reference to his own observational labour? In any case, the production of the Tychonic globes was itself a form of publication. Appeal to Tycho's allegedly secretive nature does not explain why he should have consented to their production at all, whereas reference to his desire to be credited for his astronomical work accounts for that decision extremely well indeed. The globes that the van Langrens and Blaeu would produce could propagate Tycho's reputation as an observational astronomer and, as we shall see, the Dane took particular steps to enhance their ability to do so.

Having considered the issue of secrecy, the outstanding question regarding the van Langren globes is why Arnold's visit to Hven in 1590 did not immediately result in the production of a Tychonic instrument, as Hendrik's apparently did in 1593. In his report of Tycho's life, Gassendi went on to claim that, in 1595:

various globes were made in the Low Countries according to the restoration of Tycho, but containing only the approximately eight hundred Stars which we said were contained in the first book of the *Progymnasmata*. Indeed, since those had then been printed in quarto format, Tycho distributed some copies of them, and in particular sent one to the aforementioned Amsterdam maker [Jacob] Floris [van Langren].[95]

Not only is the date incorrect, but so, most probably is the nature of the connection between the completion of Tycho's star catalogue and the production of the van Langren celestial globes that this statement seems to suggest. Perhaps, by Gassendi's time, printing had become synonymous with publication, and publication had become established as a means of establishing priority, yet this seems unlikely. Certainly, in Tycho's case, any supposed fears of plagiarism could not have been settled merely by the act of printing his stellar catalogue, especially since he did not then distribute

---

[94] See the discussion of prosthaphaeresis in Chapter 3.

[95] Gassendi 1654, 157: 'Fuere eodem anno Globi varii iuxta Tychonis restitutionem in inferiore Germania confecti; sed continentes tamen dumtaxat octingentas circiter Stellas, quas diximus in Libro primo Progymnasmatum contineri. Nempe, quia complexi illas quaterniones typis expressi iam fuerant; ideo Tycho eorum exempla aliquot iam disperserat, miseratque speciatim unum ad memoratum Florentium artificem Amsterodamensem.'

it immediately. Indeed, with completion of the *Progymnasmata* postponed until 1602, the first Tychonic star catalogues otherwise known to have been distributed were the manuscript ones circulated with presentation copies of the *Mechanica*. But there is a grain of truth in Gassendi's linkage of Tycho's work preparing the *Progymnasmata* with the issue of the celestial globes, one that explains the need for the successive visits of the van Langren brothers. When Arnold came to Uraniborg, Tycho's plan to publish his stellar positions was already formulated, but the observations and calculations that resulted in the catalogue of 777 stars had not been completed.[96] While Tycho could have provided Arnold van Langren with the newly calculated positions of some stars, perhaps even interim locations for most of them, it seems quite likely that he would have advised the Dutchman to await the completion of this work before beginning a *globus Tychonicus*. He may even have withheld his consent at this stage. So the necessity of Hendrik's subsequent visit to Hven in 1593 to the production of the van Langren celestial globe is explained simply by the chronology of the Tychonic observational project.[97]

It remains open to speculation as to whether the production of a Tychonic celestial globe was indeed the object of Arnold's trip in 1590, or a project born of it; but it seems implausible that, once at Uraniborg, the journeyman globe-maker would not have spent some of his six-week stay there inspecting the several globes in the astronomer's possession. That he was just as interested in the construction of the *Globus Magnus Orichalcicus*, as in the data it was destined to bear, is suggested by a letter that he subsequently wrote to Landgrave Wilhelm. Stating that he had studied the art of globe-making with many learned men of Italy and Germany, he offered as a gift 'a small astronomical globe I completed a few years ago, which I humbly dedicated to his Royal Majesty of Denmark'.[98] This was undoubtedly the celestial globe of the pair sent in 1589 to Christian IV. He also proposed to dedicate another globe, engraved in copper to Wilhelm; this may have been a printed terrestrial globe, larger than the companion to this instrument, which the van Langren firm published in the same year

---

[96] Thoren 1990, 285, 287, 294–295.

[97] This explanation for the lack of a Tychonic celestial globe after Arnold's visit seems more satisfactory than the invocation of his unreliable character; see van der Krogt 1993, 106.

[98] van der Krogt 1993, 105: 'kurz verrückter Jharen ich einen kleinen Globum astronomicum gefertigt, wilcher khöniglicher Maiestät in Dennemarck ich underthenig dedicirt.' Leopold 1986, 37, identified the globe dedicated to Christian IV with the first van Langren globe, and hence the globe to be dedicated to Wilhelm with the celestial globe of 1589. He thus dates the letter to 1589. That this cannot have been the case, however, is indicated by mention of Tycho; for as we have seen, Arnold van Langren's trip to Hven did not occur until 1590.

as their instrument-pair.[99] And he indicated that he would be prepared to construct a large celestial globe for the Landgrave, of six feet in diameter, like Tycho's, and also for, or according to, current observations. Whether this meant that he intended it to be based on Tycho's observational data, or to serve the same recording and calculating functions at the Kassel observatory as the *Globus Magnus Orichalcicus* did at Uraniborg, is somewhat unclear. However, the chronological arguments rehearsed above, and the statement that it was to be fashioned 'so carefully and expertly' that 'its like would not be found in Christendom' both favour the second conclusion.[100] Thus, while it is only a possibility that inspection of the *Globus Magnus Orichalcicus* prompted Arnold van Langren to propose the production of a printed Tychonic globe, it is almost certain that it induced him to offer to construct a near-duplicate metal sphere, one that would also serve as a precision working instrument.

To the best of anyone's knowledge, this globe was never commissioned. Arnold's letter could not have reached Kassel long before Wilhelm's death, and no reply to it has ever been located. In any case, the Landgrave and his astronomers already made use of a metal globe as a calculation aid, albeit one of only 0.72 metres in diameter.[101] This copper instrument, which still survives in Kassel, and which was depicted in the 1577 portrait of Wilhelm attributed to Caspar van der Borcht, was completed in 1563.[102] In the 1580s, Rothmann wrote of it that, 'By means of this globe, the Prince checks the observations and quickly seeks the true places as accurately as possible, leaving to me in the meantime the effort of the calculation and precise determination'.[103] It was this instrument that Tycho cautioned against relying upon at the beginning of 1587, after he had found errors in the globe-derived longitudes and latitudes for twenty-four stars averaging

---

[99] van der Krogt 1993, 105–106, leaves room for doubt as to whether he considers 'uf Kupfer gestochen' to mean an engraved copper globe or a standard printed globe produced from copper-plate engravings. I take the latter to be the case, as suggested by Arnold van Langren's stated intention to sell this globe.

[100] van der Krogt 1993, 105: 'Da auch e.f.G. gnediges Gefallens einen grossen Globum, allermassen Tiche Brache so de e.f.G. ohnzweifentlich bekandt einen hatt, dessen Diameter 6 Schuh hoch und nach der izigen Observation zugericht, will e.f.G. dergleichen auch einen machen und das so vleissig und meisterlich das e.f.G. sunderlichs gnediges Gefallen darab haben sollen, und deren Gleichen in der Christenheitt nit gefunden werden soll . . .'

[101] Leopold 1997, 6.

[102] Leopold 1986, 17, 56–59. Cf. von Mackensen 1988, 135 no. 18. See Figure 2.2.

[103] Leopold 1986, 57–58: 'Mittels dieses Globus prüft der Fürst die Beobachtungen und sucht schnell so genau als möglich die wahren Örter, mir unterdessen die Mühe der Ausrechnung und genauen Bestimmung überlassend.' This German translation of Rothmann's original comment in manuscripts in Kassel is dated by Leopold to 1589; cf. von Mackensen 1988, 21, where it is dated to 1586. I have not been able to check their source.

2.7 and 3.3 arc-minutes respectively.[104] That very year, perhaps in response to this criticism, work on a new calculating globe was begun at Kassel. It was to be five feet in diameter and, like Tycho's instrument, made of brass. But the construction of this globe proved difficult even for the skilled craftsmen of Wilhelm's court, and it remained unfinished at the time of the Landgrave's death.[105] Since work on the object was ongoing in 1590, Arnold's overture to Wilhelm, although shrewdly directed at the one prince whose interest in a great calculation-globe was guaranteed, would seem to have been somewhat redundant.

The letter sent by Arnold van Langren, with its juxtaposition of the proposed calculating globe and the printed globe sent to Wilhelm as a gift, and with the further contrast it conjures between each of these objects and the mechanical celestial globes of gold and silver produced at the Landgrave's court, underlines once more the variety in form and function of the globes that circulated in the early modern period. The globe he wanted to produce, modelled on Tycho's *Globus Magnus Orichalcicus*, would have been monumentally large, metal, and engraved; the ones he and his family actually constructed were much smaller, and were composed of coloured printed sheets pasted onto spheres of papier-mâché. The mechanical globes produced at Kassel were akin to the calculating globes in their individuality, their expense, and their requirement for precision manufacture, but they were more like printed globes in their size, and their limited utility for calculating and recording. They also had an interior complexity that was entirely their own. Each type of globe was, in its own way, visually striking; and each type could be deemed an appropriate offering for a prince. In other respects, however, their uses were quite different; and the printed globes, produced in larger numbers than the other two sorts, were also accessible to a much broader audience.

In the next section of this chapter, the different communicative roles of the printed Tychonic globes and the Kassel mechanical globes will be considered in detail. But before we move on to this topic, it is worth commenting on Arnold van Langren's choice of prospective patron. This is unlikely to be have been merely fortuitous. Clearly, he could have learnt of Wilhelm's astronomical activities during his stay at Uraniborg; perhaps he was led to believe by Tycho himself that the Kassel observatory required a better calculating instrument. But he might also have read about the

---

[104] *TBOO* VI, 77. Tycho tabulated the stellar coordinates derived from the globe and those produced by calculation; he did not summarise the results in this form.

[105] On this instrument see Leopold 1986, 24, 32 and 59. In 1603, Graf Simon von Lippe asked Moritz to give him the still unfinished globe; its subsequent fate is unknown.

Landgrave in Petrus Ramus' *Prooemium mathematicum* (1567) or in Heinrich Rantzau's *Catalogus* (1580), or simply heard of his astronomical work during his travels in Germany and Italy. But there is also a strong possibility that his overture to Wilhelm was suggested by a source in the Netherlands: the Leiden professor of mathematics Rodolphus Snellius, with whom Jacob Floris van Langren had collaborated in producing his celestial globe of 1586.[106] Snellius, who had studied and taught in Marburg between 1566 and 1575, and whose son Willibrord would first publish material from the astronomical manuscripts of Rothmann and the Kassel observatory, may well have met the Landgrave; he would certainly have known of his enthusiasm for astronomy.[107] The difficulty of deducing by which route Arnold van Langren first came to learn about the astronomical work in Kassel, and indeed, the very fact of his overture to the Landgrave following his visit to Hven, indicate the complexity of the traceable connections between different practitioners of astronomy and makers of astronomical instruments in the early modern period.

II PRINTING AND PRIVILEGE: BOOKS, GLOBES, AND GIFTS

Under the penalty laid down in the documents, it is decreed by the privilege of his Imperial Majesty for 30 years, of the King of France for 10 years, and of the States General for 15 years, that, without his permission, no one may copy or sell piecemeal these things concerning astronomical matters which appear by the consent and favour of the Nobleman TYCHO BRAHE. (Cartouche: the van Langren celestial globe of 1594)[108]

In 1541, two globes were published by MERCATOR at Louvain: a celestial globe and a terrestrial globe. I recently brought these home myself. However, I am informed that a few years ago the same ones were published at Cologne, in a better version and even larger, and that those maps are not to be published elsewhere. I earnestly entreat you to let me know whether this is true, and whether the maps could be acquired from the printers, if not by me, then by some fellow of superior standing. (Tobias Fischer to Tycho Brahe, 8 March 1600)[109]

Early modern globes such as the van Langren instrument in Frankfurt have much in common with printed books of similar vintage. Not only were

---

[106] See van der Krogt 1993, 92.

[107] See Leopold 1986, 32. Rodolphus Snellius was also in contact with Tycho; see Christianson 2000, 358.

[108] van der Krogt 1993, 425: 'Cautum est privilegio Caes. Maiest. ad triginta annos. & Regis Gallorum ad decennium, nec non Ordinum confoederatorum Belgii ad annos quindecim, Ne quispiam ea quae studio & consensu Nobilis viri TYCHONIS BRAHE Dani in rebus Astronomicis prodeunt, sine ipsius permissu imitetur aut divendat, sub poena in diplomatibus statuta.'

[109] *TBOO* VIII, 270.16–22. Fischer was a poet of Silesian birth; see *TBOO* VIII, 456.

the techniques of copper-plate engraving, printing, and colouring used to produce a paper globe also employed in the production of illustrated texts, but the finished instruments were frequently distributed through identical channels to other printed materials. Once bought, or otherwise acquired, globes were typically stored in the same places as texts; finding a home, as images and inventories of libraries show us, amidst the book-collections of universities, the working libraries of individuals such as Tycho and John Dee, and, frequently alongside other instruments, with the books and *objets d'art* of princely *Kunstkammern* and cabinets.[110] Indeed, with respect to production, distribution, use and display, globes were strongly associated with particular printed works: the maps, atlases and star-charts that were the other staples of the cartographic publisher. More generally, also, both producers and owners of globes used some of the same means of correcting, augmenting and modifying the instruments as could be employed for the personalisation and alteration of books: the over-pasting of corrective slips of paper, for example, or the filling of blank spaces with manuscript emendations and additions.[111] We have already seen several examples of globes furnished with dedications and acknowledgements; just as for a book, these might be in manuscript or print, and applied to an entire edition or specific to one copy. In fact, as is shown particularly clearly by the well-known example of the Behaim terrestrial globe, an artefact which predates sixteenth-century globe-printing, it is unwise to underestimate the sheer quantity of text that an early modern globe could support on its surface.[112]

A number of the stratagems used by book-publishers, printers and authors to encourage sales, to protect their investment in an edition, and to distribute their wares, were also tried by globe-makers and cartographers. These procedures are not just of interest to historians of publishing; in the case of the Tychonic celestial globes they have some bearing on the question of what the commercially produced instruments were supposed to convey to those who saw or acquired them. Both Blaeu and the van Langrens marketed their celestial globes as aids to navigation, pairing them with terrestrial globes of equal size that they also promoted as instruments suited to seafarers. In an age of exploration, when both cartographic records and commercial sailing routes were subject to rapid alteration, as

---

[110] Viellard 1973; Fucikova 1985, 49–50; Menzhausen 1985, esp. 69, 72; Babicz 1987; Sherman 1995, 33–34. Gabriel Naudé, in his *Advis pour dresser une bibliothèque* (1627), suggested that money should be spent on furnishing a library with 'mathematical instruments, globes, maps, spheres, paintings, stuffed animals, stones, and other curiosities as well of art as of nature'. See Taylor 1950, 72–73; Nelles 1997, 44.

[111] See Sherman 1995, 33–34.   [112] See Bott 1992, 745–756.

entrepreneurs and governments sought to exploit the latest discoveries, currency and accuracy were clearly important attributes of navigators' charts. Globes, however, because of their fragility and their size (too large for convenient carriage on ship, too small for accurate or easy calculation) were of limited navigational value. Much of the actual custom for them must have arisen on land, in the form of mathematicians and astronomers who acquired them for their own use, or for pedagogic purpose, and individuals or institutions who appreciated them as symbols. Yet such purchasers, particularly if they valued the association with oceanic adventure, are as likely as practising navigators to have been attracted by claims of up-to-dateness and precision. The evocation of these qualities was, it appears, one of the chief advantages that the van Langrens, Blaeu, and their competitors, saw in the use of Tycho's data and name.

One indicator of the importance of currency-claims is reflected in the date of the surviving van Langren globe, or rather in the manner in which this was marked on the instrument. While the first three digits of the year are printed, the last was entered in manuscript. Book-producers are also known to have completed years of publication in this fashion; it was one of a number of strategies for reconciling customers' preference for new material with the slow rate at which a sizeable edition could actually be sold.[113] It seems likely, therefore, that the van Langrens intended to sell their '1594' celestial globe over a number of years, and did not want to jeopardise their commercial success by dating the edition too precisely.

Working as they were with engraved copper-plates, which could be stored, reused, and even altered with relative ease, globe-makers were not necessarily committed to printing, in advance of demand, a large stock of the paper sheets used in the construction of their product. In this respect they were unlike publishers of texts, for whom it was uneconomic to let demand dictate the rate of the printing. Few could afford to tie up capital by leaving type set in formes, objects that were, in any case, bulky and hence difficult to store. Standard practice, therefore was to print as many copies of each sheet as were required to meet the demand anticipated over the period, usually of several years, in which the publisher intended to sell the edition. The large initial outlay necessary for the production of any one publication encouraged the commercially minded to develop broader networks of distribution, to promote sale by catalogue, and to contract exchanges of stock with other publishers: all strategies that improved the rate at which the initial investment was recovered.[114]

---

[113] Pantin 1988, 242.   [114] Ehrman and Pollard, 1965; Balsamo 1990, 20–25, 57–59.

Globe-makers' and cartographers' overheads and presswork costs were similar to those of other publishers, but the assemblage of globes from the printed gores required additional per-unit expenditure on labour and materials. One historian has estimated that, for an edition of instruments of the dimensions of the van Langren globes, no fewer than fifty would have had to be sold before the profit-threshold was encountered. To date, little has been done to uncover the workshop practices of early modern globe-makers, so we do not know whether the business strategy appropriate to the type-printer was carried over into the cartographer's workshop, or whether it was realised that risk could be reduced by letting demand drive production. But it probably took only about a day or so for the gores and caps of twenty globes to be printed, and constructing a globe from these and other components was perhaps another day's labour.[115] It therefore seems reasonable to suppose that the process of globe-making would have lent itself to a piecemeal approach. Consequently, it is difficult to be sure how many of a given globe were ever printed, and how many of these were actually assembled, distributed, and sold.

Blaeu's first celestial globes of 1598 were approximately the same size as the van Langrens', and, like those, clearly indicated their dependence upon Tycho's data.[116] Of these instruments only a partial set of gores has survived, although several examples of a revised version, issued in 1603, are known to be extant.[117] With celestial globes of 0.1 and 0.23 metres diameter also published by Blaeu in the intervening period, each proclaiming their use of Tycho's observations, it seems clear that the trade in these instruments was brisk.[118] Evidence of the van Langrens' business success is a little more elusive. Some time in 1598 or 1599, Jacob van Langren retired, leaving 'all the plates and instruments serving the art of globes' to Arnold and Hendrik.[119] But in 1607 or 1608, Arnold migrated to the Spanish Netherlands to escape his many debts and creditors; he was appointed 'Royal Spherographer' to the Spanish governors in 1609, but did not produce another celestial globe until 1630. Hendrik, meanwhile, went into partnership with his brother-in-law, the two of them signing an agreement in 1608 to 'prepare and deliver six hundred copies of the small globes, which are in use now, namely three hundred celestial and three hundred terrestrial globes, minus the 48 already made up and sold to their mutual profit'.[120] While the celestial globes mentioned

---

[115] van der Krogt 1993, 217–218.   [116] Warner 1979, 28; van der Krogt 1993, 492.
[117] van der Krogt 1993, 492.   [118] van der Krogt 1993, 498–499, 503–505.
[119] van der Krogt 1993, 134–135: 'alle de platen ende instrumenten tot de cunste van de Globis dienende . . .'
[120] Gemeentelijke Archiefdienst Amsterdam, Not. Arch. 208, 1 February 1608, as transcribed and translated by van der Krogt 1993, 618–619.

here are probably those produced in 1594, there is no indication of how many were distributed before this partnership began; moreover, it is unlikely that the contract was fulfilled and adhered to. But after Hendrik's death in 1648, his heirs attempted to sell the copper-plates for the globes, and for two world-maps he had also produced, a fact which suggests that that they were still considered usable. Since a single copper-plate is good for several hundred impressions, 2,000 or 3,000 at most,[121] it seems reasonable to suppose, as a conservative estimate, that somewhere between 200 and 500 copies of the van Langren celestial globe were actually published.

As in other areas of the print-trade, producers of globes suffered from, or perpetrated piratical acts.[122] In 1600, Jodocus Hondius (1563–1612) produced a set of globes 0.35 metres in diameter, with the celestial instrument of the pair advertising its reliance upon Tycho's stellar data.[123] Since this information was not then widely available, it seems likely that the star positions of this *globus Tychonicus* were actually derived from the Blaeu globe of 1598, on which Hondius also based the style of the constellations.[124] The same publisher issued another smaller globe-pair in 1601, in which the attribution was repeated; and this set was itself copied by the Milanese maker Guiseppe de Rossi in 1615. In the meantime, with the issue of the *Progymnasmata*, the Tychonic catalogue of 777 stars entered the public domain, making it very easy for cartographers to produce instruments and charts on the basis of these data. The publication of the *Tabulae Rudolphinae* in 1627, which contained the enlarged stellar catalogue of 1,000 stars, may also have helped to consolidate the practice, by then a well-established tradition amongst globe-producing cartographers of the Netherlands, of citing the Dane.[125] But with the earliest of the Tychonic globes predating the widespread distribution of Tycho's major texts, it would appear that these instruments played a significant role in propagating his reputation as an astronomer. For although *De nova stella*, the reports of visitors to Hven, and recipients of Tycho's letters and books, and the occasional remark in the printed works of other authors must collectively have engendered sufficient awareness of Tycho's astronomical competence for the van Langrens, Blaeu, and the piratical Hondius to register his name as a valuable 'brand' for their globes, these objects and the letter-book were really the only tangible evidence of the work undertaken at Uraniborg to be made commercially available during the astronomer's lifetime. It is probable that Blaeu

[121] Hind 1963, 15. For the estimate that no more than 300 globes were ever produced in a single edition, see van der Krogt 1993, 126.
[122] See Pottinger 1958, 210–237; Johns 1998b, 58–186.   [123] van der Krogt 1993, 470.
[124] van der Krogt 1993, 162.   [125] van der Krogt 1993, 475, 211–213.

4.7 Detail of the van Langren globe of 1594, clearly showing the portrait of Tycho. Photograph by Ursula Seitz-Gray, reproduced by permission of the Historisches Museum Frankfurt [HMF X 14609].

himself learnt of Tycho through the van Langrens' instrument, and Hondius through Blaeu's.

That Tycho's use of the commercial globes to represent his observational project was conscious can be inferred from the steps he took to grant legitimacy to the instruments produced by the van Langrens and Blaeu. To each maker he must have given a copy of the engraved portrait that he himself used on the first page of the *Epistolae astronomicae*. This picture of Tycho was reproduced in miniature on the first Tychonic globes (Fig. 4.7), and became a feature of the majority of seventeenth-century globes that claimed to be based on his data. In addition, the van Langren globe of 1594 was issued under several privileges that had been awarded to Tycho. Widely employed by book-publishers, privileges and charters were documents purporting to offer some protection against plagiarism and piracy, and had originally emerged from *ad hoc* attempts to reward technical innovation by granting inventors of new technology a monopoly in its commercial exploitation. As such, they were the precursors to both patent legislation and the later principle of copyright.[126] But they may have been appreciated as much for their prestige-value, and usefulness in marketing a work, as for their supposed legal function.

The application of privileges to publishing began with grants to individuals who introduced the press to a particular location. In 1469, for example, Johann Spira obtained a privilege from the Venetian Senate that granted him the exclusive right of printing in the city.[127] Subsequently, however, it became more usual to award a publisher a privilege for all the works they might produce, or, as increasingly happened, to grant a limited monopoly in the production of one text in particular. Typically, this last type of privilege was awarded for a fixed period: often for somewhere between three and ten years, but occasionally for a longer term, or one defined in some other fashion, such as the length of time it took to sell all the stock from a certain size of print-run. The penalty specified for infringing a privilege also varied from one case to another. Frequently, a fixed fine was incurred. But forfeiture of all the copies of the illegitimate publication, which might then be destroyed, was another common punishment.[128] And if an ecclesiastic was the issuing authority offenders might even face the sanction of excommunication.[129]

As a mechanism for protecting intellectual and commercial property, the early modern privilege was limited in several ways. One problem was that, having emerged as attempts to reward innovation and enterprise, privileges might be awarded for creative alteration of an existing work: the use of a new font, or layout, the provision of more sophisticated apparatus, the addition of new text or illustrations, the author's revision of his preface. A privilege did not necessarily guarantee, therefore, any rights in the reproduction of some particular body of text. Furthermore, as it was granted by a certain secular or ecclesiastical prince, parliament, civic body or university

---

[126] On printing, science, and developing notions and mechanisms of intellectual property, see Eamon 1984, 1985; Long 1991b; Rose 1993; Johns 2003.
[127] Armstrong 1990, 2.   [128] Schottenloher 1933; Grendler 1988; Armstrong 1990; Tennant 1996.
[129] Armstrong 1990, 8–9, 12.

chancellor, a privilege was only enforceable, if at all, within a limited juris-
diction. Often, it is true, a printing privilege protected against the impor-
tation of rival editions as well as local reproduction, but even this measure was only of much benefit if the publisher could expect to recoup his
investment from within the demesne of the issuing authority. In the case
of minor powers, this was unlikely unless the work was of considerable
local interest. Consequently, those involved in the production and sale
of books frequently sought to obtain broader protection for their works,
either by soliciting their privileges from the papal or Imperial chancelleries,
or by acquiring a number of them from different authorities. Both of these
strategies were adopted by Tycho for his own publications, and both were
carried over to the globe constructed by the van Langren family.

The application of printing privileges to globes was not new. Johannes
Schöner obtained an eight-year Imperial privilege for a globe-pair and
accompanying manuals in 1515; another such privilege, for five years, was
awarded for a terrestrial globe by Gaspard van der Heyden, and the manual by Franciscus Monachus.[130] A ten-year Imperial charter, subsequently
extended, was awarded to Gemma Frisius and Gaspard van der Heyden
in 1531; the terms of this privilege were printed on their globe-pair of
1536/1537.[131] In 1541, Gerard Mercator received an Imperial privilege for
the production of a terrestrial globe, and these credentials were likewise
displayed on the instrument's surface.[132] It is likely that in each of these
later cases the globe-publisher consciously set out to follow the example
of his immediate predecessor. But with the emergence of the northern
provinces of the Netherlands as the centre of globe-production at the
end of the sixteenth century, the States General replaced the Imperial
Chancellery as the most immediate and relevant privilege-issuing authority.
It was from this body that the van Langrens obtained a ten-year charter for
their globes in 1592, extended for a further ten years in 1596.[133] In theory,
the celestial globe of 1594 was protected by this privilege, as well as by those
that were awarded to Tycho.

The initial van Langren petition for a globe-privilege may have been a
somewhat belated response to the production in England of a globe-pair
that borrowed from the van Langren instruments of 1589.[134] The engraver of
these instruments, published by Emery Molyneux, was Jodocus Hondius;
and in 1597 he, having since moved to Amsterdam, obtained a ten-year

---

[130] van der Krogt 1993, 31, 40–47.
[131] van der Krogt 1993, 49–50, 57, 410–412, 575.   [132] van der Krogt 1993, 413–415.
[133] van der Krogt 1993, 576–578.   [134] See van der Krogt 1993, 108–115.

privilege from the States General for a new terrestrial instrument. The award was contested by the van Langrens, who claimed that it infringed their own extended charter, but Hondius' rights were confirmed on the grounds that, although he had indeed adopted features of the van Langren models, his globe was enough of an improvement on theirs to be considered a new and separate enterprise. As evidenced by the fact that no further globe-privileges were sought by cartographers working in the northern provinces for almost a hundred years, this ruling effectively undermined the privilege as a means of preventing globe-piracy.

Even before the efficacy of globe-privileges had been tested in this way and found to be wanting, the significance of printing privileges in general would seem to have broadened. While the full text of a privilege was sometimes displayed at the front or rear of a volume, it became increasingly common during the sixteenth century for printers, when they had the option to do so, simply to place the words *cum privilegio* on the title-page of a book. By not specifying the identity of the issuing authority, this may have left some room for doubt as to the legal extension of the privilege's authority. But whether or not either strategy deterred would-be pirates, we can assume that advertising a privilege imbued a text with degree of respectability and legitimacy; not least because, in many regions, the process of awarding a privilege was coupled to a system of licensing intended to check the publication of works expressing dangerous political or theological sentiments.[135] Works not protected by privilege might therefore be suspected not just of being piratical, but also somewhat questionable in respect of their content. Thus, by publishing his *De astronomicis hypothesibus* with the words 'without any privilege' displayed on the title-page, Ursus was deliberately choosing to emphasise its unauthorised nature, perhaps with a view to dissociating the work from his patrons.[136] Equally, the privileges of Tycho cited on the van Langren globe, awarded by the States General, the King of France, and the Emperor, were supposed to ratify both the instrument and the Tychonic observational project in general.

When he applied for his first printing privilege, a ten-year Imperial charter awarded in 1586, Tycho used a similar form of words to those employed by commercial book-producers. He had expended such time and effort in carrying out his astronomical work, he wrote, and would incur further costs in preparing his publications, that he did not want these burdens to have been assumed in vain, as would be the case if anything 'printed here

---

[135] Pottinger 1958, 54–81; Evans 1975, 29–30; Grendler 1975, 1988; Richardson 1999, 38–46.
[136] See also Jardine and Launert 2005, 90, 94.

on my press were immediately printed in Germany or elsewhere'.[137] But he added that if a privilege were affixed to the front of his works about celestial phenomena, these would, by virtue of the Imperial authority, be adjudged the product of his own labours and expense that much more readily.[138] A further indication of Tycho's approach to privileges is provided by the fact that, after the French crown had issued him with a splendid-looking ten-year charter in 1588, he sought to have his rather dowdy Imperial privilege redrawn in a more impressive form.[139] Clearly the document itself, as well as the text, was something to display. In fact, Tycho hoped to acquire the visible affirmation of a panoply of powers: his fifteen-year privilege from the States General obtained in 1594, almost certainly with the aid of the van Langrens, complemented the Danish royal privilege awarded the same year, and there were also attempts to obtain protection from the monarchs of England, Spain, and Scotland, the Venetian Senate, and, as we saw in Chapter 3, the Spanish-controlled Netherlands.[140] Interestingly, just as Archbishop Ernst of Cologne informed Tycho that the last would be difficult to obtain for works sight unseen, he was told by his agent in Prague that he was the recipient of an Imperial privilege, 'such as no one else has', for 'privileges are not usually granted to anyone unless a manuscript copy has first been deposited in the Chancellery'.[141] In fact, both Ursus and Rothmann were also granted privileges for works they had yet to complete.[142]

Since Tycho's privileges were essentially those of a publisher, awarded to him for the astronomical and philosophical works printed on his own presses, their application to the content of the van Langren globe represented a profound, and perhaps unwarranted, reinterpretation in favour of their recognition of intellectual property.[143] But the impossibility of enforcing an injunction against the reproduction of improved determinations of stellar coordinates, rather than the reproduction of a printed document or globe bearing that information, suggests that the status and legitimacy conferred by the privileges, not the protection they afforded, was the motivating factor. Within the United Provinces, at least, the globe was independently

[137] *TBOO* VII, 95.41–96.15; *TBOO* XIV, 32.   [138] *TBOO* VII, 96.24–28.

[139] *TBOO* VII, 217.21–218.23. Through the agency of Hagecius and Kurtz, Tycho acquired his 'more splendid' privilege, which he received sometime between April and August of 1590. See *TBOO* VII, 243.35–38, 256.12–14. He later sought and obtained another emendation to his Imperial privilege; it was probably on this occasion that the term of ten years featured in the original document was extended to the thirty years advertised by the van Langrens. See *TBOO* VII, 347.11–12, 348.14–24; *TBOO* II, 8–10.

[140] *TBOO* VII, 122.31–34, 141.11–18, 217.25–35, 262.31–37, 282.26–29, 284.26–35, 304.1–15, 306.30–33, 330.7–13, 344.24–26, 354.20–22, 377.29–31, 382.36–42; *TBOO* VIII, 260.20–28.

[141] *TBOO* VII, 103.18–24.

[142] Launert 1999, 352–355; Granada 2002, 219–223.   [143] *TBOO* II, 9.31–39; XIV, 33–34.

covered by the van Langrens' privilege, but by having it issued under his privileges as well, Tycho strengthened the object's association with himself.

Evidence for the success of this strategy comes from the pen of Kepler. In May 1599, a few weeks before he was expelled from the Styrian city of Graz, he wrote to the Bavarian chancellor Herwart von Hohenburg about an example he had seen there. Perhaps prompted by the mention of the instruments in the preface of the *Epistolae astronomicae*, Hohenburg appears to have asked Kepler if he knew where he could procure one. The astronomer replied that:

Since Tycho enjoys the privileges of many kings and provinces, a certain resident of Amsterdam, whose name escapes me, sent his son to him in Denmark, in order to copy down the eight hundred stars in Tycho's restored catalogue. The father of he who was sent, having received the indulgence of the privilege from Tycho, undertook to have these engraved in copper, with the remaining 200 northern stars added from the ancient catalogues. One example, now made into the form of a globe, equipped with a horizon, meridian, and everything necessary for the globe's use, was brought to Graz from Holland, along with other painted and very well-appointed goods by a certain member of the Mercator family of Duisburg who resides in Nuremberg. A terrestrial globe equipped in the same way and illustrated with pictures was paired with it. At first he valued both at 50 Thaler and said he would not sell them separately. And although in the end they could have been bought for 40 florins, there was no one in this time of great confusion who would purchase them. The diameter of the globe equalled a standard foot. When that man, a merchant [Mercator] by both name and nature, departed, he left them here with a certain David Heldius, who indeed has kept them until now. But they have been made the property of another. For tomorrow he is sending them to Vienna to another merchant, to whom they were long ago sold by his master. And indeed that Viennese man buys so as to sell again; in this way the price will increase. But although I know that, I am unable to prevent them from being taken to Vienna, since I am not clear about your wishes.[144]

Several things are apparent from Kepler's report. One of them is the manner in which Kepler read the globe: he registered and recounted those details marked on the instrument that related to Tycho's involvement in its construction, yet was unable to recall the name of its maker. And except to mention the size of the globe and its price, and to confirm that it was indeed mounted as one would have expected, Kepler had very little to say about the form of the instrument. It is perhaps worth remembering that he and Herwart were probably unusual amongst the purchasers and potential purchasers of the globe in having independent evidence of the

---

[144] *KGW* XIII, 341.77–101.

Danish astronomer's observational project. Many others who saw the globe must have encountered Tycho's name and features on the instrument for the first time, or something very close to it.

Kepler's letter to Herwart also provides some insight into the commercial distribution of the van Langren globe. Tycho's own comments in the *Epistolae astronomicae* and the *Progymnasmata* may be read as endorsements of the printed *globi Tychonici*, but it is clear that the astronomer had no involvement in their sale. In 1600, when Tobias Fischer (1559–1616) wrote to enquire about the globes of Mercator, Tycho said that these had now been superseded, and suggested that he look instead for the instruments by the van Langrens or Blaeu.[145] If Fischer were to visit him in Bohemia, he wrote, then he would happily show him an example; but he did not know if they were being shipped to Frankfurt.[146] In fact, this comment must be taken as indicating that Tycho was unaware whether they were still being sold there; in October of 1594, Aslachus had informed him that the van Langren instruments were available at the book-fairs, to the combined pleasure and dismay of the masses. The pleasure was caused by their skilful and clever construction, the dismay by their price: 22 thalers for the celestial globe along with its terrestrial partner.[147] Perhaps this report inclined Tycho to approve Blaeu's plan for another globe when he visited Hven a year or so later.[148] But in any case, Tycho's remark to Fischer also implies that he understood the globes were not to be found for sale in Prague; presumably he had received his copies whilst still in Denmark, and transported them to the Imperial Court along with his observing apparatus. Through Kepler's letter, however, we can trace the movement of one of the globes from the Netherlands to Graz, presumably via Frankfurt and Nuremberg, and then on to Vienna, where it was to be sold again, at a considerable mark-up. Ironically, the man responsible was apparently one of the relatives of Gerard Mercator of Duisburg, the cartographer whose celestial globes Tycho claimed had been superseded by these instruments. This Mercator had paired the instrument with the terrestrial globe of another maker, and seemed determined to sell these mismatched objects together.[149]

[145] *TBOO* VIII, 322.31–41.

[146] *TBOO* VIII, 322.41–323.4. Fischer had written from Swidnica in Silesia; see *TBOO* VIII, 270.9–10. Tycho's comment about Frankfurt therefore reflects the role of the fairs as a major centre for the distribution of books and printed matter to sites across Europe, not his correspondent's location.

[147] *TBOO* VII, 368.10–14.

[148] It must certainly have been the source for his complaint to Pontanus, in 1595, that he had heard nothing from globe-maker, and had received no globe from him; and that he seemed to be excessively greedy. See *TBOO* VII, 371.35–38.

[149] Kepler noted in the margin of his letter that the terrestrial globe was, 'Published in 1589 by a certain maker of Antwerp, whose name, if I remember correctly, is Rosenberg'. See *KGW* XIII, 341.88–90.

That Tycho possessed several of the printed *globi Tychonici* is indicated by his presentation in 1600 of three globes, probably all by Blaeu, to various notables: Rudolph Corraducius, the Imperial Prochancellor; Matthias, Archduke of Austria; and Christian II, the Elector of Saxony. For a man so strongly associated with the construction of instruments, Tycho seems to have given relatively few away to others: in addition to the globes, he is known to have presented a horizontal sundial and a pocket watch to Rantzau; and, somewhat reluctantly, a mechanical statue of Mercury to Heinrich Julius, the Duke of Braunschweig.[150] When, in 1590, Hagecius and Kurtz were looking to acquire the instruments necessary to make accurate observations in Prague, Tycho was unable to meet Kurtz's request that he be the one to supply them. The problem was partly their size, which would have made transportation difficult, partly the fact that each of his instruments had been the work of multiple craftsmen, some of them having taken five or six artisans to construct and up to three years to perfect. Once Uraniborg was fully furnished with observing instruments, Tycho had dismissed most of his mechanics, retaining only those one or two necessary to maintain them.[151] In place of the real thing, therefore, Tycho sent, as Hagecius had suggested, several scale models made from, 'glued paper or little wooden sticks', even though he expressed his doubts about the possibility of using these as templates for the full-size constructions.[152] That Hagecius and Kurtz succeeded, nevertheless, in having at least one observing instrument made that Tycho was willing to acknowledge as Tychonic was due to the involvement of Jost Bürgi, the Kassel instrument-maker, from whom they commissioned a sextant.[153] And when Tycho himself came to Prague, some sort of automaton or mechanical device in his possession was offered to Rudolf II after it was spied in his carriage as he drove up to the palace; but the Emperor decided to have this instrument copied by his own craftsmen, rather than appropriate the original.[154] The presentation of the Blaeu globes to Corraducius, Matthias and Christian are perhaps, therefore, with the exception of another model to be discussed later in this chapter, the

---

[150] For the sundial sent to Rantzau, see *TBOO* VII, 124.36–37, 385.35–386.27. The pocket watch is on display in the Danish National Museum, Copenhagen, Inv. Nr. D 125. Regarding the statue, see *TBOO* VI, 287.19–32; Thoren 1990, 335.

[151] *TBOO* VII, 273.12–24.

[152] *TBOO* VII, 257.22–24. Hagecius had asked for such models of the sextant and the armillaries, claiming that a quadrant would be easy to construct. See *TBOO* VII, 244.6–11.

[153] *TBOO* VII, 357.6–9, 360.13–23.

[154] *TBOO* VIII, 166.11–25; Thoren 1990, 413. It is likely that the device was a form of waywiser, and had been constructed by Peter Jachenow, an instrument-maker whom Tycho recommended to Hagecius in 1591; see Christianson 2000, 293–295.

only occasions on which Tycho gave away an instrument closely associated with his own astronomical project.

Tycho said little about his reasons for presenting globes to the Prochancellor and the Archduke, but his desire to establish and maintain good relations with the Imperial official and the Imperial heir is not hard to comprehend. In both cases, the globe was given at the same time as copies of his manuscript catalogue of stars and his printed *Mechanica*.[155] Tycho was, however, a little more forthcoming about the spirit of reciprocation in which he sent the globe to Christian of Saxony, and more explicit about the relationship between the globe and the catalogue. Thus, writing to the Elector on 6 May 1600, Tycho thanked him for a gold chain bearing his portrait, and for the hospitality he had offered during the astronomer's journey from Denmark to Bohemia. At first he claimed that he could offer no recompense for the Elector's benevolence, writing that:

Since I am, I confess, unable to manage anything which would in turn please so great a Prince, I can only humbly promise that as long as I live, I shall always be mindful of that most merciful goodwill and kindness towards me which I experienced in his company, and shall speak of it among others.[156]

But he added that, so as not to show himself utterly unable to perform his duty, he would send by means of Johannes Jessenius, who had recently visited him in Prague:

a book of the accurate restoration of the fixed stars, the product of my study and labour over many years; and with it a celestial globe, on which, by the labour of a certain skilful Dutchman on the island of my former residence, Uraniborg, in the Kingdom of Denmark, our fatherland, the fixed stars are represented according to this same rectification carried out by us, as adroitly and accurately as the size of this artificial sphere allows.[157]

The supposition that the text to be sent with a globe was a manuscript copy of the Tychonic stellar catalogue is confirmed by the letter that Tycho wrote to Jessenius the following day; he was sending it, he said, in case the Elector wished to have an even larger globe constructed.[158] In the light of Tycho's acknowledgement of the chief limitation of the instrument, that it was only as accurate as its size permitted, Tycho's explanation could be construed as no more than a form of words to justify the presentation with the globe of what he considered to be a gift of greater value. It is likely, however, that in offering both the catalogue and the suggestion, he was

[155] *TBOO* VIII, 259.5–7, 309.11–18. From Tycho's letter to Rosencrantz of 3 June 1600, it is evident that the globe presented to the Archduke was of Blaeu's manufacture. See *TBOO* VIII, 332.1–6.
[156] *TBOO* VIII, 313.20–24.   [157] *TBOO* VIII, 313.24–39.   [158] *TBOO* VIII, 316.20–30.

conscious of the fact that the Saxon court had, during the second half of the sixteenth century, become one of the most important sites of princely instrument-making and collecting across the whole of Europe.

Although appreciation of scientific instrumentation was widespread throughout the European courts, it was often the result of a *pro forma* interest in instruments rather than a genuine enthusiasm for their practical application. In general, such objects were appreciated for their preciousness, the ingenuity of their construction, and their contribution to the illusion of polymathic princely erudition.[159] On this basis, courtly commissions sustained the finest instrument-makers and horologists, many of whom worked, like Tycho's 'clever craftsman', in the Imperial cities of Augsburg and Nuremberg. In addition, it was not uncommon for princes to sponsor their own craftsmen, selecting them from the ranks of the commercially trained; Tobias Volckmer of Braunschweig, for example, began his career as a goldsmith in Salzburg, and was subsequently called to the Bavarian court, where he served for thirty years or more.[160] But the actual acquaintance of the nobility with the techniques and technology of such disciplines as astronomy, geodesy and navigation was, in the majority of cases, quite superficial. It was not so much that it was beneath the dignity of aristocrats to engage in practical activities: turning was widely promoted as a craft particularly suited to the study of princes, and often on the grounds that it gave rulers an insight into the operation of the mechanical arts.[161] Moreover, the economic and military value of the mathematical disciplines was becoming increasingly apparent to many rulers, especially as mathematicians themselves found it advantageous to promote their work in terms of its utility.[162] But it was rare for a noble to have any genuine affinity for the study of such subjects. And of those exceptional individuals whose interests and enthusiasm led them to assemble a truly outstanding collection of scientific instruments, we can single out three: Wilhelm IV, by virtue of the extraordinary precision with which his planetary-clocks and mechanical globes were constructed, and the interdependence of the instrument-building activity and the observational work at Kassel; Emperor Rudolf II, whose appetite for luxury goods, *artificialia* in the Mannerist mode, and occult devices of all descriptions was matched, if at all, only by the resources at his disposal to engineer their acquisition; and Christian II's grandfather, Elector August

---

[159] See Olmi 1985, 12; Scheicher 1985, 34–35; Seelig 1985; daCosta Kaufmann 1994; Bredekamp 1995; Leopold 1995; Turner 1995a; Jardine 1996, 410–412.
[160] Zinner 1979, 574–577; Miniati and Rudan 1981.
[161] Maurice 1985, 8; Dreier 1985, 106; Menzhausen 1985, 71.
[162] Bennett and Johnston 1996, 9–16; Neal 1999.

of Saxony, whose *Kunstkammer* was especially remarkable in its orientation towards practical technology.[163]

The collections and courts of these three princes may be viewed as repositories of expertise rendered mutually accessible through the bonds of kinship and the culture of the gift. Like the majority of his contemporaries, Elector August collected in a number of areas: he enlarged the Saxon holdings of coins, medals and silverware, and established an armoury, an anatomical cabinet, a treasury and an outstanding library.[164] His *Kunstkammer*, however was quite extraordinary. At his death, it numbered around 10,000 items, of which approximately 7,400 were practical tools drawn from such diverse fields as gunnery, locksmithery, cabinet-making, gardening and surgery, and a further 950 were mathematical or scientific instruments. The 100 or so quadrants, armillary spheres, astrolabes, globes, astronomical clocks and similar objects, testify to August's active interest in astronomy, time-keeping, and astrology.[165] The *Kunstkammer*, rather than the library proper, also contained over 250 texts: described as 'astronomy-', astrology-', geometry-', perspective-', arithmetic- and other craft-books',[166] these, along with the objects, were made available to artists and craftsmen, particularly those working on the commissions of the Elector. But, fittingly for the ruler of the state whose fortunes were largely bound up with its mining and metal-working industries, August's greatest interest was reserved for surveying and geodesy. Both Wilhelm IV and Rudolf II benefited from the Elector's cultivation of geodetic technology: in 1568, the Landgrave solicited surveying instruments from August, and later a waywiser, a speciality of the Saxon Court (Fig. 4.8); the Emperor received one of these instruments as a gift in 1583.[167]

August, for his part, tended to consult Wilhelm on the subject of astronomy. It was through him that Wilhelm's attention was first drawn to the new star of 1572; he was one of a number of ruling princes who sent the Landgrave reports on this celestial prodigy, and requested his opinion of it. In 1576, August asked for Wilhelm's assessment of a horoscope concerning the Polish royal succession; the Landgrave had little time for most astrological prognostication, as his response made clear, but he was not above attempting to exploit August's credulity in an attempt to bring about the

---

[163] On Rudolf, see Fucikova 1985, 1997a; Trevor-Roper 1991, 79–115; Evans 1997; Findlen 1997; Gouk 1997. On August, see Menzhausen 1985; Schillinger 1990.

[164] Watanabe-O'Kelly 2002, 71–99.   [165] Menzhausen 1985.

[166] Menzhausen 1985, 69: 'An astronomischen, astrologischen, geometrischen, perspectivischen, arithmetischen und anderen Kunstbüchern.'

[167] Schillinger 1990, 279 n. 12, 283 n. 16; Dolz 1993, 92–93.

4.8 A gilt-brass waywiser produced by the instrument-maker Christoph Trechsler, who served the Saxon court from at least 1571 until his death in 1624. It is signed and dated 'C. T. 1584'. The 1587 inventory of the Dresden *Kunstkammer* listed twelve such instruments, which measure distances traversed for the purpose of surveying. In 1583, one was sent to Emperor Rudolf II as a gift; Landgrave Wilhelm IV also received one. By permission of the Staatliche Kunstsammlungen Dresden, Mathematisch-Physickalische Salon [Inv. Nr. C.III.a.4].

release of Caspar Peucer, whom the Elector had imprisoned as a crypto-Calvinist in 1570.[168] August also received from Wilhelm one of the planetary clocks constructed by Eberhard Baldewein, Hans Bucher and Hermann Diepel; requested by the Elector shortly after the completion of the first

---

[168] Kolb 1977; Moran 1978, 31; Moran 1981; Leopold 1986, 17–19, 29–30.

such clock in 1562, it was not itself finished until 1568 (Fig. 4.9). Thus Wilhelm's request for geodetic equipment, mentioned above, could well have been an attempt to capitalise on the obligation incurred by the presentation of this object.

Delivery and installation of the Saxon astronomical clock were overseen by Baldewein himself. It seems that, while in Dresden, the Hessen *mechanicus* saw a celestial globe, by the Nuremberg instrument-maker Christopher Heiden (1526–1574), that may have inspired the series of mechanical globes subsequently produced for the court of Kassel.[169] The first of these, completed by Baldewein in 1574, took forward some elements from earlier globes produced in Kassel, but also, in what appears to be a genuine case of the transfer of technology, adopted the same method as Heiden in order to represent the motion of the Sun on the ecliptic without an unsightly external drive. This was a problem that continued to defy the ingenuity of other craftsmen (Fig. 4.10).[170] Placement of the drive-mechanism within the globe-sphere solved the aesthetic difficulty at the expense of a practical one, since the solar movement then bisected the sphere, making it (almost) impossible to impart a diurnal motion to via a stationary mechanism. Heiden's solution, and subsequently Baldewein's, was to have the inner driving-assembly rotate with the globe. This was a strategy that limited the overall size of the instrument, but allowed a reasonably accurate reproduction of the annual solar motion.[171] Although Wilhelm had intended to present Maximilian II with a prettier copy of this first mechanical celestial globe produced at his court, the Emperor preferred to take immediate possession of the instrument. Remaining in the possession of the Imperial family, it was later the inspiration and technical model for a similar object, one that was commissioned from the clock-maker Gerhard Emmoser by Rudolf II.[172]

At Kassel, the production of mechanical celestial globes continued, first under the supervision of Baldewein, then under Bürgi.[173] The Landgrave sent several of these instruments to other princes, and the last such presentation is recorded in the Hven–Kassel correspondence. In 1592, Tycho wrote to the Landgrave to report the death of his goldsmith Johannes Crol, who had 'for some years rendered me careful labour in certain things

[169] Leopold 1986, 35, 72–75; Leopold 1997, 6. On Heiden, see Zinner 1979, 369–371.
[170] For examples of instruments with external drives, see Maurice and Mayr 1980, 301; Dolz 1993, 90–91.
[171] Leopold 1986, 75; Leopold 1997, 6.
[172] Zinner 1979, 303–304; Maurice and Mayr 1980, 292–293; Leopold 1986, 104–111; Leopold 1997, 6–7.
[173] Leopold 1986, 1997.

4.9 The planetary clock by Baldewein, Bucher and Diepel, manufactured at Marburg and Kassel between 1563 and 1568 for Elector August of Saxony. The near faces are the planetary dial for Mars (above) and the *primum mobile* in the form of an astrolabe (below), with thirty-nine star-pointers on the rete. Between the two is a small dial for the quarter-hours and weekdays. Partially visible are the planetary dials for Saturn and Jupiter; dials for Venus, Mercury, the Moon and a calendar dial are obscured. The exterior of the clock is of gilt-brass and silver with enamel ornaments. A celestial globe is mounted on the top. By permission of the Staatliche Kunstsammlungen Dresden, Mathematisch-Physickalische Salon [Inv. Nr. D.IV.d.4].

4.10 A gilt mechanical celestial globe, with terrestrial globe below and armillary sphere above, produced by Georg Roll and Johannes Reinhold in Augsburg in 1586. Unlike the mechanical globes produced at Kassel, the motions of the Sun and Moon are represented on the surface of the celestial globe by figures that move along externally mounted iron bands. By permission of the Staatliche Kunstsammlungen Dresden, Mathematisch-Physickalische Salon [Inv. Nr. E.II.2].

pertaining to the fabric of my instruments'.[174] He was seeking assistance, from Bürgi rather than from Wilhelm, in finding someone 'who knows how to construct and handle clocks and automata' to be his replacement.[175] Wilhelm replied that Bürgi would do what he could to find such a person on his way from Kassel to Prague. For, he explained:

his Imperial Majesty wrote to us most graciously a few days ago, and requested that we send to him our clock-maker Jost Bürgi, with the globe which he recently made, and into which he has brought the motion of the planets. Since we could not deny his Imperial Majesty this, we decided to send the aforesaid clock-maker to attend him at once, but said that he should return to us promptly; so that we trust he will be back with us again by Michaelmas.[176]

Sadly, at the end of August, a month before the Feast of St Michael, Wilhelm himself passed away. But shortly before he expired, he was gratified to receive Rudolf's letter of thanks for the Bürgi instrument, written in the Emperor's own hand – a fact of which much was made when the whole incident was recounted in Treutler's funeral oration. Wilhelm's cultivation of the horological and astronomical expertise that procured him such success in the competitive field of princely gift-exchange was treated as one of the great accomplishments of his life.[177]

Two other outcomes of Bürgi's visit to Prague deserve our attention. The first is another instance of the transfer of horological technology. In constructing his own mechanical celestial globes, Bürgi had replaced the Heiden-Baldewein construction of the internal drive, which was highly limiting with respect to size, with a delicate arrangement which allowed the movement to remain stationary within the globe whilst keeping the two hemispheres together. But in Prague he saw the Emmoser globe of Rudolf II, in which an alternative solution was adopted: the majority of the movement remained stationary, but the drive for the Sun rotated with the case of the instrument. After returning to Kassel, Bürgi constructed another mechanical globe that utilised the very same arrangement. In this way, therefore, Baldewein's instruments both led to a second series of mechanical globes constructed at Kassel, and inspired the production in Prague of an instrument that, in its turn, contributed to a third generation.[178] Such interactions are of some significance to the historian of astronomy, since horological engineering had an important role to play in the development of both observational and theoretical branches of the discipline. Although Tycho had experimented with clocks and found them somewhat unreliable,

---

[174] *TBOO* VI, 297.36–41.   [175] *TBOO* VI, 297.31–35.   [176] *TBOO* VI, 301.5–14.
[177] Treutlerus 1592, 84–85.   [178] Leopold 1986, 31, 36; Leopold 1997, 7.

the Kassel observatory, with its access to superior clock-makers, made extensive use of precision time-keeping instruments.[179] Moreover, as we shall
soon see, its horological engineers, Bürgi in particular, made their own
contributions to contemporary debates concerning the true and the possible systems of the world.

Bürgi's trip is also significant because it paved the way for his eventual
entry into Imperial service. He visited Prague a second time near the end
of 1596, and once more in 1604 when, as the envoy of Moritz, he again
presented the Emperor with some instruments that he had constructed.[180]
On this occasion, Rudolf determined to appoint him instrument-maker
to the Imperial Court, and in 1605 he was retrospectively released from
Hessen employ. From then until 1631, he lived and prospered in Prague,
establishing a workshop in the Hradcany, the city quarter surrounding the
palace, and serving three emperors. Despite the attribution to Bürgi of a
pair of Tychonic sextants made in Prague *c.*1600, the dates of his known
sojourns in the city do not coincide with those of Tycho himself, who arrived
in Prague in 1599 and died there in 1601.[181] But he was to meet and work with
Kepler, collaborating with him in observational astronomy, mathematics,
and pump technology.[182] As striking a conjunction of expertise as the more
famous partnership of Kepler and Tycho, this is just one of the many
alliances of instrument-making skills and academic learning that the desire
to possess tokens of wealth and erudition helped to foster at sites across
Europe.

That Wilhelm had Baldewein and Bürgi accompany the instruments
which he presented to August and Rudolf might be taken to indicate the
measure of his respect for these technicians.[183] But in Chapter 3, we saw
a comparable situation, when Rothmann anticipated that Wilhelm would
want him to personally present his work on the observations of the fixed
stars to Frederick II. Such incidents suggest that the noble sense of ownership extended beyond the products of courtly patronage to the scholars
and artisans: talented men were, in some respects, just as collectible as
any of their work. In fact, patrons and clients undoubtedly colluded in
representing one another as reflections of their own importance, making
their relationships visible via the exchanges of goods and favours through
which, to a large extent, they were formally constituted. This helps to

---

[179] von Bertele 1988; von Mackensen 1988, 135–137.   [180] Leopold 1986, 31–32.
[181] Sima 1993, esp. 450–451. One of these instruments could actually have been the sextant that Bürgi
made for Kurtz and Hagecius reported seeing at the end of 1593.
[182] Leopold 1986, 33; Caspar 1993, 164–165.   [183] Leopold 1986, 30.

explain the role of the personal portrait, a class of object that has featured several times in this chapter. Not only did it help to define an individual's sense of self (one thinks here of the portraits of Wilhelm and Sabine with the Kassel astronomical instruments, and of Tycho's mural quadrant) but, conferred on another, it offered confirmation of a binding connection between patron and client.[184] In this manner, Tycho endorsed the products of his collaboration with the van Langrens and Blaeu, while the King of Denmark and the Elector of Saxony appointed or reappointed themselves as patrons of Tycho. Christian of Saxony was, it is true, not the great patron of Tycho that Frederick II and Rudolf II had been, but casual alliances of the sort that operated between him and the astronomer, and which operated on essentially the same principles of courtesy and reciprocation as the longer-term ones, were also an important feature of the cultural landscape. At all levels, clients and patrons were promiscuous in forming multiple relationships. It was this sensibility, as much as the desire to protect intellectual property, which informed the use made of privileges by an individual like Tycho.

III MOVING HEAVEN AND EARTH: MODELS OF CELESTIAL MOTION

To describe the evolutions in the dance of these same celestial gods, their juxtapositions, the counter-revolutions of their circles relative to one another, and their advances; to tell which of them come into line with another at their conjunctions, and which in opposition; and in what order they pass in front of or behind one another; and at what periods of time they are severally hidden from our sight and, reappearing again, send to men who cannot calculate panicky fears and signs of things to come – to describe all this without visible models of them, would be labour spent vainly. (Plato, *Timaeus* 40C.-D)[185]

I feel that very many things which pertain to the inequality of the year were developed from imperfect or false observations; for it is not necessary for that celestial motion to have been perpetually confined within the same limits, and by such complex hypotheses as those of COPERNICUS in ascribing a motion to the Earth – which, there is no doubt, are false, and perhaps do not even satisfy the manifest appearances. For this reason I wanted an instrument that would present to view that whole fantasy of COPERNICUS; which, I confess, I could never understand from a simple sketch. (Heinrich Brucaeus to Tycho Brahe, 12 June 1584)[186]

---

[184] See also *TBOO* VI, 269.6–8; daCosta Kaufmann 1993, 100–135; Christianson 2000, 99–102, 113–118.
[185] Adapted from Cornford 1937, 135.   [186] *TBOO* VII, 85.20–28.

Most of the instruments discussed in this chapter functioned as moving models of celestial motion, one way or another. The majority of the objects considered have been celestial globes and even a non-mechanical globe will readily display, if it is appropriately mounted, the rotation of the *primum mobile* which accounts in the Ptolemaic scheme for day and night and the diurnal motion of the stars. Some historians, it is true, have objected to the application of the term model to instruments of this kind. Static globes and planispheric astrolabes were, it has been pointed out, described and promoted by early modern scholars as mathematical problem-solving devices rather than mobile representations of the heavens.[187] Yet the evidence also suggests that such instruments were understood as working in virtue of their embodiment of mathematical models, and the notion that these mathematical models enjoyed *some* kind of correspondence to physical reality was seriously entertained.[188] In this way, in fact, a period notion of representation might extend from devices of these kinds to others that it is now much less tempting to think of as celestial models: Robert Recorde (*c.*1512–1558), for example, in his *The castle of knowledge* (1556), grouped a number of instruments with astrolabes and globes, including quadrants, staves and dials, as parts or representations of the celestial sphere.[189] At the same time, early modern scholars quite naturally saw continuities between static instruments and self-moving machines that further blurred the distinction between the calculation of celestial motions and their visual representation. As we have seen, sophisticated mechanical globes not only reproduced the diurnal motion of the heavens, but also portrayed the movements of the Sun and the Moon with the aid of figures and dials. Even more elaborate devices, ranging in size from the Strasbourg cathedral clock to table-top instruments, displayed in addition the motions of the planets. Such instruments could function in the same way as hand-operated analogue calculators, but tended to show the current position of celestial bodies rather than the underlying geometry of the mathematical models. Unsurprisingly, given the expense and effort that it took to produce them, their production was mandated by courtly and municipal cultures of display, not the requirements of mathematical utility. Modelling and representing the cosmos better describe their primary function than calculating, therefore, even if the result was not quite the legendary sphere of Archimedes to which they were often compared.[190]

---

[187] Bennett 2002.   [188] See Mosley 2006.   [189] Recorde 1556, 35.
[190] Rose 1975, 7–9; King and Millburn 1978, *passim*; Poulle 1980, 495–498; Haber 1981, 10; Simms 1995, 53–55; Granada, Hamel and von Mackensen 2003, 61; Mosley 2006.

Among the most elaborate celestial models of the age were the planetary clocks produced at Landgrave Wilhelm's behest. Still extant in Kassel and Dresden are those that Baldewein and Bucher manufactured in 1560–1562 and 1563–1568. These clocks were not only furnished with a celestial globe, mounted on the instrument, that displayed the annual motion of the Sun. They were also equipped with eight separate dials: a calendar dial, a dial for the *primum mobile* (in the form of an astrolabe for the appropriate latitude), and individual dials for the Moon and each of the planets. On the planetary dials, the courses of the planets around the Earth were depicted in analogue, according to a physically realised arrangement of eccentric deferent and epicycle. Three indicators allowed the latitude of the planet, and its true and mean place on the ecliptic, to be read off from the circular scales of these discs.[191] When operating, therefore, the clocks displayed the planets' motions in real time, effectively calculating their locations according to a particular set of planetary hypotheses.

The astronomical hypotheses represented by the Baldewein instruments were the Ptolemaic ones. The design of the dials was derived by the mathematician Andreas Schöner from the planetary volvelles contained in Peter Apian's *Astronomicum Caesareum*. The Bürgi instrument given to Rudolf II, which Wilhelm described as a globe representing the courses of the planets, was probably a similar combination of mechanical globe and astronomical clock, and was most likely based on the same mathematics as the earlier Kassel models. But with a range of candidate world-systems available at the end of the sixteenth century, instrument-makers were not restricted to mechanical instantiations of Ptolemaic hypotheses. At about the same time as he was constructing his planetary globe, Bürgi produced a table-clock that displayed the lunar motions according to the theory of Copernicus (Fig. 4.11).[192] The Copernican model of the moon was, it is true, considered superior to the Ptolemaic scheme even by scholars who discounted the possibility that the Copernican scheme could represent the true form of the cosmos.[193] Bürgi's use of it does not, therefore, necessarily imply any commitment to a particular cosmology. Nevertheless, the instrument presented to Rudolf in 1592 could have showed the celestial motions according to heliocentric hypotheses.

The extent to which the existence of such extravagant models of celestial motion stimulated cosmological discussion in the court environments in which they were displayed is rather unclear. But other astronomical models,

---

[191] Leopold 1986, 60–71; von Mackensen 1988, 118–121; Dolz 1993, 106–109.
[192] Leopold 1986, 158–173.   [193] See Neugebauer and Swerdlow 1984, 193–194.

4.11 (a) Table-clock produced by Jost Bürgi, *c.*1591, with a case by the goldsmith Hans Jakob Emck. The instrument includes an horary dial, a dial for the days of the week with planetary rulers, and (on the inner side of the hinged lid) a dial with calendar, lunar phase indicator, solar indicator, and dials for the lunar epicycles according to the Copernican model of lunar motion.

some of them produced at the same sites, certainly did play a role in the presentation and evaluation of world-systems at this period, and did feature in the attendant cosmological debates and disputes over priority. While the existence of these instruments has previously been noted, they have, until recently, been accorded little significance. This neglect is due in part to the nature of the evidence available: with few such devices surviving, speculation about their character and importance has to proceed on the basis of very little information. But the privileging, by historians, of intellectual culture over material culture, also bears some responsibility for the lack of interest in these objects. In fact, as we shall see, there are grounds for believing that instrument-building provided scholars with another route to the geoheliocentric planetary hypotheses promoted by Tycho, Ursus and others. The need to adapt Copernican models to a form that would allow analogue calculation of geocentric celestial coordinates also resulted

4.11 (b) A detail from the case, showing Copernicus with a diagram of his world-system. The entire clock displays a miniature visual history of astronomy and geometry, with further images of Abraham, Thales, Euclid, Archimedes, Hipparchus, Ptolemy, and Alfonso X of Castile, all cast in silver. By permission of the Astronomisch-Physikalisches Kabinett, Staatliche Museen Kassel [Inv. Nr. U 24].

in geoheliocentric devices. This need could be generated by a desire to satisfy courtly appetites for magnificent clocks and machines, but it could also stem, perhaps, from more straightforwardly pedagogic concerns.

That mechanical models of planetary motion possessed an important cognitive function had been recognised for a very long time. The passage in Plato's *Timaeus* that pointed out their indispensability to the study of the heavens was known throughout the Middle Ages, and may have provided the intellectual justification, if any were needed, for the construction of the volvelles, equatoria, demonstrational armillaries and other forms of model developed prior to the early modern period.[194] Like globes, these instruments varied in their precise function and form, but were generally promoted for their pedagogic value, labour-saving functions and, in the case of the more elaborate examples, for their ability to excite admiration and wonder. Alongside appreciation of the didactic role and captivating appearance of such instruments, however, it is also possible to find concern about their capacity to mislead. The fifth-century author Martianus Capella, in his *De Nuptiis Philologiae et Mercurii*, had *Astronomia* herself express the opinion that a bronze armillary sphere was not an authoritative guide to the workings of the cosmos, since features peculiar to it might be mistaken for characteristics of the universe.[195] And this criticism was probably derived from the report of Marcus Varro's similar comments in the *Noctes Atticae* of Aulus Gellius (*c.*130–180).[196] Such misgivings identify specific instances of a practical epistemological problem inevitably faced by users of any analogical model, physical or conceptual: that of articulating and enforcing the distinction between valid and invalid inferences from the characteristics of the model to those of the thing being modelled.[197] Frequently, the erosion of boundaries of this sort proved intellectually fruitful; this was the route taken by the early modern astronomers who argued that their mathematical models of the cosmos represented the physical reality. Often, however, failure to differentiate the limits of the analogy led only to confusion. Medieval astronomical models and diagrams appear to have induced misconceptions on more than one occasion,[198] and with the advent of new cosmological systems in the sixteenth century, the difficulties inherent in the use of astronomical models seem to have been exacerbated rather than improved.

A particular problem for the representation of the Copernican hypotheses, with the mean Sun as the centre of motion, was that celestial positions were always given with reference to the terrestrial observer.[199] Consequently, while instruments that enabled mechanical calculation of the Copernican planetary positions could be constructed, these might differ significantly from models intended to represent the fundamental principles of heliocentricity. In the first case, the Copernican world-system would have to

---

[194] See King and Millburn 1978, 1–61; Lindberg 1979; Poulle 1980; Gingerich 1993; Dekker 1999, 3–12.
[195] Burge, Johnson and Stahl 1971–1977, II, 319; Willis 1983, 309–310.
[196] Rolfe 1927–1928, I, 267–269.
[197] For a philosophical treatment of this issue, see Hesse 1966.
[198] For one example, see Eastwood 1981; cf. Eastwood 2000.
[199] For a brief discussion of this issue in relation to pedagogic concerns, see Johnson 1953, 288.

be geometrically transformed, or 'inverted', so as to produce a device from which the coordinates of the planetary positions could be directly read or inferred. The resulting instrument, perhaps taking the form of an equatorium, volvelle, or planetary clock-dial, would look geocentric on first inspection, even though it employed heliocentric parameters and assumptions.[200] A good mathematical analogue of the Copernican system, it would be a poor physical representation. A model of the second type, however, which displayed the principles of heliocentricity, would need to show the rotation of the Earth and planets around the Sun. Although it could also be constructed as an equatorium or volvelle, it might instead have a form more like that of an orrery.[201] Capable in principle of displaying the relative planetary positions with great fidelity, it might, in practice, be produced with much less emphasis on the mathematical details, even to the extent of omitting the epicyclic refinements. But despite being a poor representation of the complex geometry of the Copernican system, such a model would do more than the former type of instrument to foster appreciation of it as a candidate cosmology.

Prior to the promulgation of his own world-system in 1588, Tycho displayed some enthusiasm for modelling planetary hypotheses by means of an instrument. Thus, in a letter of September 1576 to Petrus Severinus (1542–1602), he concluded a description of the three motions Copernicus had attributed to the Earth by commenting that:

all these things, as they were ingeniously devised by that very clever artificer, can only be understood with difficulty unless a machine of the sort in which they are represented visually is constructed. But this, and many other things, God aiding our efforts, I shall do sometime, when I am able to enjoy a philosophical respite away from the noise of the rabble; and I shall then weigh more carefully those things which are still wanted in the motions.[202]

Eight years later, however, writing to Brucaeus, Tycho's reservations about these instruments were such that he denied ever having contemplated producing a model of the Copernican cosmos:

You ask that I should send to you some instrument of the Copernican hypotheses, such as I once saw at SCHRECKENFUCHS'; which I would willingly do, if I had prepared such a thing. For it has never entered my mind to prepare instruments of such a sort, which only display the findings of others, since they can readily be

---

[200] See North 1969, 418–421.

[201] The term 'orrery' is anachronistic, but useful in specifying a certain type of planetary instrument. See King and Millburn 1978, 150–167, esp. 154.

[202] *TBOO* VII, 40.14–20.

understood from a simple sketch, and the case, as to whether the matter is or is not thus constituted, remains before the judge. For this reason every effort ought rather to be expended in building instruments with which the heavenly motions are established accurately, so that through these the findings of our predecessors can be evaluated and tested to the limits of the truth.[203]

The shift in Tycho's thinking between 1576 and 1584, sufficiently powerful to lead him to forget or deny what he had previously thought, is perhaps explained by the more critical stance towards the Copernican system that had emerged as he was increasingly called upon to articulate his philosophical objections to heliocentrism and, at the same time, found confirmation of the astronomical shortcomings of *De revolutionibus* in the results of his own observations.[204] Stating that Copernicus himself had only advanced his hypotheses in order to save the appearances, Tycho suggested that it was of little benefit to waste time and effort constructing a model of the Copernican scheme when there was still work to be done laying the astronomical foundation for the determination of the actual world-system.[205] Somewhat defensively, Brucaeus hastened to agree. He justified his request by explaining that he had not wanted the model because he thought the heliocentric system was the true one, but because he found it difficult to understand with just a diagram to help him.[206] He desired an instrument that could play the key cognitive role that the younger Tycho – and Plato before him – had acknowledged.

Tycho presented specific objections to the form of planetary model for which his correspondent had asked. Erasmus Oswald Schreckenfuchs (1511–1579), professor at the University of Freiburg, had possessed a device that he claimed could be used to illustrate either the Copernican or the Ptolemaic world-system. But, wrote Tycho, 'when I afterwards considered the matter, I realised that this construction did not satisfy the scheme of COPERNICUS. For he [Schreckenfuchs] thought that the same machine, according as it was variously inverted, could represent both the Ptolemaic and the Copernican hypotheses; which is clearly impossible.'[207] Tycho's point, it seems, was that since the Copernican system required a much larger interval between the centre of the universe and the sphere of the fixed stars than the Ptolemaic, then even if a geocentric model could be rearranged so as to position the Sun at the centre, rather than the Earth, the resulting instrument would not constitute an adequate representation of the heliocentric world-system.

---

[203] *TBOO* VII, 79.22–31. The disparity between the views expressed in the letter to Severinus and those contained in this letter has previously been remarked upon by Moesgaard 1972a, 37, 46–47.
[204] Gingerich and Voelkel 1998, 1–9.   [205] *TBOO* VII, 80.41–81.3.
[206] *TBOO* VII, 85.20–28.   [207] *TBOO* VII, 79.35–39.

Indeed the problem of achieving proportionality would, he indicated, make it difficult to construct any model whatsoever of the Copernican hypotheses:

And it will not readily be allowed to portray all the motions of the Copernican invention mechanically in a single machine which would agree with itself in every respect, especially because of the incomparable distance of the eighth sphere to the orb of the Earth; even though the speculation of Copernicus about the symmetry of the world is far more beautiful, and would require fewer circles, than the Ptolemaic scheme.[208]

This comment is particularly significant in the light of the emphasis placed on cosmological proportionality in the refutation of Copernicus that appeared in the *Epistolae astronomicae*. It suggests that thinking about a mechanical representation of planetary hypotheses may have helped to crystallise one of Tycho's key objections to the Copernican world-system. And, even if that was not in fact the case, it indicates that Tycho's judgement of Schreckenfuchs' model was predicated on it functioning as more than just an indicative depiction of the planetary system. To be properly analogous, scalar as well as angular quantities had to be properly represented.

It has been suggested that Schreckenfuchs' belief that his instrument depicted both Copernican and Ptolemaic world-systems would have been objectionable not only on the grounds of the scalar difference between the two cosmologies, but also because the model would actually have portrayed the heliocentric and geoheliocentric arrangements.[209] In combination with an earlier assertion, that Tycho acquired the device of the Freiburg professor in the mid-1570s, this raises the possibility that the instrument played some important role in the evolution of the Tychonic world-system.[210] But neither claim, as we shall see, is strongly supported by the evidence. Let us deal first with the issue of Tycho's ownership of the planetary model. When, in 1575, Hieronymus Wolf and Paul Hainzel wrote to Tycho about acquiring an instrument for him from Schreckenfuchs, they described the object as a 'Ptolemaic Armillary'.[211] A note sent to Wolf by Schreckenfuchs, and forwarded to Tycho, spoke of a *sphaera materialis*, a designation that could have applied to either an armillary sphere or a globe. 'As pertains to the setting up of this *sphaera materialis*', Schreckenfuchs wrote, 'a certain ingenuity will be necessary, lest the colours and numbers on it are rubbed

---

[208] *TBOO* VII, 79.39–80.2. [209] Gingerich and Voelkel 1998, 8–9.
[210] Zinner 1979, 530; Zinner 1988, 291, 294; Thoren 1990, 30 n. 51.
[211] Wolf, who claimed not to know anything about the object, wrote of an *armilla* and a *machina*; Hainzel of *Armillae Ptolemaicae*. See *TBOO* VII, 15.24–26, 18.10–16, 19.8–13.

off'.[212] This caution, reminiscent of Pratensis' comments about the difficulties of producing the mechanical globe commissioned by Tycho, is not,
to be sure, inconsistent with an instrument having the form of a mechanical armillary showing the planetary motions. Hainzel, however, added that
Copernicus had described the device in book two, chapter fourteen, of *De
revolutionibus*, and that both its construction and use had been treated by
Regiomontanus. Indeed, he had decided to transcribe the latter's account
and forward it to Tycho, just in case, he said, his correspondent did not own
a copy already.[213] These comments make it clear that Hainzel thought the
instrument was an observational armillary sphere of the zodiacal configuration originally described in Ptolemy's *Syntaxis*.[214] Given that he went on
to advise Tycho of an improvement to this kind of observing instrument,
it seems unlikely that he could have been mistaken about the nature of
the object.[215] So the identification of the instrument sent to Tycho with
the planetary model described in the letter to Brucaeus is almost certainly
erroneous.

The claim that Schreckenfuchs' instrument might have depicted the
heliocentric and geoheliocentric hypotheses is somewhat more plausible.
The proponents of this idea suggest that the model was 'some sort of
orrery', a description that most readily suggests a physical analogue of the
Copernican world-system. If we suppose a simple machine capable of representing the order and centres of motion of the heavenly bodies, then
representation of the geocentric and heliocentric systems might be accomplished by interchange of components representing the Earth, with Moon
attached, and the Sun, a procedure that would certainly account for Tycho's
criticism that it failed to accommodate the scalar difference of these two
accounts of the cosmos. But a model of this kind could not be altered
in the same way so as to display the geoheliocentric system, with its two
distinct centres of motion. It is however, possible that a model intended to
display the Copernican world-system might inadvertently represent a semi-
Tychonic arrangement. One such model was designed by Johannes Kepler,
although never constructed, as an illustration of the scheme he described in
his *Mysterium cosmographicum*. Kepler's plans for this instrument, commissioned by Duke Friedrich of Württemberg, changed over time. The central
thesis of his text, that the planetary intervals of the (heliocentric) cosmos
were determined by the nested Platonic solids, was one that did not require

[212] *TBOO* VII, 18.26–28.
[213] *TBOO* VII, 19.13–14. The work to which Hainzel refers is the *Scripta mathematici M. Ioannis Regiomontani* (1543); see Bennett and Bertoloni Meli 1994, 30–31.
[214] In *Syntaxis* V.1. See Dreyer 1890, 81–82.   [215] *TBOO* VII, 20.16–26.

a mechanical model to represent it.[216] His first design therefore, was not any form of planetarium, but rather a static model that showed the relations between the solids and the spheres of the planets, and dispensed drinks as part of an elaborate court amusement.[217] In a letter of 1598 to Michael Maestlin, however, Kepler described a more complex device, by which motion was to be imparted to the models of the planets. And in the scheme he described, the Earth and the Sun would actually have moved one around the other: the Earth, like the rest of the planets, was to rotate about the Sun; but it was to do it in such a way that it always remained at the centre of the outermost sphere of the cosmological model, with respect to which all the other planets, and the Sun as well, would appear to move eccentrically. This arrangement, undoubtedly designed to preserve the astrologically and calculationally important positional relationship between the Earth and the sphere of fixed stars, represented a geoheliocentric scheme at least as well as it did the system of Copernicus. But this was something that Kepler himself seems not to have realised, and certainly never intended.[218]

A similar effect would also be produced through an equatorium or dial of the sort described above, geometrically analogous to the heliocentric world-system, but actually presenting to view the 'inverted' Copernican arrangement. Had the Schreckenfuchs instrument been this kind of device, or a model like Kepler's, then its owner's belief that it displayed the Ptolemaic and Copernican hypotheses would indeed be erroneous. But in the former case, the instrument could hardly be said to contribute the sort of understanding of heliocentrism that a simple sketch might provide, as Tycho had suggested. This might be thought to be evidence in favour of an instrument of the same type as Kepler's. However, an instrument of that form would have misrepresented the Copernican system not by presenting an arrangement that was inconsistent with the expanded scale of the universe, but by employing a scheme which would seem to have rendered such expansion unnecessary. This does not seem to accord with Tycho's criticism of the Schreckenfuchs model. Moreover, for either type of instrument, the change from geoheliocentric to heliocentric representation, and vice versa, would require not a mechanical inversion, that is to say an adjustment of the model's configuration, but merely a cognitive one. The instrument itself would remain the same, but at one moment one would choose to see it displaying the Sun in motion about the Earth, at the next the Earth about the Sun. And this does not agree at all with Tycho's description of the instrument in question. Considering his reports and his criticisms with care, it seems much more likely that the machine was a very modest device supposed to represent the fundamental differences of the Ptolemaic and Copernican systems of the world.[219] And if this machine took the form of a representational equatorium, then the transformation between one system and the other might even have been a literal inversion. On one side, the instrument could have depicted the Earth at the centre of the universe, and on the other, the Sun.

The evidence does not, therefore, readily support the idea that an instrument displaying the geoheliocentric system was known to Tycho prior to 1588, let alone that it made any contribution to his adoption of this arrangement as a serious cosmological theory. But Tycho was subsequently to perceive a role for such an instrument in conveying the principles of his system. Having received notice of the new hypotheses through the agency of Heinrich Rantzau, Caspar Peucer and Georg Rollenhagen both produced objections that were forwarded to Tycho by the Holstein Governor. Seemingly having seen only the diagram of the Tychonic system, Peucer objected that the new arrangement required a great quantity of empty (and therefore superfluous) space in the cosmos.[220] In combination with Rollenhagen's misapprehension, that the new hypotheses did not preclude an eventual collision of Mars and the Sun, this comment evidently convinced Tycho once more of the value of a demonstrational model.[221] As he wrote to Rantzau:

since the disclosure and explanation of my hypotheses is not easily perceived by anyone from a diagram alone, I have had constructed from brass a certain mechanical instrument, in which the courses of the planets with respect to the Sun, and of the Sun and the Moon around the Earth, can be revolved by hand and understood more swiftly; through which also those doubts of both PEUCER and ROLLENHAGEN

---

[219] Gingerich and Voelkel 1998, 9, state that the machine was made to depict Copernicus' hypotheses. But while this is possible, the evidence does not entirely justify such an unequivocal assertion. Tycho's description of the instrument, allows us only to claim that the instrument was intended to depict *either* the Ptolemaic system, *or* the Copernican system, and Schreckenfuchs believed that it displayed the other as well – or that it was always intended to display both arrangements. See also Mosley 2004.

[220] *TBOO* VII, 125.24–35. The translation of *delineatio* as 'exposition' in Rosen 1986, 24, might encourage the view that Peucer and others were forwarded the entirety of the account of the Tychonic world-system contained in the work. However, since Tycho later wrote that Rantzau had extracted and distributed the *delineatio* from the book, while disregarding the brief explanation that accompanied it, the term must here be interpreted as referring to the diagram alone. See *TBOO* VII, 131.14–23.

[221] *TBOO* VII, 125.35–42.

are very readily dispelled, just as the presenter of these [documents] (provided that his judicious head can comprehend the holy objects of Urania with sufficient care and hold them in memory, which I do indeed wish, but can scarcely hope for) will show you more fully. For I am sending to you, through him, the archetype of this equatorium; which I request you receive in good part.[222]

While neither this instrument, nor any diagram of it, is known to have survived, it seems clear from Tycho's brief description that it was a relatively simple hand-operated device, perhaps designed to display no more than the range of planetary configurations possible under the geoheliocentric scheme. If indeed it took the form of an equatorium, rather than a model like an orrery, then it might have been no more complex than Tycho's cosmological diagram rendered in a sequence of rotating brass discs on a circular base. Such an instrument would have sufficed to demonstrate to Rollenhagen that as the Sun moved around the Earth it always remained at the centre of Mars' orbit, thereby making plain that the point of intersection of the courses of the two bodies, where Rollenhagen feared they would collide, was only an artefact of the diagram of the system. And such an instrument would also have sufficed to indicate to Peucer that the space shown as empty on the sketch would come to be occupied as the solar motion proceeded, with every part of the universe being traversed by the planets as they rotated in their circumsolar course. If construction of the instrument was indeed prompted by the need to address these concerns, then Tycho could have had little time to produce a machine that was any more complex.

Tycho's description of what he was sending to Rantzau as the archetype (*Idea*) of this instrument is somewhat ambiguous; it is compatible with him having forwarded only its design to the Governor of Holstein. In fact, from a letter of Tycho's written a little while later, we learn that his correspondent had received 'that instrument by which our hypotheses of the celestial revolutions are more conveniently displayed'.[223] And in May 1589, Peucer informed Tycho that he had, 'through the agency of Lord Rantzau, received the brass *ichnograph* and construction of your hypotheses', adding that the instrument had helped him to understand his system much better.[224] Apparently therefore, it was the instrument itself that Tycho sent to Rantzau; with good reason, of course, for while it remained on Hven, it could not convey to those who had failed to understand his diagram the way that the Tychonic world-system actually functioned. But the astronomer seems not to have built further copies of this model, either

[222] *TBOO* VII, 126.26–35.   [223] *TBOO* VII, 385.35–386.1.   [224] *TBOO* VII, 191.15–19.

for distribution to others or display at Uraniborg. Although we will never know for sure why that should be so, several reasons for Tycho's restraint would seem to be suggested by the character of the other instruments on Hven and his remarks to Brucaeus. One possibility is that the planetary model was just too simple for his taste and for that of the courtly visitors whom he might have hoped to impress. In the form suggested for it, the instrument would not have been very striking, and Tycho may well have wished to represent his hypotheses in a much more elaborate device. Indeed, his later request for Bürgi's assistance in finding a new instrument-maker who knew 'how to construct and handle clocks and automata' might have been prompted by the thought that such expertise would allow him to construct cosmological models as splendid as the Kassel planetary clocks. At the same time, however, Tycho might have considered it inappropriate, even shameful, to commission something so elaborate before he could complete the empirical determination of the parameters that would allow the construction of an accurate and truly Tychonic cosmological model. Like his unfulfilled promise to present a fuller account of his world-system, therefore, Tycho's apparent failure to promote his planetary hypotheses by means of instruments intended for presentation or display could well have been a consequence of the difficulties he encountered in completing his astronomical programme as he had envisioned it.

This explanation finds some support in remarks Tycho made in the *Mechanica* concerning a universal astrolabe. He had, he declared, obtained the brass plates for such an instrument some years before, but had been kept from completing it by the press of other occupations; in any case, his failure to do so had caused him no regret prior to his own observational determination of the star positions to be marked on it. Now that these investigations were complete, the instrument could be easily finished whenever time allowed.[225] But Tycho's thoughts on the value of representing his world-system in the form of a demonstrational model may also have been shaped by news of Ursus' planetary machine. A description of this instrument was contained in the Dithmarschener's *Fundamentum astronomicum* of 1588. Ursus claimed that he had thought up his 'new, and true, and natural hypotheses of the motions of the planets or heavenly bodies' almost three years previously, 'whilst in a certain corner of the kingdom of Poland'.[226] Afterwards, however:

---

[225] *TBOO* V, 99.9–41.
[226] Ursus 1588, 37r: 'novas nostras, nec non veras ac naturales Hypotheses motuum planetarum, seu corporum Mundanorum quas ante elapsum iam fere triennium in extremo quodam angulo amplißimi regni Poloniæ excogitavimus . . .'

we presented them to the Most Illustrious Prince of Hesse, whose best artisan Jost Bürgi of Switzerland constructed the presented hypotheses from brass. And on the basis of these hypotheses any individual man, in his house, and at very slight expense, may have such an Astrarium, as they call them, constructed: which not only precisely enough indicates and shows just the mean or equal motions of the planets, as certain instruments or clocks, more stately and expensive than skilful, are accustomed to attempt (and that barely, and not even barely), but also the true and apparent motions of all the celestial bodies; indeed far more precisely than all the hundreds and thousands of all the calculated tables. For that these have been constructed and sewn together from false and invented hypotheses, and therefore (as also the very observations demonstrate and teach) are quite false, we shall prove in its own time in our *Astronomia*.[227]

These are claims that would certainly have raised readers' expectations regarding the capabilities of a planetary model based on geoheliocentric hypotheses, and perhaps set Tycho a goal that he knew he could not accomplish. But the diagram that Ursus included in the text, and that purported to show the arrangement of wheels required for the model of his world-system, would almost certainly have been inadequate for anyone hoping to construct a device such as he had described.[228]

In fact, according to Rothmann, Ursus had arrived at Kassel with little idea about how to represent his world-system mechanically, and had not known what parameters to specify.[229] This was probably because, like Tycho, he saw geoheliocentrism as a conceptual development, and had not anticipated that Wilhelm would commission a precision instrument to represent it. Rothmann reported that when the model was finally constructed, with values derived from Copernicus' *De revolutionibus*, both Wilhelm and Bürgi had realised at once, 'especially from Mars', that Ursus' hypotheses could not be the true ones. Probably this was because, having been

---

[227] Ursus 1588, 37r: 'posteaque Illustrissimo Principi Hassiae obtulimus, cuius summus artifex Iustus Byrgi Helvetius easdem oblatas Hypotheses ex orichalco extruxit. Earumque Hypothesium adminiculo quivis homo privatus, suae domi, minimoque aere impenso tale instrumentum Astrarium, ut vocant, extrui curare potest, quod non solum modo medios seu aequales planetarum motus, ut pomposa quaedam ac sumptuosa magis quam artificiosa instrumenta vel Horologia duntaxat (idque vix ac ne vix quidem) factitare solent: sed etiam veros ac apparentes motus omnium corporum Mundanorum satis exacte indicat atque commonstrat: imo longe exactius, quam omnes omnium tabularum supputatarum centuriae atque myriades: Eas namque ex falsis & fictitiis Hypothesibus extructas atque consutas, ideoque (ut & ipsae observationes convincunt atque docent) & falsissimas esse, suo tempore in nostra Astronomia demonstrabimus.'

[228] See Ursus 1588, 40v. Leopold 1986, 190–192, has produced a conceptual reconstruction of Ursus' instrument on the basis of the diagram and description; but as he points out, there are clearly errors in the display of the arrangement.

[229] *TBOO* VI, 362 n. to 183.13: 'Nec homo ille sciebat rationes et quantitates diametrorum et eccentricitatum sed haec ex Copernico desumebat Horologiopaeus noster, qui satis coram Illustrissimo Principe ex me audiebat, quomodo res tractanda esset.'

built on the basis of Copernican parameters, the instrument displayed the shortcomings in Ursus' diagram of the cosmos (Fig. 4.12). This showed the path of Mars around the Sun entirely enclosing that of the Sun around the Earth when, like Tycho's, it should have shown the orbits intersecting. But Rothmann's own criticisms of the Ursine instrument concentrated on its lack of mathematical sophistication. The hypotheses, he stated, 'assumed only an eccentric of the Sun; and on that they led around the eccentrics of the rest of the Planets, with the theory of the equants omitted – or, if one is to speak following Copernicus, with the epicycles omitted entirely'.[230] Since his model did not incorporate any epicyclic refinements, Ursus' claims to have designed an instrument that could reproduce the true and mean motions of the planets 'far more precisely' than astronomical tables was certainly hollow.

When Ursus left Kassel, perhaps in some disgrace, he took his planetary model with him.[231] But for some time, Tycho may have laboured under the misapprehension that a splendid Ursine model remained in the possession of Wilhelm. Rothmann's full account of the machine constructed for Ursus appears in the draft of a letter preserved at Kassel. In the version of the letter printed by Tycho, however, Rothmann stated only that: 'I was also going to write about that Dithmarschener Ursus, and about that system of the world stolen from you, but the courier hastens, so I shall put it off until another time'.[232] In the draft letter, Rothmann had gone on to declare that, after Ursus' departure from Kassel, he had collaborated with Bürgi to produce for the Landgrave a planetary model that accurately represented the motions in accordance with the parameters of Copernicus, but from a geocentric perspective. It was this model of the 'inverted' Copernican hypotheses that Rothmann had mentioned in a previous letter. However, in the light of his reading of the *Fundamentum astronomicum*, Tycho had naturally, but mistakenly, interpreted this instrument as the product of Ursus' perfidy.[233] Apparently lacking the explanation that Rothmann had intended to send, this misconception was probably not corrected until the Hessen *mathematicus* visited Uraniborg himself; that is to say, not before August 1590.

Actually, Rothmann's earlier letter was not so very vague about the nature of the model in the Landgrave's possession. Having just received a copy of the *De recentioribus phaenomenis*, Rothmann had wondered if the Tychonic

[230] *TBOO* VI, 362 n. to 183.13: 'Assumebant enim tantum eccentriculum Solis, inque eo eccentricos reliquorum Planetarum circumducebant, omissa aequantium ratione, seu, si secundum Copernicum loquendum est, omissis prorsus illis circellis.'
[231] *TBOO* VI, 362 n. to 183.13. [232] *TBOO* VI, 183.13–15. [233] *TBOO* VI, 179.14–25.

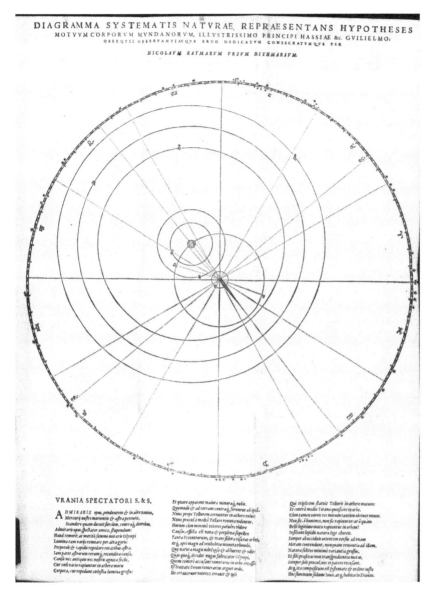

4.12 The diagram of Ursus' world-system, as published in his *Fundamentum Astronomicum* (Strasbourg, 1588). The Earth stands at the centre, and is orbited by the Sun, about which revolve all the other planets. Unlike the diagram of Tycho's geoheliocentric system, however, while the orbits of Mercury and Venus are shown intersecting with that of the Sun, the orbit of Mars is not. Note the dedication to Landgrave Wilhelm IV. By permission of the British Library [BL 8561.c.56].

hypotheses were simply the Copernican ones accommodated to the geocentric perspective, as he himself had shown in his (unpublished) *Elementa
astronomica*.[234] Following this method, he wrote, the Landgrave had in the
previous year, 1587:

> had an automaton constructed, of amazing smallness, but showing the motions of
> all the planets. For it is a disk, flat on both sides, of scarcely 6 inches in diameter,
> on one face of which the Theory of the Moon with its epicycles and the *caput
> draconis* is captured, but on the other the Theories of the Sun and all the other
> planets; and all the motions are shown not only as true in the longitude of their
> anomalies and centres, but also in the latitude of the three superior planets.[235]

Perhaps because of his preoccupation with Ursus' 'theft', Tycho seems not
to have realised that Rothmann's query about the relationship between the
new hypotheses and the 'inverted' Copernican system did not follow from a
misconstrual of his and Ursus' intentions, but indicated a previous interest
in geoheliocentrism that took no account of its potential as a candidate
cosmology. Rothmann stated that in his *Astronomical Elements*, a work that
was probably composed prior to his arrival in Kassel, he had not advanced
the hypotheses of Copernicus, since these were so difficult that complete
comprehension of them eluded even such supposed masters of the art as
Michael Maestlin.[236] Instead, he had provided greatly simplified planetary
models.[237] In addition, however:

> So as not to neglect clever men, and those who already understood the general
> forms of each motion, or who wished to construct automata, I attached at the end,
> as a corollary, the inversion of Copernicus which I spoke of above; so that in this
> way I prepared students for the reading and easier understanding of Copernicus.[238]

Although it is possible that Rothmann's consideration for those wishing
to construct a planetary model based on Copernican parameters was retrospectively generated as a result of his experiences in the service of the
Landgrave, this application for the inverted Copernican hypotheses is in
some ways more plausible than the claim that they were advanced for a
pedagogic purpose. But whichever was the chief reason for Rothmann's
labour, it would appear that, having accomplished a geoheliocentric transformation of the Copernican scheme several years prior to its realisation in
the models of Bürgi, he was first led to consider (and reject) the arrangement as a cosmological system when presented with the claims of Tycho
and Ursus. If Wittich's geometric manipulations of the Copernican system

[234] *TBOO* VI, 156.40–157.4.   [235] *TBOO* VI, 157.8–16.
[236] *TBOO* VI, 160.8–13; also 361 n. to 183.13.   [237] *TBOO* VI, 160.13–18.   [238] *TBOO* VI, 160.18–22.

were pursued for similar reasons to Rothmann's, then he too may have been working within a purely mathematical tradition: one that had application to the construction of planetary models for the purpose of display and calculation, but was not all motivated by the concerns of cosmology.[239] And Tycho, for his part, might not have perceived that his system would be so readily identified with the geometrical inversion of Copernicus until informed by this letter of Rothmann that a mathematician other than Wittich had been pursuing the same enterprise as his erstwhile collaborator. But once alerted to this possibility, any doubts he may have had about presenting a model of the 'Tychonic' world-system that relied upon the Copernican planetary parameters, rather than the ones he was striving to arrive at, would surely have been strongly reinforced.

Tycho may also have become disillusioned with planetary models by the reaction of the Kassel astronomers to his world-system, which, as he thought, they had already seen realised in material form. Rothmann reported by letter that:

when I showed our Most Illustrious Prince, in your second book on comets [i.e. the *De recentioribus phaenomenis*], that hypothesis of yours, he recognised it as soon as he saw it, inasmuch as it was known well enough to him from the planetarium: Good God, he joked, that circle of the Sun ought to be far more sturdy than brass, since it can drag so many planets with it.[240]

And Rothmann himself, even though he was entirely in agreement with Tycho about the dissolution of the celestial spheres, was moved to ask the astronomer an embarrassing question:

who would ever believe, that the centre of the greater epicycle of the Sun is provided with so much virtue that it can drag all the planets after it, indeed pull them down from their spheres and raise them again, when they are connected by no corporeal and gripping substance?[241]

Faced with the prospect of an analogy being made between a mechanical instrument and mechanical cosmos, Tycho, whose world-system depended upon the elimination of physical apparatus in the heavens and the attribution of volitional motion to the celestial bodies, may have supposed that any instrument he produced was going to do more harm than good to his cause. With his correspondence and rivalry with Kassel providing so many reasons for him to revert to and stick fast by his words to Brucaeus, that his effort as an astronomer should be directed to activities other than the building of models of hypotheses, it does not seem surprising that Tycho

[239] See Goulding 1995. [240] *TBOO* VI, 158.21–26. [241] *TBOO* VI, 158.18–21.

made no mention of any such instrument in the *Mechanica's* pages, nor indeed, that according to the *Synopsis* of 1591, it was a diagram of his world-system, not an automaton, that had pride of place in Stjerneborg's gallery of astronomical accomplishments.[242]

But perhaps Tycho was too pessimistic. His cause was not advanced at Kassel as a result of the production there of a model of the 'inverted' Copernican hypotheses. Elsewhere, however, the work of mathematically transforming the heliocentric world-system in order to produce such objects, and occasionally instances of such instruments themselves, may have assisted the reception and adoption of Tychonic and semi-Tychonic world-systems. After his move to Prague, for example, Bürgi constructed an astronomical clock that displayed the Copernican planetary motions and positions according to the geoheliocentric inversion.[243] And while Bürgi is, in many respects, an atypical figure, it remains a possibility that evidence of the design and production of similar objects by other individuals will emerge in the future. Previously unknown instruments of the period continue to be brought to the attention of historians and, as we have seen, documentary evidence of an object may survive when the instrument itself is no longer available. Although some of this material has been examined, it has not always been interpreted with care and skill. And the archives, libraries and museums of the world contain considerable quantities of material that have yet to receive the attention they require. Concerning the place of instruments in the culture and history of early modern astronomy, there is still a great deal of work to be done.

IV CONCLUSION

I agree with you, that one ought not to expend the same labour on those instruments which expose to view the fancies of another, as on those which reveal the celestial motions to us: although it is the mark of a learned man that he wishes to explain the inventions of others, to understand these, and to present them to view as much as is possible. (Brucaeus to Tycho, 12 June 1584)[244]

In recent years, studies of early modern approaches to the production of knowledge have emphasised the role played by the collection and display of material objects. 'Through the possession of objects one physically acquired knowledge, and through their display, one symbolically acquired the honour and reputation that all men of learning cultivated'.[245] As a description

---

[242] *TBOO* VI, 275.22–276.25.   [243] Maurice 1980, 97–101; Maurice and Mayr 1980, 222–223.
[244] *TBOO* VII, 85.28–32.   [245] Findlen 1994, 3.

of the 'scientific' culture of early modern Europe, this applies just as much to astronomy and cosmology as to a field like natural history. As we have seen, the craft of the instrument-maker was not only essential to the observational reform of astronomy, which did so much to further the integration of mathematical and physical conceptions of the heavens. It was also crucial to the expression of astronomical knowledge in ways that made it accessible to novices and individuals lacking mathematical expertise, and that were fundamental to its representation in material forms that could be owned and appreciated at the highest levels of early modern society. The value invested in such objects was one motivation to engage in the geometrical manipulation of the heliocentric world-system that produced geoheliocentric schemes of the cosmos.

Display was a function intrinsic to certain instruments of astronomy. Globes, armillaries, and models of planetary motion were all constructed so as to convey astronomical knowledge and concepts in a visual way. But the representational value of these objects was enhanced and extended through decorations and inscriptions which made them vehicles for the dissemination of a broader range of messages concerning status, expertise, patronage, and wealth. At sites such as Hven, Kassel, and Prague, observing and calculating instruments were employed as objects of display as well as functional devices. In part, this was possible because the size and fabric of the apparatus lent them a certain majesty and grandeur. But conscious effort could also be made to enhance their symbolic qualities. The Atlas-mount of Tycho's great observing armillary discussed in the introduction is one exemplary instance of a working instrument that was furnished with emblematic significance just as carefully as its sights were fitted, and its scale was divided. Tycho's great mural quadrant was another: decorated by no less than three artists, it became an emblem of the Tychonic astronomical and alchemical projects, a function that its image frequently serves in textbook reproductions. Many of the other instruments at Uraniborg were also furnished with allegorical figures and the portraits of famous astronomers. One small quadrant even bore an emblem that spoke of the value of philosophical study, particularly study of the heavens, as a route to a form of immortality – although it also explicitly acknowledged the greater importance of securing Christian salvation.[246] At Kassel, however, the same instrument-makers who were put to work constructing highly decorative clocks and globes freighted with symbolic meaning seem to have produced

---

[246] *TBOO* V, 14.3–32; *TBOO* VI, 284.10–285.28; Segonds 1994. For an example of a later instrument with a similar motif, see Turner 1979.

observing instruments that, although they were undeniably attractive, were quite plain by comparison.

This was not the only difference between the use made of instruments by Tycho and Wilhelm. Opportunistically, perhaps, the Danish astronomer allied himself with commercial instrument-makers so as to make visible his observational labours to a broad constituency of globe-users and purchasers. But Wilhelm's mechanical globes, although they drew on the work of the Kassel observatory, were intended only for the most privileged, and were as much instruments of state diplomacy as devices for the promotion of any individual. In the funeral oration of the Landgrave, his accomplishments as a ruler were merged with his achievements as an astronomer, the transition between the two aided by the motif of Atlas as *Coelifer*. At a fundamental level, however, Tycho and Wilhelm were similar in both finding a role for instruments in the cultivation of a lasting reputation; one, moreover, that extended beyond the use of observing apparatus to obtain astronomical data of unprecedented accuracy.

Representations of instruments could serve some of the same functions as the objects themselves. Lacking the resources or the inclination to give away the actual apparatus of Uraniborg, Tycho conceived the idea of distributing engravings and descriptions of his instrumental technology at an early stage in the life of his observatory. He enacted this plan in the range of ways discussed in this book: within his correspondence and manuscript *Synopsis*, through publication of the *Epistolae astronomicae*, and via production of the *Mechanica*. The policy of describing one's instruments, and the Tychonic mode of representation, were both adopted by subsequent astronomers as a means of establishing the credentials of their observational programmes, the task that Tycho had at one point intended the account of his instruments to accomplish.[247] As important for the Danish astronomer, however, was the fact that, by dedicating the *Mechanica* to Rudolf II, he could symbolically present the Emperor with his entire astronomical project. The eventual outcome of this presentation was the work that posthumously glorified both Rudolf and Tycho: the *Tabulae Rudolphinae*, edited by Kepler.

Not all communication via instruments was deliberate on the part of their producers. Instruments, as well as books, and frequently the combination of an instrument and accompanying manual, could in theory be protected against copying by means of a privilege. But the limitations of the privilege system were such that it seems likely that acquisition of these documents was motivated at least as much by the prestige they conferred

---

[247] See Chapman 1984; van Helden and Winkler 1993.

upon the practitioner, and his productions, as by their effectiveness in deterring plagiarists and pirates. Thus Bürgi, who came from a craft tradition, was not readily persuaded of the merits of making his knowledge available to others, despite his ready access to the privileges issued by the Imperial Chancellery.[248] Tycho, who was certainly no less aware of the dangers of others 'stealing' one's productions, but who was partially driven by pursuit of the credit and honour that could only come from some form of publication, was necessarily less reticent, notwithstanding his failure to release his texts for sale as soon as they were printed. To some extent, however, the culture of gift-exchange and display that governed diplomatic and scholarly relations could overcome artisanal secrecy.[249] For as instruments embodied some of the knowledge involved in their construction, their movement and exhibition had the potential to distribute designs and techniques of production. At many early modern courts, collections of instruments were a resource untapped either for this purpose or for the pursuit of astronomical knowledge. But at a site like Dresden, where objects in the *Kunstkammer* were made available to craftsmen to study, this potential was recognised and exploited.

The early modern courts helped to sustain and develop instrument-making skills that were used for observational astronomy at just a handful of locations. But the relationship between the courtly aesthetic and the commercial instrument-making trade was reciprocal and complex.[250] With the emergence of professional instrument-makers, the production of instruments of increasing quality raised the expectations of the elite consumers, and stimulated their appetite for just this kind of commodity. And a drive to emulate the elite helped broaden the base of the instrument-making trade by creating a demand for instruments that were cheaper and simpler. In turn, this led to larger workshops and greater concentrations of expertise in particular urban environments, providing further opportunities for propagating craft techniques and developing the skills of the talented. Thus was born a virtuous circle of consumption that stimulated production that stimulated consumption. But given the range of ways in which instruments

---

[248] Bürgi obtained an Imperial privilege in 1602, for example, awarded for his surveying instrument and the book that he proposed to write describing it; but the publication was not forthcoming. Likewise he hesitated over publishing his mathematical innovations; leading Kepler to complain in the *Tabulae Rudolphinae* that he was a 'homo cunctator, & secretorum suorum custos . . .' See Leopold 1986, 31, 33; von Mackensen 1988, 33–34.

[249] There has been some debate about the extent to which artisanal traditions fostered secrecy and philosophical ones openness. See McMullin 1985; Long 2001. Iliffe 1992, points out the extent to which scholars had to be aware of modes of knowledge transmission.

[250] See Moran 1978, 150–157; Bedini 1980, 19–26; Chandler and Vincent 1980; Groiss 1980; Maurice 1980.

assisted in the production of, or embodied, such heterogeneous astronomical commodities as prognostications, letters, tables, books, world-systems, and practitioners, and the interrelationships between each of these goods, it is tempting to interpret the broader role played by the court setting in assisting the transformation of astronomy as the function of an analogous cycle of mutually reinforcing demand for, and supply of, 'heaven-bearers' of almost every description.

CHAPTER 5

Concluding remarks

Anyone who has passed through Frankfort at the time of the Fair, may say that in the same act he had been to Nuremberg. And not only the citizens of Nuremberg, but also those of Augsberg, Ulm, Strasburg, Brunswick, and of many other cities are in the habit of sending there each for himself a specimen of his handicraft.

Henri Estienne, *Francofordiense Emporium sive Francofordienses Nundinae* (1574)[1]

Instruments, letters, and books were all fundamental to the practice of astronomy in the early modern period. But instruments were not just devices for acquiring knowledge of the heavens, correspondence was not simply a means of maintaining contact with one's colleagues, and books were not only a way of presenting the final outcome of one's labours. Instruments and letters, as well as books, publicised astronomical endeavours, represented new theories, and made results available to others. Letters and books were important carriers of information about the celestial realm, and were read as part of the process of generating new knowledge. And the movement of books and instruments supplemented the role played by correspondence in establishing and sustaining relations with practitioners in other locations. Moreover, instruments, books and letters encompassed one another in a variety of complex and subtle ways. Letters were reproduced in books, or composed *de novo* within them. Books described the construction and use of instruments, or contained instruments of paper. Instruments advertised texts and prestigious documents, and were in their turn advertised through them. And letters assessed, and evaluated, both books and instruments.

Viewed within the early modern culture of display as embodiments of knowledge and labour, all of these objects were invested with a prestige value that was not necessarily proportional to their importance for the

[1] As translated by Thompson 1911, 163–165.

advancement of astronomy as an academic subject. By encouraging the production and distribution of all manner of goods of astronomy, this feature of the early modern cultural landscape helped to foster the skills required for the transformation of the subject, at the same time as it engendered a competitive economy of credit that shaped the behaviour of a wide range of practitioners. The dispute between Ursus and Tycho was one product of the courtly commodification of early modern astronomy; the manner in which the subject was patronised by Wilhelm IV was another.

In the early modern period, the movements of letters, books and instruments were necessarily dependent upon the movement of people. In Chapter 2, I described the wide range of individuals employed as letter-couriers by Tycho and his correspondents. Merchants, noblemen, and students such as Petrus Jacobi and Gellius Sascerides, were all pressed into service. Sascerides also played a role in the distribution of copies of Tycho's *De recentioribus phaenomenis* and, as shown in the Appendix, Tycho made use of his future son-in-law Franz Tengnagel to courier presentation copies of the *Mechanica* and his manuscript star-catalogue. A broader network for the movement of second-hand books, manuscripts, and gift-copies, that supplemented and exploited the structures of the commercial book-trade, functioned along the same *ad hoc* lines as the correspondence, making use of both serendipitous encounters with travellers, and the seasonal migrations of men of commerce. And in Chapter 4, we saw not only the commercial distribution of instruments, and their movement with individuals such as Eberhard Baldewein and Jost Bürgi, but also the fact of Tycho's own relocation to Prague with the instruments of his Danish observatory. Thus, as auxiliaries to the communication of ideas and expertise through material objects, individuals and their movements have appeared throughout *Bearing the Heavens*.

In passing, however, and in the last chapter more than just in passing, we have also noted the communications and collaborations that the movements of individuals enabled in and of themselves. The journeys made by Jacobi and Sascerides brought not only letters and books to Kassel, but also their knowledge of Tycho's instruments, methods and theories – knowledge that they were able to impart in discussion with Wilhelm and Rothmann. Having travelled onwards to Italy, Sascerides represented the Tychonic astronomical project there as well, assisting Magini with the production of a Tychonic sextant, and with the planning of certain observations. The correspondence between Hven and Kassel also resulted in Rothmann's trip to Uraniborg; and although there is reason to be cautious about the account of his meeting with Tycho given in the *Epistolae astronomicae*, we can be certain that the two astronomers found much to discuss. One topic of conversation was the visits of Paul Wittich and Ursus to their observatories, and the consequences of the disclosures and interventions of these highly mobile individuals for their respective astronomical projects.[2] In considering the production of globes and planetary models, we have noted the significance of the journeys of commercial instrument-makers, such as the van Langrens and Blaeu, and equally those of court instrument-makers like Baldewein and Bürgi. And we have seen, moreover, how sites such as Hven, Kassel, Dresden, and Prague, functioned as centres of expertise at which collaborations and forms of cooperation that transcended the social stratification distinguishing prince and nobleman from mathematician and mechanic, were made possible and encouraged. Not unexpectedly, perhaps, it has therefore become clear that the peregrinations and relocations of various practitioners had an importance extending beyond their role in assisting the distribution of objects. Appealing as it might be to imagine Tycho scurrying across Germany with the *Globus Magnus Orichalcicus* strapped to his back, we should not forget that he was described as bearing the heavens *in* his head, not *upon* it.

Of the many ways in which it would be possible to develop and extend the picture of late sixteenth-century astronomy contained in this book, therefore, one would be to pay more attention to two intertwined topics: the movements of individuals involved in the study of astronomy, and the significance of the sites between which they migrated. To some extent, construction of such an account would be relatively straightforward. One way of proceeding would be to anatomise the movements of individuals according to their cause and their outcome. As we have already seen, individuals made their way from one site to another for a variety of reasons. Sometimes they travelled to accomplish a specific objective, sometimes their relocation was prompted by unfortunate and unforeseen events, and sometimes movement was natural to the trajectory of their life and career. In the first category, for example, can be placed Elias Olsen Morsing's missions to Frauenburg and Königsberg to quantify the error in the latitudes of these sites as determined by Copernicus and Reinhold. In this case the outcome was information important for the interpretation of *De revolutionibus* and the *Tabulae Prutenicae*. Motion of this sort, with the purpose of re-evaluating the legacy of previous scholars, continued to be an important feature of astronomical practice, although another kind of motion popular amongst later astronomers, travel with the intent of observing

---

[2] *TBOO* VII, 323.5–324.4.

specific celestial phenomena, was not common in this period. Other forms of purposeful journey, with different kinds of outcome, include those made in order to present a book or other object; those prompted by the desire to consult a particular individual; those governed by the need to make a certain acquisition, for example a book or an instrument; and those with the intent of exploiting a specific resource, such as a library or printing press.

The second category of movement encompasses not only Tycho's relocation to Prague towards the end of his life, and Kepler's expulsion from Graz, which encouraged his entry into Imperial service, but also the migration of talented craftsmen and cartographers from the Spanish Netherlands, and Arnold van Langren's flight from the United Provinces to escape his creditors. The troubles of the times, the vagaries of patronage, and difficult personal circumstances, could all make it inadvisable or impossible to remain in one place. The times were such, indeed, that once one has noticed this phenomenon it is not hard to find other equally striking examples: the employment of Michael Maestlin at Tübingen followed Philipp Apian's refusal to subscribe to the Formula of Concord; the wanderings of Giordano Bruno, which unhappily ended in his execution for heresy, were made prudent by the highly heterodox nature of his philosophy; and John Dee was expelled from Prague after representations made by the Papal Nuncio to the Emperor.[3] The fact that such events and outcomes were contingent on the unique combination of local personalities and interests teaches an important lesson about the extent to which particular social structures and cultural conventions were both significant and enabling. Patronage, for example, sometimes protected individuals against charges of impropriety and animosity caused by confessional differences. Unlike Dee, Tycho seems to have remained untouched by the enmity of Catholic residents of Prague, in particular that of the Capuchins he was accused of instructing Rudolf to banish; and Kepler, perhaps possessing Jesuit support, or an ally at court, was for some time exempted from the 1598 dismissal from Styria of Lutheran schoolmasters.[4] But Galileo appears to have miscalculated the degree to which his flair as a client could secure a tolerant reception for his advocacy of heliocentrism; and Caspar Peucer's service at the court of Elector August of Saxony, and cordial relations with Landgrave Wilhelm, neither prevented him from being imprisoned for crypto-Calvinism, nor secured his early release from confinement.[5] At some sites, for certain periods of time, particular practitioners were liberated by

---

[3] *DSB* IX, 167–170; Yates 1964; Dibner 1967; Clulee 1988, 225.
[4] Thoren 1990, 445; Caspar 1993, 79–80.   [5] Biagioli 1993, 313–352; Leopold 1986, 19, 30.

patronage from the constraints imposed by the custodians of orthodoxy, allowed to transcend the hierarchical disciplinary distinctions of the universities, and given the opportunity to pursue expensively ambitious astronomical projects. But this liberty, in effect the freedom to remain and practice as one willed, where one willed, was far from universal.

The third category of motion provides further reason to qualify statements about, for example, the particular importance to astronomy of the early modern court. Mobility between sites of practice and types of such was normal to the career path of many practitioners, so individuals frequently brought to one location skills, and ideas, obtained at another. Princes inclined to support instrument-making could draw on the pool of talented men trained in commercial workshops; in similar fashion, individuals of a mind to patronise astronomy had recourse to the university-taught medics and mathematicians of early modern Europe. With the formative experiences of so many practitioners occurring at a site other than the one at which they were employed, the necessity of examining the nature of their intellectual heritage and vocational training is surely not in question. It was no coincidence that Rothmann and Tycho shared a belief in the importance of the providential ordering of the cosmos as a guide and motivation for astronomical study; even if they internalised this doctrine in subtly different ways, it was the common legacy of their education in the Philippist schools and universities of Reformed northern Europe. Astronomical work was encouraged, and enabled, by the *series* of environments through which individuals travelled.

In addition to studying the sequential propagation of knowledge and skills, it is important to take equal account of expertise that was simultaneously accessible in two or more locations. Certain princes exerted a strong influence over the educational institutions in their domain, thereby making the expertise located in these foundations available to their courts. It was Wilhelm IV's control of the University of Marburg, for example, that enabled the emplacement there of Victorin Schönfeld, and his collaboration with the Landgrave mentioned in the second of our chapters. It is notable, moreover, that when Rothmann considered himself accused of impiety by Tycho, he responded in such a way as to suggest that the theologians of the university were also readily available for his consultation: he asserted that both Wilhelm and his theologians had read his account accommodating scripture to the theory of Copernicus, and had found nothing in it that was at all worthy of censure.[6] Courts also shared expertise with the commercial urban environment – for although princely support of workers

---

[6] *TBOO* VI, 181.31–36. See also Moran 1978, 80.

without guild membership, and the consequent distinction of court artisans from town craftsmen, might be a cause of tension, it was nevertheless
possible for individuals to hold a respected position in both of these realms.
Christoph Trechsler (*c.*1550–1624), who served at the Dresden court, was
not initially accepted into the town's guild of locksmiths, spur-makers,
nailsmiths, clock-makers and gunsmiths; from 1595 onwards, however, he
was working for the town council at the same time as he was employed in
the Electoral *Kunstkammer.*[7]

Assessing the cause and outcome of the movements of a particular individual is a task that can be accomplished through the biographical mode
of writing. The importance of such studies as the foundation for investigating communication is illustrated by the extent to which they have been
cited in each of the chapters of *Bearing the Heavens.* But the examples presented here suggest that bringing more of this material to the fore would
supplement and qualify our account of the constitution and operation of
astronomical communities in several ways that are important. The same can
be said of studies that explore the forms of collaboration, practical expertise and scholarly traditions characteristic of particular locations. Thus,
although the terminology of the *domesticus* and the topography of Tycho's
observatory both indicate that the operation on Hven was more closely
modelled on the traditions of the noble household, and princely court,
than the contemporary academic environment, accounts of Uraniborg as
an early modern research institute offer an analogy useful for drawing out
the fact of a concentration of individuals working together to advance a
particular field of study. It is important too, to recognise that collaboration with Tycho was often an intermediary stage in the academic career
of his assistants: Longomontanus, Pontanus, and others, went on to distinguish themselves as holders of university positions.[8] Studies of Kassel
and Prague have drawn out the relationship between the various scholarly
activities pursued there, and their subordination to the imperative of man-
ifesting princely status and power.[9] There is much to be gained as well
from accounts of the discourses and debates current at universities such as
Wittenberg and Tübingen.[10] And analyses of Nuremberg and Augsburg as
key sites of the commercial instrument-making trade, of Frankfurt as the
location of expertise and materials for the early modern book-trade, and
of these places and other urban locales as communications centres crucial

---

[7] Schillinger 1990, 283–288.　　[8] As shown by Thoren 1985; Christianson 2000.
[9] Moran 1978, 1980, 1981, 1982; Leopold 1986; Evans 1997; Fucikova 1997b.
[10] Westman 1975a; Hofmann 1982; Kusukawa 1995; Methuen 1998.

for enabling individuals to treat with one another, and objects to circulate, need to be pursued, and utilised, with equal vigour.[11]

Yet if the further integration of such work into *Bearing the Heavens* would further our understanding of its topic, the converse, that this study sheds much light on the role of particular people and locations, should also be apparent. The bridging of the geographical interval between individuals, and of the intellectual gaps between different institutional environments, was an important consequence of the production and circulation of the letters, books, and instruments that I have described and considered. Tycho maintained correspondence with university professors and schoolmasters, court physicians and *mathematici*, nobles and patricians. He read and assessed the books produced by a similar range of individuals; he drew on the practical experience and skills of commercial printers and technicians. Study of communicative practices deepens our understanding of the career of individuals, as well as the manner in which knowledge was produced at sites like Uraniborg and Kassel. It also suggests that we can, and should, extend our research beyond the familiar boundaries of traditional grand narratives such as the 'Astronomical' or 'Copernican Revolution'.[12] Narratives of this kind promote the study of a sequence of great individuals – e.g., from Copernicus to Newton, via Tycho and Kepler. But they tend to distinguish between these individuals and their contexts in ways that unduly constrain. As a reason for studying the astronomical communities of the early modern period, the fact that they gave rise to and nourished the likes of Tycho and Kepler will only take us so far.

To anyone wishing to develop the topic of this book, therefore, I would recommend pursuing a historical study restricted neither by the *termini* of an individual's career, nor by the horizons of a single location. Such a work would seek to draw out the extent and depth of contacts between astronomical practitioners of many varieties, displaying multiple connections of the type identified with respect to Arnold van Langren and the question of how he acquired his knowledge of Wilhelm. This was a question the possible answers to which, it will be recalled, involved links to, and between, such figures as Tycho, Petrus Ramus, Heinrich Rantzau, and Rodolphus Snellius. But in shifting focus from individual people and places to the objects and events that united them – from the bones and organs of the astronomical community, we might say, to the blood vessels and connective tissue of this metaphorical body – it is not clear that the traditional narrative or analytic

[11] E.g. Thompson 1911; Bobinger 1954; Strauss 1976; Groiss 1980; Gouk 1988; Smith 1983; Lindner 1987; Seelig 1995; von Stromer 1997.
[12] Kuhn 1957; Koyré 1973.

forms are the most useful way of proceeding. One way forward might be to exploit the possibilities presented by the new media of our age, producing hypertextually linked narratives capable of encompassing transitions in practice, the evolution of ideas, the overlapping successions of individual practitioners, and the changing significance of particular locations. The product of many individuals' efforts, such a project could offer a new way for historians themselves to collaborate in the construction of accounts of the past.

For some time now, historians of sixteenth-century astronomy have tended – when they have chosen to consider groups of practitioners rather than a sequence of singular individuals – to focus on the community comprised by those who read, understood and responded to Copernicus. I do not wish to deny the value of this work or the standard of scholarship that pursuing it has required.[13] But it should now be apparent that the group identified by concentrating on this particular activity is somewhat problematic. Following Ursus' translation of *De revolutionibus* for Bürgi, this extraordinary autodidact, instrument-maker, observer, and mathematician could be considered to be a member of this group, whereas it is less obvious that the enterprise was one that engaged Landgrave Wilhelm IV.[14] But were one to adopt Rothmann's position, then almost all university professors of mathematics, even Michael Maestlin, could be excluded from this community on the grounds that they possessed insufficient understanding of Copernican planetary theory. The identification of other shared practices, beliefs and aspirations provides a way of differentiating a range of overlapping communities engaged in the study and representation of the heavens. Having utilised correspondence as a key resource in writing this book, it seems quite natural to want to constitute an astronomical community in terms of the participation of individuals in the exchange of letters on the subject. But we have also seen such different motivations for astronomical work as the pursuit of contemporary credit and posterior fame, the commercial exploitation of astronomical knowledge, and the desire to appreciate the providential construction of the cosmos. We have encountered shared and divergent practices of maintaining control over intellectual property, of establishing and sustaining links with other practitioners and supporters, and of producing, representing, and interpreting, astronomical data and cosmological theories. And we have seen the various ways in which astronomical knowledge was constructed from, and combined with, the doctrines and skills of a number of other fields: not just mathematics and

---

[13] E.g. Jarrell 1989; Gingerich 2002a.   [14] See Hamel 1998, 111–173.

philosophy, but also theology, alchemy, horology, and cartography. One lesson of such diversity is that, as we reorient the focus of historical investigation so as to consider the distribution and evolution of the full range of practices and theories implicated in the study of the heavens, we shall arrive at a deeper understanding of the culture of early modern astronomy in all its variety. Much of this book has been about Tycho Brahe, a figure about whom there is still more to be written; but, in the final analysis, he was only one astronomer among many, even if he was described as a prince. We need therefore to press on with studying the astronomy of the early modern period via a history of communication; a history in which we consider both who was communicating with whom, and how, as well as what it was that they said.[15] For the history of communication must be a part, and an important one, of the history of science as a practice. Only a history that encompasses the transmission and evolution of techniques and technologies, as well as the sharing and evaluation of data and ideas, can claim to represent the *culture* of science, and hence account for what is taken to be its product, knowledge of the world.

[15] On 'history of communication', see Darnton 2000.

Appendix: Known and presumed owners of Tycho's works prior to 1602

I SCHOLARS

Recipient	DRP	EA	AIM	Catalogus
Jacques Bongars		(1597)		
Heinrich Brucaeus	1588			
Joachim Camerarius the Younger	1590	1590		
David Chytraeus		1597		
John Craig	1588/89			
John Dee	(1590)			
Thomas Digges	(1590)			
Andreas Dudith	1588			
David Fabricius		1597	1598	
Galileo Galilei		(1599)		
Johannes Grynaeus			(1598)	
Thaddaeus Hagecius	1588	1590	1599	
Johannes Jessenius			(1600)	
Johannes Kepler		(1599)/1600	1599/1600	
Jakob Kurtz	(1589)	1590		
Duncan Liddel	*	*		
Michael Maestlin	1588			
Giovanni Antonio Magini	1589/90	1600	1598/99	1598/99
Paul Melissus			1599/1600	
Jacob Monaw		1598	*	
Caspar Peucer	1588	1590		
Vincenzo Pinelli	(1589)	(1590)	(1600)	
Johannes Praetorius	1588			
Heinrich Rantzau	1588	*	1598	
Christian Hansen Riber		*		
Georg Rollenhagen	1588		1598	
Helisaeus Röslin	(1588)	(1597)		
Christoph Rothmann	1588			
Gellius Sascerides	1588			
Thomas Savile	1590	(*)		

(cont.)

Recipient	*DRP*	*EA*	*AIM*	*Catalogus*
Joseph Scaliger			1598	1598
Bartholomaeus Scultetus	1588	(1600)		
Friedrich Taubmann			1599	
Nicolai Reymers Baer = Ursus		(1597)		
Matthias Wacker von Wackenfels	1599			

Note: Columns give the year of significant dates or dates when information about the text's presentation and ownership is recorded; if no date is known an asterisk is entered. Copies of the *Epistolae astronomicae* presented in 1590 were partial publications of the correspondence. Brackets indicate something other than a straightforward presentation of a text by Tycho. For details, see below.

The division of owners into scholars and aristocrats is somewhat arbitrary, but of some value in indicating Tycho's likely motive for presenting a copy of the text to a given individual.

Abbreviations: *DRP* represents *De recentioribus phaenomenis*, *EA* the *Epistolae astronomicae*, and *AIM* the *Astronomiae instauratae mechanica*. An asterisk after *AIM* indicates that the manuscript *Catalogus* was also included in one or more of the references. The locations of extant copies, when known, are given in parentheses. Other abbreviations are *d.* for date despatched, *a.* for date arrived, *c.* for date confirmation of arrival/ acquisition provided, *pl.* date pledged, *l.* date loaned, *p.* date presented. All despatches and presentations, etc., were by Tycho unless otherwise indicated.

Bongars: *EA*, *c.* 30 March 1597, probably purchased at Frankfurt, not presented; see *TBOO* VII, 383.8–9.

Brucaeus: *DRP*, *c.* 20 September 1588. *TBOO* VII, 142.4–7.

Camerarius: *DRP*, *c.* and *EA*, *pl.* 21 October 1590. This date refers to a letter of Tycho to Camerarius, replying to one sent 'around the vernal Equinox of this year' that had evidently acknowledged receipt of *De recentioribus phaenomenis*. See *TBOO* VII, 276.6–277.10.

Chytraeus: *EA*, *c.* 16 June 1597. See *TBOO* VIII, 3.21–24.

Craig: *DRP* (Edinburgh University Library), *d.* 2 November 1588, *c.* May 1589. See Nørlind 1970, 122; *TBOO* VII, 175.10–17.

Dee and Digges: *DRP*, *d.* 1 December 1590. Tycho sent two copies to Thomas Savile, and asked that he show them to Dee and Digges so that they might form an opinion of the text's contents. He did not state that ownership of these copies was to be transferred to them. *TBOO* VII, 284.15–16, 284.37–285.3.

Dudith: *DRP* (Kungliga Biblioteket, Stockholm), *d.* by 17 August 1588. In his letters to Scultetus and Hagecius of this date, Tycho asked both to greet Dudith on his behalf, and declared himself eager to know the bishop's opinion of his work. See *TBOO* VII, 123.11–16, 124.11–14; Nørlind 1970, 123.

Fabricius: *EA*, *d.* before June 1597. According to his letter to Kepler of 14 January 1605, Fabricius was sent this copy before Tycho left Denmark. See *KGW* XV, 121.242–243. *AIM* (Kiel Universitätsbibliothek), *p.* 1598. See Nørlind 1970, 288.

Galilei: *EA*, *d.* 1599. This is to follow Nørlind 1970, 236; he assumes that Franz Tengnagel presented Galileo with a copy of the letter-book in Padua, when he referred to him instead of Vincenzo Pinelli. But Tycho's later letter to Galileo indicates only that Tengnagel and Galileo discussed the work together; Galileo could have purchased it himself. See *TBOO* VIII, 226.20–33, 311.20–40.

Grynaeus: *AIM* (Danmarks Natur og Laege Videnskabelige Bibliotek, Copenhagen), *d.* 7 June 1598. This copy was destined for Tycho's nephew Otto Brahe, but the astronomer instructed the courier that Grynaeus should be permitted to examine it *en route*. See *TBOO* VIII, 70.27–37; Nørlind 1970, 288. Grynaeus was a professor at Basle.

Hagecius: *DRP*, *d.* 3 May 1588, *a.* July 1588. See *TBOO* VII, 120.19–22, 145.14–15. *EA* partial copy, *d.* 23 February 1590. See *TBOO* VII, 225.41–226.4. *AIM* (British Library), *p.* 14 January 1599. See Nørlind 1970, 289.

Jessenius: *AIM* (Oslo Deichmanske Bibliotek), *d.* 7 May 1600. This copy was to be presented to Samuel Mosbach. Tycho suggested that Jessenius complete the dedication (he was uncertain of Mosbach's title), and have the book bound. See Nørlind 1970, 289; *TBOO* VIII, 316.34–39. Jessenius was a professor of medicine at Wittenberg.

Kepler: *EA*, *l.* 29 August 1599 by Herwart von Hohenburg, according to a letter of this date; however Kepler subsequently complained that he had not received it. See *KGW* XIV, 61.80–81, 75.518–522. *EA* and *AIM*, *d.* 9

December 1599, *c.* 12 July 1600. See *TBOO* VIII, 207.36–39; *KGW* XIV, 93.170–172, 130.100–101.

Kurtz: *DRP, l.* 1589 by Hagecius. See *TBOO* VII, 201.24–31. *EA* partial copy, *d.* 23 February 1590, via Hagecius. See *TBOO* VII, 225.41–226.4.

Liddel: *DRP* (Aberdeen University Library), *EA* (Aberdeen University Library). Dates of presentation and despatch are unknown, but Liddel is thought to have visited Hven in June 1587. See Wightman 1962, II, 42–44.

Maestlin: *DRP* (British Library), *d.* 14 May 1588, *a.* 15 August 1588. See Gingerich and Westman 1988, 60.

Magini: *DRP, d.* autumn 1589, via Gellius Sascerides, *c.* 13 September 1590. Favaro 1886, 193; *TBOO* V, 125.8–12. *EA* partial copy, *d.* 1590, *c.* 4 March 1600. *TBOO* VIII, 252.14–18. *AIM**, *d.* 28 November 1598, *c.* 4 November 1599. This copy was presented to Magini by Franz Tengnagel. See *TBOO* VIII, 123.33–37, 190.10–22.

Melissus: *AIM, d.* early 1599, *c.* 5 March 1600. *TBOO* VIII, 248.29–33, 267.32–37.

Monaw: *EA, c.* 7 March 1598. *TBOO* VIII, 30.4–7. *AIM* (Wroclaw Biblioteka Uniwersytecka). Date of presentation/despatch unknown. See Nørlind 1970, 291.

Peucer: *DRP, d.* 13 September 1588. *TBOO* VII, 131.23–25. *EA* partial copy, *d.* 1590. *TBOO* VII, 131.23–25.

Pinelli: *DRP, l.* before February 1589 by Joachim Camerarius. See Nørlind 1970, 126; *TBOO* VII, 276.14–22. *EA, pl.* 21 October 1590. The pledge was made to Camerarius. Nørlind 1970, 236, notes later evidence which indicates that the Paduan did acquire the work: he mentioned it in correspondence with the Leiden botanist Carolus Clusius (1526–1609) in 1599. *AIM*, possible *l.* after 3 January 1600. On this date Tycho suggested that Pinelli ask the Venetian patrician Giovanni Francesco Sagredo if he wished to see the work; I have assumed that this is a reference to one of the two copies sent to the Doge and Senate, rather than an indication that Sagredo had received a copy of his own. See *TBOO* VIII, 228.22–32. On Pinelli, see Grendler 1980.

Praetorius: *DRP* (Wroclaw Biblioteka Uniwersytecka), *d.* 18 September 1588. See Nørlind 1970, 124.

Rantzau: *DRP, d.* 13 September 1588. On this date, Tycho sent Rantzau two *further* copies of the work, making the number despatched nine in total. See *TBOO* VII, 127.9–12. *EA, p.* 1596? Since Rantzau sent one to Ernst of Cologne, I consider it unlikely that he was not also provided with a copy. *AIM, p.* 1598? Since Tycho completed the *Mechanica* as a guest of Rantzau's, I am assuming that he presented his host with a copy.

Riber: *EA* (Lunds Universitet Bibliotek). Nørlind 1970, 233.

Röslin: *DRP, l.* December 1588 by Maestlin. See Granada 2002, 284–288. *EA, d. c.* 30 March 1597, by Bongars, not Tycho. See *TBOO* VII, 383.8–9.

Rollenhagen: *DRP, d. c.* 13 September 1588. Rollenhagen was sent the sketch of the new world-system excerpted from the text by Rantzau, and it seems probable that, like Peucer, he was subsequently sent the complete work. *AIM, d.* or *p.* 1598. See Nørlind 1970, 286.

Rothmann: *DRP, d.* June 1588. *TBOO* VI, 132.36–38. *EA*, c. 5 April 1600, in a letter from Tycho to Rollenhagen.

Sascerides: *DRP* (Det Kongelige Bibliotek, Copenhagen), *p.* 20 March 1588. Nørlind 1970, 122.

Savile: *DRP* (Bodleian Library), *d.* 1 December 1590. Tycho sent two copies of the work to Savile, and asked him to solicit the opinion of others regarding it, particularly Dee and Thomas Digges. See *TBOO* VII, 284.15–16, 285.1–3; Nørlind 1970, 123. *EA* (Bodleian Library). Ownership of this copy has been assigned to Thomas Savile, but he died in 1593. It is possible that Tycho despatched a copy to Savile in ignorance of this fact, or sent one to his elder brother Henry. Nørlind 1970, 233.

Scaliger: *AIM** (Leiden Universiteitsbibliotheek), *d.* 23 August 1598. *TBOO* VII, 107.8–10; Nørlind 1970, 288, 295.

Scultetus: *DRP, d.* May 1588. *TBOO* VII, 123.27–31. *EA, d.* by 7 January 1600? Tycho wrote to Scultetus in 1600, to enquire about the possibility of the printers in Görlitz finishing his *Progymnasmata* and *De recentioribus phaenomenis* (by printing his refutation of John Craig), as well as his second book of *Epistolae astronomicae*. He stated that the mathematician undoubtedly had a copy of his first book of letters, although whether this was because he had been given one was left unclear. See *TBOO* VIII, 235.9–39.

Taubmann: *AIM, p.* during 1599. Nørlind 1970, 286. Taubmann was professor of poetry at Wittenberg.

Ursus: *EA, c.* 1 June 1597, in a letter of complaint to the Emperor. See Jardine and Launert 2005, 90–92.

Wacker von Wackenfels: *DRP* (Wroclaw Biblioteka Uniwersytecka), *p.* July 1599. See Nørlind 1970, 124. Wacker von Wackenfels is better known to historians of science as the dedicatee of Kepler's *Strena seu de Nive Sexangula* (1611).

II PRINCES, ARISTOCRATS, AND CHANCELLORS

Recipient	EA	AIM	Catalogus
Otto Steenson Brahe, Tycho's nephew		1598	
Christian II, Elector of Saxony		1600?	1600
Christian IV, King of Denmark	1596	1598	1598
Rudolph Corraducius, Imperial Prochancellor		1600	1600
Ernst, Archbishop-Elector of Cologne	1597	1598	1598
Ferdinand, Archduke of Austria		*	
Ferdinand II de' Medici, Grand Duke of Tuscany		1598/99	1598/99
Friedrich Wilhelm, Duke of Saxe-Altenburg		*	*
Christian Friis, Chancellor of Denmark	1596		
Anton II Fugger, Augsburg patrician	(*)		
Philip Eduard Fugger, Augsburg patrician		1600	
Marino Grimani, Doge of Venice		1598	1598
Heinrich Julius, Duke of Braunschweig-Wolfenbüttel		*	*
Johann Georg Herwart von Hohenburg, Chancellor of Bavaria	(1599)	1599	1599
Johann Adolf, Duke of Holstein-Gottorp	1596	1598	1598
Sdenko Popel von Lobkowitz, Chancellor of Bohemia		1600	
Matthias, Archduke of Austria		1600	1600
Maurice of Nassau, Stadtholder of the United Netherlands		1598	1598
Maximilian I and Wilhelm V, Dukes of Bavaria		1598	1598
Moritz, Landgrave of Hesse-Kassel	1597	1600	
Samuel Mosbach, Chancellor of Saxe-Altenburg		1600	
Otto II, Duke of Braunschweig-Harburg		*	
Oldrich Desiderius Pruskovsky z Pruskova, Bohemian noble		1601	
Wolf Dietrich von Raitenau, Archbishop of Salzburg		*	*
Holger Rosenkrantz, Danish noble		1598	
Petr Vok z Rozmberka, Bohemian noble		1600	
Rudolf II, Holy Roman Emperor		1599	1599
Johann Sitsch von Stübendorf, Bishop of Wroclaw		*	
Ulrich, Duke of Mecklenburg.		1598	1598
The Venetian Senate		1598	1598
Jan Zbynek z Hazmburk, Bohemian nobleman		*	

Note: Abbreviations as before.

Brahe: *AIM* (Danmarks Natur og Laege Videnskabelige Bibliotek, Copenhagen), *d.* 7 June 1598, via Johannes Grynaeus. See Nørlind (1970), 288, where this copy is described as extant in the Det Kongelige Bibliotek, Copenhagen; formerly a branch of this library, the *DNLB* became an independent library in 1986.

Christian II, Elector of Saxony: *AIM**, *d.* 6 May 1600. Unusually, the *Mechanica* itself is not mentioned as being sent, although a copy was sent via the same intermediary, Johannes Jessenius, to be presented to the Chancellor of Saxe-Altenburg. See *TBOO* VIII, 313.24–49, 316.25–30.

Christian IV, King of Denmark: *EA*, *d.* December 1596. Nørlind 1970, 234–235. *AIM**, *d.* 8 May 1598, via Holger Rosenkrantz.

Corraducius: *AIM** (Österreichische Nationalbibliothek, Vienna), *d.* 8 March 1600. See *TBOO* VIII, 259.5–7; Nørlind 1970, 291, 296.

Ernst, Archbishop-Elector of Cologne: *EA*, *c.* 13 January 1597, transmitted via Rantzau. See *TBOO* XIV, 310.24–28. *AIM**, *c.* 25 June 1598, presented by Franz Tengnagel. *TBOO* VIII, 71.28–36, 80.15–18.

Ferdinand, Archduke of Austria: *AIM* (Det Kongelige Bibliotek, Copenhagen). This copy is not listed in Nørlind 1970, 288. However inspection suggests that it has either been repaired, resulting in a reversal of the orientation of the manuscript dedication with respect to the title-page, or tampered with, so that a gift-copy presented to one individual now contains a dedication to another.

Ferdinand II, Grand Duke of Tuscany: *AIM** (Det Kongelige Bibliotek, Copenhagen), *d.* 8 November 1598, *c.* June/July 1599. *TBOO* VIII, 119.3–7, 156.14–17; Nørlind 1970, 288, 295.

Friedrich Wilhelm, Duke of Saxe-Altenburg: *AIM** (Forschungsbibliothek, Gotha). See Nørlind 1970, 287–288.

Friis: *EA*, *d.* 31 December 1596. *TBOO* XIV, 100.36–44.

Anton Fugger: *EA* (University of Texas, Austin). The stamped arms on the cover bear the date 1586, but this must be a feature of the stamp alone. It seems likely that this copy was purchased rather than presented. See Nørlind 1970, 233.

Philip Eduard Fugger: *AIM*, *d.* September 1600. The copy was presented by Herwart von Hohenburg on Tycho's behalf. See *TBOO* VIII, 354.19–22.

Grimani: *AIM** (Bodleian Library, Oxford), November 1598. This copy was taken to England by Sir Henry Wotton (1568–1639). See Nørlind 1970, 289–290, 295.

Heinrich Julius, Duke of Braunschweig-Wolfenbüttel: *AIM** (Niedersächsische Staats- und Universitäts Bibliothek Göttingen). See Nørlind 1970, 288, 295.

Herwart von Hohenburg: *EA, c.* 23 July 1599. Nørlind 1970, 235, suggests that Herwart von Hohenburg purchased his copy at the Frankfurt bookfairs. However, Herwart also indicated that he had read the *De recentioribus phaenomenis*, and where he obtained this remains unclear. See *TBOO* VIII, 157.34–46. *AIM** (Universitätsbibliothek, Munich), *pl.* 31 August 1599. Tycho promised to send three copies on this date, two of them for the Bavarian dukes Maximilian and Wilhelm, but said that, at that precise moment, he lacked the colourist required to complete them. See *TBOO* VIII, 162.25–29; Nørlind 1970, 289.

Johann Adolf, Duke of Holstein-Gottorp: *EA* (Det Kongelige Bibliotek, Copenhagen), *d./p.* before 29 August 1596. Nørlind 1970, 232–233, proposes this *terminus ante quem* on the grounds that the extant copy is dedicated to Johann Adolf as Bishop of Bremen, a post that the duke then ceded to his brother Johann Friedrich. *AIM** (*Catalogus* only extant, Det Kongelige Bibliotek, Copenhagen), *d./p.* before 10 August 1598. See *TBOO* VIII, 89.5–29; Nørlind 1970, 295.

Popel von Lobkowitz: *AIM, p.* 8 August 1600. This copy was extant in Cambrai before World War II. It may have been despatched to the Chancellor in May 1600, and subsequently furnished with Tycho's manuscript dedication. On both points, see Nørlind 1970, 295.

Matthias, Archduke of Austria: *AIM** (Österreichische Nationalbibliothek, Vienna), *d.* 2 May 1600. *TBOO* VIII, 309.11–18; Nørlind 1970, 291, 296.

Maurice, Stadtholder of the United Netherlands: *AIM**(Bibliothèque nationale, Paris), *d.* May 1598, *c.* 6 July 1598. *TBOO* VIII, 67.10–18; Nørlind 1970, 290, 296.

Maximilian I and Wilhelm V, Dukes of Bavaria: *AIM** x 2, *pl.* 31 August 1599, in a letter to Herwart von Hohenburg.

Moritz, Landgrave of Hesse-Kassel: *EA*, by June 1597. *TBOO* XIV, 107.44–48. *AIM, d.* 10 November 1600. *TBOO* VIII, 387.41–388.18.

Mosbach: *AIM* (Oslo Deichmanske Bibliotek), *d.* 7 May 1600, via Jessenius. See Nørlind 1970, 289.

Otto II, Duke of Braunschweig-Harburg: *AIM* (National Library of Russia, St Petersburg). Christianson 2000, 227; Nørlind 1970, 288.

Pruskovsky z Pruskova: *AIM* (Narodni Muzeum, Prague), *d.* March 12, 1601. See Hadravova, Hadrava and Shackelford, 1996, xii–xiii.

von Raitenau: *AIM** (Bibliothèque nationale, Paris). Nørlind 1970, 290, 296.

Rosenkrantz: *AIM*, *d.* 8 May 1598. Tycho first indicated that this copy was for Rosenkrantz, but then said that it should be presented to King Christian; since the king did not much like astronomy, he thought that Rosenkrantz would probably be allowed to keep it. But if the king did accept the gift, then Rosenkrantz would be sent another copy. It seems, however that Rosenkrantz was reluctant to make the presentation, and Tycho later promised him that his second copy would also be coloured. The question of whether Rosenkrantz had or had not presented the text continued to be an issue, and eventually Tycho turned for others to assistance. See *TBOO* VIII, 64.23–25, 66.31–40, 88.37–89.29, 135.9–17; Christianson 2000, 225, 344–346.

Vok z Rozmberka: *AIM* (Det Kongelige Bibliotek, Copenhagen), *p.* 18 September 1600. Nørlind 1970, 288.

Rudolf II, Holy Roman Emperor: *AIM**, *p.* before 9 September 1599. See *TBOO* VIII, 165.41–166.11.

Sitsch von Stübendorf: *AIM* (Herzogin Anna Amelia Bibliothek, Weimar). Nørlind 1970, 290.

Ulrich, Duke of Mecklenburg: *AIM* (Universitätsbibliothek Rostock), *d.* before 10 August 1598; *(Forschungsbibliothek, Gotha), *d.* 14 September 1598. Nørlind 1970, 290 and 295.

Venetian Senate: *AIM** (Biblioteca Nazionale Marciana, Venice), *d.* before November 1598, via Franz Tengnagel, assuming that it was despatched at the same time as the copy to Grimani. Nørlind 1970, 290–291, 296.

Zbynek z Hazmburk: *AIM* (Strahovksy Klaster, Prague). Hadravova, Hadrava and Shackelford 1996, xiii–xiv.

References

Agrimi, J. and C. Crisciani, 1994. *Les Consilia Médicaux*, Typologie des sources du moyen âge occidental 69, Turnhout: Brepols.

Aiton, E. 1977. 'Johannes Kepler and the *Mysterium Cosmographicum*', *Sudhoffs Archiv* 61, 173–194.

1981. 'Celestial Spheres and Circles', *History of Science* 19, 75–114.

1987. 'Peurbach's *Theoricae novae planetarum*: A Translation with Commentary', *Osiris* 3, new series, 5–43.

Allen, D. C. 1966. *The Star-Crossed Renaissance: The Quarrel About Astrology and its Influence in England*, London: Frank Cass and Co. Ltd.

Allen, E. J. 1972. *Post and Courier Service in the Diplomacy of Early Modern Europe*, The Hague: Martinus Nijhoff.

Ami, A. and V. de Michele and A. Morandotti, 1985. 'Towards a History of Collecting in Milan in the Late Renaissance and Baroque Periods', in Impey and MacGregor 1985, 24–28.

Anderson, R. and J. Bennett, and W. Ryan, eds. 1993. *Making Instruments Count: Essays on Historical Scientific Instruments Presented to Gerard L'Estrange Turner*, Aldershot: Variorum.

Anglo, S. 1990. 'Humanism and the Court Arts', in A. Goodman and A. Mackay, eds., *The Impact of Humanism on Western Europe*, London: Longman, 66–98.

Apian, P. 1540. *Astronomicum Caesareum*, Ingolstadt: P. Apian.

Armstrong, E. 1990. *Before Copyright: The French Book-privilege System 1498–1526*, Cambridge: Cambridge University Press.

Ashworth, W. 1990. 'Natural History and the Emblematic World-view', in Lindberg and Westman 1990, 303–332.

Aslachus, C. 1597. *De natura caeli libri triplicis libelli tres*, Siegen: n. p.

Baade, W. 1945. 'β Cassiopeiae as a Supernova of Type I', *Astrophysical Journal* 102, 309–317.

Babicz, J. 1987. 'The Celestial and Terrestrial Globes of the Vatican Library, Dating from 1477, and their Maker Donnus Nicolaus Germanus (*c.*1420–*c.*1490)', *Der Globusfreund* 35/37, 155–166.

Baldini, U. 1983. 'Christoph Clavius and the Scientific Scene in Rome', in G. Coyne, M. Hoskin and O. Pedersen, eds., *The Gregorian Reform of the Calendar: Proceedings of the Vatican Conference to Commemorate its 400th Anniversary*. Vatican City: Specola Vaticana, 137–169.

Balsamo, L. 1990. *Bibliography: A History of a Tradition*, Berkeley: Bernard M. Rosenthal.

Barker, P. 1985. 'Jean Pena (1528–1558) and Stoic Physics in the Sixteenth Century', in R. Epp, ed., *Recovering the Stoics*, The Southern Journal of Philosophy Supplement 23, Memphis: Memphis State University, 93–107.

    1991. 'Stoic Contributions to Modern Science', in M. Osler, ed., *Atoms, Pneuma and Tranquility. Epicurean and Stoic Themes in European Thought*, Cambridge: Cambridge University Press, 135–154.

    1993. 'The Optical Theory of Comets from Apian to Kepler', *Physis* 30, 1–25.

    1997. 'Kepler's Epistemology', in D. Di Lisca, E. Kessler, and C. Methuen, eds., *Method and Order in Renaissance Philosophy of Nature. The Aristotle Commentary Tradition*, Aldershot: Ashgate, 355–368.

    1999. 'Copernicus and the Critics of Ptolemy', *Journal for the History of Astronomy* 30, 343–358.

    2000. 'The Role of Religion in the Lutheran Response to Copernicus', in M. Osler, ed., *Rethinking the Scientific Revolution*, New York: Cambridge University Press, 59–88.

    2004. 'How Rothmann Changed his Mind', *Centaurus* 46, 41–57.

Barker, P. and B. Goldstein, 1984. 'Is Seventeenth-Century Physics Indebted to the Stoics?', *Centaurus* 23, 148–164.

    1988. 'The Role of Comets in the Copernican Revolution', *Studies in History and Philosophy of Science* 19, 299–319.

    1995. 'The Role of Rothmann in the Dissolution of the Celestial Spheres', *British Journal for the History of Science* 28, 385–403.

    1998. 'Realism and Instrumentalism in Sixteenth-Century Astronomy: A Reappraisal', *Perspectives on Science* 6, 232–258.

    2001. 'Theological Foundations of Kepler's Astronomy', *Osiris* 16, 88–113.

Basore, J. trans., 1928–1935. *Seneca. Moral Essays*. Loeb Classical Library, Cambridge MA: Harvard University Press. 3 vols.

Baumgartner, F. 1988. 'Galileo's French Correspondents', *Annals of Science* 45, 169–182.

Beaver, D. 1970. 'Bernard Walther: Innovator in Astronomical Observation', *Journal for the History of Astronomy* 1, 39–43.

Bedini, S. 1980. 'The Mechanical Clock and the Scientific Revolution', in Maurice and Mayr 1980, 19–26.

    1985. *Clockwork Cosmos: Bernardo Facini and the Farnese Planisferologio*, Vatican City: Biblioteca Apostolica Vaticana.

Benedetti, G. 1585. *Diversarum Speculationum Mathematicarum & Physicarum Liber*, Turin: Haeres Nicolai Bevilaquae.

Bennett, J. 2002. 'Geometry in Context in the Sixteenth Century: The View From the Museum', *Early Science and Medicine* 7, 214–230.

Bennett, J. and D. Bertoloni Meli, 1994. *Sphaera Mundi. Astronomy Books in the Whipple Museum 1478–1600*, Cambridge: Whipple Museum of the History of Science.

Bennett, J. and S. Johnston, 1996. *The Geometry of War, 1500–1700*, Oxford: Museum of the History of Science.

Biagioli, M. 1990. 'Galileo the Emblem Maker', *Isis* 81, 230–258.

1993. *Galileo, Courtier: The Practice of Science in the Culture of Absolutism*, Chicago: University of Chicago Press.

1996. 'Playing with the Evidence', *Early Science and Medicine* 1, 70–105.

Bietenholtz, P. 1971. *Basle and France in the Sixteenth Century: The Basle Humanists and Printers in their Contacts with Francophone Culture*, Geneva: Librairie Droz.

1977. 'Erasmus and the German Public 1518–1520: The Authorized and Unauthorized Circulation of his Correspondence', *Sixteenth Century Journal* 8.2, 61–78.

Björnbo, A. A., ed., 1907. 'Ioannis Verneri De Triangulis Sphaericis Libri Quattuor', *Abhandlungen zur Geschichte der Mathematischen Wissenschaften* 24, 1–184.

Black, R. 1982. 'Ancients and Moderns in the Renaissance: Rhetoric and History in Accolti's *Dialogue on the Pre-eminence of Men of his Own Time*', *Journal of the History of Ideas* 43, 3–32.

Blackmore, H. 1976. *The Armouries of the Tower of London. Volume I: Ordnance*, London: HMSO.

Blackwell, R. 1991. *Galileo, Bellarmine and the Bible*, Notre Dame: University of Notre Dame Press.

Blaeu, J. 1662. *Atlas Maior*, Amsterdam: J. Blaeu. 12 vols.

Blair, A. 1990. 'Tycho Brahe's Critique of Copernicus and the Copernican System', *Journal for the History of Ideas* 51, 355–377.

Blair, R. and T. Faulkner, and N. Kiessling, eds., 1989–1994. *Robert Burton's The Anatomy of Melancholy*, Oxford: Clarendon Press. 3 vols.

Blum, R. 1958–1960. 'Vor- und Frühgeschichte der nationalen Allgemeinbibliographie', *Archiv für Geschichte des Buchwesens* 2, 233–303.

Boas Hall, M. 1965. 'Oldenburg and the Art of Scientific Communication', *British Journal of the History of Science* 2, 277–290.

Bobinger, M. 1954. *Christoph Schissler der Ältere und der Jüngere*, Augsburg and Basle: Verlag Die Brigg.

Bots, H. and F. Waquet, 1997. *La République des Lettres*, Paris: Belin De Boeck.

Bott, G., ed., 1992. *Focus Behaim Globus*, Nuremberg: Germanischen Nationalmuseum. 2 vols.

Brahe, T. 1573. *De nova et nullius aevi memoria prius visa stella*, Copenhagen: L. Benedictus.

1588. *De mundi aetherei recentioribus phaenomenis liber secundus*, Uraniborg: C. Weida.

1596. *Epistolarum astronomicarum liber primus*, Uraniborg: Ex officina Typographica Authoris.

1598. *Astronomiae instauratae mechanica*, Wandesburg: Propria Authoris Typographia.

1602. *Astronomiae instauratae progymnasmata*, Uraniborg and Prague: n.p.

Brandolinus, A. 1549. *De ratione scribendi libri tres . . . Adiecti sunt, Io. Ludovici Vivis, D. Erasmi Roterodami, Conradi Celtis, Christophi Hegendorphini, De Conscribendis epistolis libelli*, Basle: J. Oporinus.

Braunmuller, A. R. 1981. 'Editing Elizabethan Letters', *Text: Transactions of the Society for Textual Scholarship* 1, 185–199.

Bredekamp, H. 1995. *The Lure of Antiquity and the Cult of the Machine: The Kunstkammer and the Evolution of Nature, Art and Technology*, Princeton: Markus Wiener Publishers. Trans. A. Brown.

Brotóns, V. N., ed., 1981. *Jerónimo Muñoz. Libro del Nuevo Cometa. Littera ad Bartholomaeum Reisacherum. Summa del Prognostico del Cometa*, Valencia: Artes Gráficas Soler.

Brotton, J. 1997. *Trading Territories: Mapping the Early Modern World*, London: Reaktion Books.

Brown, E., ed. and trans., 1990. *Regiomontanus, His Life and Work*, Amsterdam: Elsevier.

Brown, J. 1979. 'Guild Organisation and the Instrument-Making Trade, 1550–1830', *Annals of Science* 36, 1–34.

Bud, R. and D. Warner, eds., 1998. *Instruments of Science. An Historical Enyclopaedia*, New York and London: Garland Publishing Inc.

Burckhardt, F., ed., 1887. *Aus Tycho Brahes Briefwechsel*, Basle: Schultze'sche Universitäts-Buchdruckerei.

Burge, E. and R. Johnson, R. and W. Stahl, 1971–1977. *Martianus Capella and the Seven Liberal Arts*, New York: Columbia University Press. 2 vols.

Burke, P. 1969. *The Renaissance Sense of the Past*, London: Edward Arnold.

2000. *A Social History of Knowledge: From Gutenberg to Diderot*, Cambridge: Polity Press.

Burmeister, K. H. 1967–1968. *Georg Joachim Rhetikus 1514–1574. Eine Bio-Bibliographie*, Wiesbaden: Guido Pressler Verlag. 3 vols.

Burnett, C., ed. and trans., 1998. *Adelard of Bath, Conversations with his Nephew: On the Same and the Different, Questions on Natural Science, and On Birds*, Cambridge: Cambridge University Press.

Cairns, H. and E. Hamilton, trans., 1963. *The Collected Dialogues of Plato, including the Letters*, Princeton: Princeton University Press. 2nd corrected printing.

Camargo, M. 1991. *Ars Dictaminis, Ars Dictandi*, Typologie des sources du moyen âge occidental 60, Turnhout: Brepols.

Capp, B. 1979. *Astrology and the Popular Press: English Almanacks 1500–1800*, London: Faber and Faber.

Cardanus, H. 1551. *De subtilitate*, Nuremberg: J. Petreius.

1557. *De rerum varietate*, Basle: H. Petri.

Caspar, M. 1993. *Kepler*, New York: Dover Publications. Trans. C. Doris Hellman. With a New Introduction and References by Owen Gingerich.

Chamber, J. 1601. *A Treatise Against Iudicial Astrologie*, London: John Harison.

Chandler, B. and C. Vincent, 1980. 'To Finance a Clock. An Example of Patronage in the 16th Century', in Maurice and Mayr 1980, 103–113.

Chapman, A. 1979. 'Astrological Medicine', in C. Webster, ed., *Health, Medicine and Mortality in the Sixteenth Century*, Cambridge: Cambridge University Press, 275–300.

    1983a. 'A Study of the Accuracy of Scale Graduations on a Group of European Astrolabes', *Annals of Science* 40, 473–488.

    1983b. 'The Accuracy of Angular Measuring Instruments Used in Astronomy between 1500 and 1850', *Journal for the History of Astronomy* 14, 133–137.

    1983c. 'The Design and Accuracy of some Observatory Instruments of the Seventeenth Century', *Annals of Science* 40, 457–471.

    1984. 'Tycho Brahe in China: The Jesuit Mission to Peking and the Iconography of European Instrument-Making Processes', *Annals of Science* 41, 417–443.

    1989. 'Tycho Brahe: Instrument Designer, Observer, and Mechanician', *Journal of the British Astronomical Association* 99.2, 70–77.

    1993. 'Scientific Instruments and Industrial Innovation: The Achievement of Jesse Ramsden', in Anderson, Bennett and Ryan 1993, 418–430.

    1994. 'Reconstructing the Angle-measuring Instruments of Pierre Gassendi', in Hackmann and Turner 1994, 103–116.

Chartier, R. 1988. *Cultural History: Between Practices and Representations*, Cambridge: Polity Press.

    1989. 'Texts, Printings, Readings', in L. Hunt, ed., *The New Cultural History*, Berkeley: University of California Press, 154–175.

    1995. *Forms and Meanings: Texts, Performances, and Audiences from Codex to Computer*, Philadelphia: University of Pennsylvania Press.

Christianson, J. 1967. 'Tycho Brahe at the University of Copenhagen, 1559–1562', *Isis* 58, 198–203.

    1968. 'Tycho Brahe's Cosmology from the Astrologia of 1591', *Isis* 59, 312–318.

    1970. 'Tycho Brahe's Facts of Life', *Fund og Forskning i Det Kongelige Biblioteks Samlinger* 17, 21–28.

    1979. 'Tycho Brahe's German Treatise on the Comet of 1577: A Study in Science and Politics', *Isis* 70, 110–140.

    2000. *On Tycho's Island. Tycho Brahe and His Assistants, 1570–1601*, Cambridge: Cambridge University Press.

Christianson, J. and A. Hadravova, P. Hadrava and M. Solc, eds., 2002. *Tycho Brahe and Prague: Crossroads of European Science*, Thun and Frankfurt am Main: Harri Deutsch.

Clarke, A. 1985. 'Giovanni Antonio Magini (1555–1617) and Late Renaissance Astrology', unpublished Ph.D. thesis, University of London.

Clavius, C. 1585. *In sphaeram Iohannis de Sacro Bosco commentarius*, Rome: D. Basa. 3rd edition.

    1593. *Astrolabium*, Rome: Typographiana Gabiana.

    1607. *In sphaeram Iohannis de Sacro Bosco commentarius*, Lyons: Q. Hug. a Porta. 5th edition.

Clough, C. H. 1976. 'The Cult of Antiquity: Letters and Letter Collections', in C. H. Clough, ed., *Cultural Aspects of the Italian Renaissance: Essays in*

Honour of Paul Oskar Kristeller, Manchester: Manchester University Press, 33–67.

Clulee, N. 1988. *John Dee's Natural Philosophy: Between Science and Religion*, London and New York: Routledge.

Collijn, I. 1933. 'Neue Beitrage zur Geschichte der Bibliothek des Heinrich Rantzau', *Zentralblatt für Bibliothekswesen* 50, 111–120.

Collins, H. 1985. *Changing Order: Replication and Induction in Scientific Practice*, London: Sage.

Constable, G. 1976. *Letters and Letter Collections*, Typologie des sources du moyen âge occidental 17, Turnholt: Brepols.

Coote, C. H. 1888. *Johann Schöner, Professor of Mathematics at Nuremberg*, London: Henry Stevens and Son. Trans. Henry Stevens.

Copenhaver, B. 1990. 'Natural Magic, Hermetism, and Occultism in Early Science', in Lindberg and Westman 1990, 261–301.

Copenhaver, B., trans., 2002. *Polydore Vergil. On Discovery*, The I Tatti Renaissance Library, Cambridge MA and London: Harvard University Press.

Copernicus, N. 1543. *De revolutionibus orbium coelestium*, Nuremberg: Johannes Petreius.

Corcoran, T., trans., 1971–1972. *Seneca. Naturales Quaestiones*, Loeb Classical Library, Cambridge MA: Harvard University Press. 2 vols.

Cornford, F. 1937. *Plato's Cosmology*, London: Kegan Paul.

Cornford, F. and P. Wicksteed, trans., 1929–1934. *Aristotle. The Physics*, Loeb Classical Library, Cambridge MA: Harvard University Press. 2 vols.

Costil, P. 1935. *André Dudith, humaniste hongrois*, Paris: Les Belles Lettres.

Crosland, M. 1962. *Historical Studies in the Language of Chemistry*, London: Heinemann.

Cuningham, W. 1559. *The Cosmographicall Glasse*, London: John Daye.

Czartoryski, P. 1978. 'The Library of Copernicus', *Studia Copernicana* 16, 355–396.

daCosta Kaufmann, T. 1993. *The Mastery of Nature: Aspects of Art, Science, and Humanism in the Renaissance*, Princeton: Princeton University Press.

1994. 'From Treasury to Museum: The Collections of the Austrian Habsburgs', in R. Cardinal and J. Elsner, eds., *The Cultures of Collecting*, London: Reaktion Books, 136–154.

Dall' Olmo, U. 1980. 'Latin Terminology Relating to Aurorae, Comets, Meteors and Novae', *Journal for the History of Astronomy* 11, 10–27.

D'Amico, J. 1988a. 'Manuscripts', in Schmitt and Skinner 1988, 11–24.

1988b. *Theory and Practice in Renaissance Textual Criticism: Beatus Rhenanus Between Conjecture and History*, Berkeley: University of California Press.

Darnton, R. 2000. 'An Early Information Society: News and the Media in Eighteenth-Century Paris', *American Historical Review* 105, 1–35.

Daston, L. and K. Park, 1998. *Wonders and the Order of Nature, 1150–1750*, New York: Zone Books.

Davis, N. 1961. 'Sixteenth-century Arithmetics on the Business Life', *Journal for the History of Ideas* 21, 18–48.

1983. 'Beyond the Market: Books as Gifts in Sixteenth-Century France', *Transactions of the Royal Historical Society* 33, ser. 5, 69–88.

Dear, P. 1988. *Mersenne and the Learning of the Schools*, Ithaca: Cornell University Press.

1995. *Discipline and Experience: The Mathematical Way in the Scientific Revolution*, Chicago: University of Chicago Press.

Dekker, E. 1993. 'Epact Tables on Instruments: Their Definition and Use', *Annals of Science* 50, 303–324.

2002. 'The Doctrine of the Sphere: A Forgotten Chapter in the History of Globes', *Globe Studies* 49/50, 25–44.

Dekker, E. ed., 1999. *Globes at Greenwich: A Catalogue of the Globes and Armillary Spheres in the National Maritime Museum, Greenwich*, Oxford: Oxford University Press.

Dekker, E. and G. L'E. Turner, 1993. 'An Astrolabe Attributed to Gerard Mercator *c*.1570', *Annals of Science* 50, 403–443.

Delisle, C. 2004. 'The Letter: Private Text or Public Place? The Mattioli–Gesner Controversy about the *aconitum primum*', *Gesnerus* 61, 161–176.

Dewald, J. 1996. *The European Nobility 1400–1800*, Cambridge: Cambridge University Press.

Dibner, B. 1967. 'Of Martyrs, Books and Science', in Lehman-Haupt 1967, 168–182.

Dick, W. and J. Hamel, eds., 1999. *Beiträge zur Astronomiegeschichte*, Acta historica astronomiae 2, Thun and Frankfurt am Main: Harri Deutsch.

Diesner, P. 1935. 'Leben und Streben des elsässischen Arztes Helisaeus Röslin 1544–1616)', *Elsass-Lotharingisches Jahrbuch* 14, 115–141.

Dolz, W. *et al.* 1993. *Uhren-Globen wissenschaftliche Instrumente*, Staatlicher Mathematisch-Physikalischer Salon Dresden, Dresden: Karl M. Lipp Verlag.

Donahue, W. 1975. 'The Solid Planetary Spheres in Post-Copernican Natural Philosophy', in Westman 1975c, 244–284.

1981. *The Dissolution of the Celestial Spheres 1595–1650*, New York: Arno Press.

Donahue, W., trans., 2000. *Johannes Kepler. Optics. Paralipomena to Witelo and Optical Part of Astronomy*, Santa Fe: Green Lion Press.

Doppelmayr, J. G. 1730. *Historische Nachricht von den Nürnbergischen Mathematicis und Künstlern*, Nuremberg: P. C. Monath.

Drake, S. 1970. 'Early Science and the Printed Book: The Spread of Science Beyond the University', *Renaissance and Reformation* 6, 43–52.

1978. *Galileo at Work. His Scientific Biography*, Chicago: University of Chicago Press.

Drake, S. trans., 1953. *Galileo. Dialogue Concerning the Two Chief World Systems*, Berkeley and Los Angeles: University of California Press.

1957. *Discoveries and Opinions of Galileo*, New York: Doubleday Anchor Books.

Dreier, F. 1985. 'The *Kunstkammer* of the Hessian Landgraves in Kassel', in Impey and MacGregor 1985, 102–109.

Dreyer, J. 1890. *Tycho Brahe: A Picture of Scientific Life and Work in the Sixteenth Century*, Edinburgh: Adam and Charles Black.

    1916. 'On Tycho Brahe's Manual of Trigonometry', *The Observatory* 39, 127–131.

    1917. 'On Tycho Brahe's Catalogue of Stars', *The Observatory* 40, 229–233.

    1924. 'Address', *Monthly Notices of the Royal Astronomical Society* 84, 298–305.

Duncan, A., trans., 1981. *Johannes Kepler. The Secret of the Universe*, New York: Abaris Books.

Durling, R. 1980. 'Konrad Gesner's Briefwechsel', in Krafft and Schmitz 1980, 101–111.

Eamon, W. 1984. 'Arcana Disclosed: The Advent of Printing, the Books of Secret Tradition and the Development of Experimental Science in the Sixteenth Century', *History of Science* 22, 111–150.

    1985. 'From the Secrets of Nature to Public Knowledge: The Origins of the Concept of Openness in Science', *Minerva* 23, 321–347.

    1991. 'Court, Academy and Printing House: Patronage and Scientific Careers in Late Renaissance Italy', in Moran 1991, 25–50.

Eastwood, B. 1981, 'The Diagram *Spera Celestis* in the *Hortus Deliciarum:* A Confused Amalgam from the Astronomies of Pliny and Martianus Capella', *Annali dell' Istituto e Museo di Storia della Scienza di Firenze* 6.1, 177–186.

    2000. 'Astronomical Images and Planetary Theory in Carolingian Studies of Martianus Capella', *Journal for the History of Astronomy* 31, 1–28.

Eckhardt, W. 1976. 'Erasmus Habermel – Zur Biographie des Instrumentmachers Kaiser Rudolfs II', *Jahrbucher der Hamburger Kunstsammlungen* 21, 55–92.

    1977. 'Erasmus und Josua Habermel – Kunstgeschichtliche Anmerkungen zu der Beiden Instrumentenmacher', *Jahrbucher der Hamburger Kunstsammlungen* 22, 12–74.

Edwards, M. 1994. *Printing, Propaganda and Martin Luther*, Berkeley and Los Angeles: University of California Press.

Ehrmann, A. and G. Pollard, 1965. *The Distribution of Books by Catalogue from the Invention of Printing to A.D. 1800. Based on Material in the Broxbourne Library*, Cambridge: The Roxburghe Club.

Eisenstein, E. 1979. *The Printing Press as an Agent of Change: Communications and Cultural Transformations in Early Modern Europe*, Cambridge: Cambridge University Press. 2 vols.

Elsmann, T. and H. Lietz and S. Pettke, eds., 1991. *Nathan Chytraeus 1543–1593. Ein Humanist in Rostock und Bremen. Quellen und Studien*, Bremen: Edition Temmen.

Evans, J. 1998. *The History and Practice of Ancient Astronomy*, Oxford: Oxford University Press.

Evans, R. 1975. *The Wechel Presses: Humanism and Calvinism in Central Europe 1572–1627*, Past and Present Supplement 2, Oxford: Seacourt Press.

    1984. 'Rantzau and Welser: Aspects of Later German Humanism', *History of European Ideas* 5, 257–272.

1997. *Rudolf II and His World: A Study in Intellectual History 1576–1612*, London: Thames and Hudson. Corrected 2nd edition.

Ezell, M. 1999. *Social Authorship and the Advent of Print*, Baltimore: Johns Hopkins University Press.

Fabian, B., ed., 1972–2001. *Die Messkataloge Georg Willers*, Hildesheim and New York: Georg Olms Verlag. 5 vols.

Fairclough, H., trans., 1967–1969. *Virgil. Eclogues; Georgics; Aeneid*, Loeb Classical Library, Cambridge MA: Harvard University Press. 2 vols.

Falk, T. and R. Abler, 1985. 'Intercommunications Technologies: The Development of Postal Services in Sweden', *Geografiska Annaler. Series B, Human Geography* 67, 21–28.

Favaro, A., ed., 1886. *Carteggio inedito di Tichone Brahe, Giovanni Keplero e di altri celebri astronomi e matematici dei secoli XVI. e XVII. con G. A. Magini*, Bologna: Nicola Zanichelli.

Febvre, L. and H.-J. Martin, 1984. *The Coming of the Book: The Impact of Printing 1450–1800*, London and New York: Verso. Trans. D. Gerard.

Feingold, M., ed., *Jesuit Science and the Republic of Letters*, Cambridge MA: MIT Press.

Field, J. 1996. 'European Astronomy in the First Millennium: The Archaeological Record', in Walker 1996, 110–122.

Figala, K. 1972. 'Tycho Brahes Elixier', *Annals of Science* 28, 139–176.

Findlen, P. 1991. 'The Economy of Scientific Exchange in Early Modern Italy', in Moran 1991, 5–24.

1994. *Possessing Nature: Museums, Collecting and Scientific Culture in Early Modern Italy*, Chicago: University of Chicago Press.

1997. 'Cabinets, Collecting and Natural Philosophy', in Fucikova 1997b, 209–219.

1999. 'The Formation of a Scientific Community: Natural History in Sixteenth-Century Italy', in A. Grafton and N. Siraisi, eds., *Natural Particulars: Nature and the Disciplines in Renaissance Europe*, Cambridge MA: MIT Press, 369–400.

Fiorentino, F. and F. Tocco *et al.*, cds., 1879–1891. *Jordani Bruni Nolani opera latine conscripta*. Naples: Morano/Florence: Le Monnier. 3 vols.

Fleet, B., trans., 1989. *Simplicius. On Aristotle's Physics 2*, Ithaca: Cornell University Press.

Folkerts, M. 1996. 'Johannes Praetorius (1537–1616) – ein bedeutender Mathematiker und Astronom des 16. Jahrhunderts', in J. Dauben *et al.*, eds., *History of Mathematics: States of the Art: Flores Quadrivii Studies in Honor of Christoph J. Scriba*, San Diego: Academic Press, 146–169.

Forbes, E. and L. Murdin and F. Willmoth, eds., 1995–2001. *The Correspondence of John Flamsteed, the First Astronomer Royal*, Bristol and Philadelphia: The Institute of Physics. 3 vols.

Fracastoro, G. 1538. *Homocentrica. Eiusdem de causis criticorum dierum per ea quae in nobis sunt*, Venice: n. p.

Frasca-Spada, M. and N. Jardine, eds., 2000. *Books and the Sciences in History*, Cambridge: Cambridge University Press.

Friis, F. R., ed., 1875. *Breve og Aktstykker angaaende Tyge Brahe og hans Slaegtninge samlede og udgivne*, Copenhagen: Andr. Fred. Høst and Søns.

1876–1886. *Tychonis Brahe et ad eum Doctorum Virorum Epistolae ab Anno 1568 ad Annum 1587*, Copenhagen: G. E. Gad.

1896. *Epistolae quas per annos a 1596 ad 1601 Tycho Brahe et Oligerus Rosenkrantzius inter se dederunt*, Copenhagen: Typis Martii Truelsen.

1900–1909. *Tychonis Brahe et ad eum Doctorum Virorum Epistolae ab Anno 1588*, Copenhagen: Graebe.

Fucikova, E. 1985. 'The Collection of Rudolf II at Prague: Cabinet of Curiosities or Scientific Museum', in Impey and MacGregor 1985, 47–53.

1997a. 'Prague Castle under Rudolf II, His Predecessors and Successors', in Fucikova 1997b, 2–71.

Fucikova, E. *et al.*, eds., 1997b. *Rudolf II and Prague: The Imperial Court and Residential City as the Cultural and Spiritual Heart of Central Europe*, London: Thames and Hudson.

Fuhrmann, H. 1959. 'Heinrich Rantzaus römische Korrespondenten', *Archiv für Kulturgeschichte* 41, 63–89.

Fumaroli, M. 1978. 'Genèse de l'épistolographie classique', *Revue d'Histoire littéraire de la France* 78, 886–905.

Garberson, E. 1993. 'Bibliotheca Windhagiana. A Seventeenth-Century Austrian Library and Its Decoration', *Journal of the History of Collections* 3, 109–128.

Gaskell, P. 1972. *A New Introduction to Bibliography*, Oxford: Clarendon Press.

Gassendi, P. 1654. *Tychonis Brahei Equitis Dani, Astronomorum Coryphaei Vita. Accessit Nicolai Copernici, Georgii Peurbachii and Ioannis Regiomontani Astronomorum celebrium Vita*, Paris: M. Dupuis.

1658. *Epistolae, quibus accesserunt clarissimorum quorumdam ad ipsum epistolae, et responsa*, Lyon: L. Anisson and J. B. Devenet.

Gatti, H. 1999. *Giordano Bruno and Renaissance Science*, Ithaca: Cornell University Press.

Gebele, E. 1927. 'Auf den Spuren Einer Verschollenen Bibliothek', *Zentralblatt für Bibliothekswesen* 44, 550–553.

Gemma, C. 1575. *De Naturae Divinae Characterismis*, Antwerp: C. Plantin. 2 vols. 1578. *De prodigiosa specie naturaque cometae*, Antwerp: C. Plantin.

Gerlo, A. 1971. 'The *Opus de Conscribendis Epistolis* of Erasmus and the Tradition of the *Ars Epistolica*', in R. Bolgar, ed., *Classical Influences on European Culture A.D. 500–1500*, Cambridge: Cambridge University Press, 103–114.

Gesner, C. 1551–1558. *Historiae animalium libri IV*, Zurich: C. Froschoverus. 1563. *Thierbuch*, Zurich: C. Froschower.

Gilmont, J.-F., ed., 1998. *The Reformation and the Book*, Aldershot: Ashgate.

Gingerich, O. 1971a. 'Apianus' *Astronomicum Caesareum* and its Leipzig Facsimile', *Journal for the History of Astronomy* 2, 168–177.

1971b. 'Johannes Kepler and the Rudolphine Tables', *Sky and Telescope* 42, 328–333.

1973a. 'The Role of Erasmus Reinhold and the Prutenic Tables in the Dissemination of Copernican Theory', *Studia Copernicana* 6, 43–63 and 123–125.

1973b. 'From Copernicus to Kepler: Heliocentrism as Model and as Reality', *Proceedings of the American Philosophical Society* 117, 513–522.

1978. 'Early Copernican Ephemerides', *Studia Copernicana* 16, 403–417.

1987. 'The Alfonsine Tables in the Age of Printing', in M. Comes, R. Puig, and Julio Samsó, eds., *De astronomia Alphonsi Regis*, Barcelona: University of Barcelona, 89–95.

1990. 'Alfonso the Tenth as a Patron of Astronomy', in F. Marques-Villanuev and C. A. Veja, eds., *Alfonso X of Castile the Learned King (1221–1284)*, Cambridge MA: Harvard University Department of Romance Languages and Literatures, 30–45.

1993. 'Astronomical Paper Instruments with Moving Parts', in Anderson, Bennett, and Ryan 1993, 63–74.

2002a. *An Annotated Census of Copernicus' De Revolutionibus (Nuremberg, 1543 and Basel, 1566)*, Leiden: Brill.

2002b. 'Recent Notes on Tycho Brahe's Library', in Christianson and Hadravova et al., 2002, 323–328.

Gingerich, O. and J. Voelkel, 1998. 'Tycho Brahe's Copernican Campaign', *Journal for the History of Astronomy* 29, 1–34.

Gingerich, O. and R. Westman, 1988. 'The Wittich Connection', *Transactions of the American Philosophical Society* 78.7.

Glasemann, R. 1999. *Erde, Sonne, Mond and Sterne: Globen, Sonnenuhren und astronomische Instrumente im Historischen Museum*, Frankfurt am Main: Waldemar Kramer.

Goldgar, A. 1995. *Impolite Learning: Conduct and Community in the Republic of Letters 1680–1750*, New Haven: Yale University Press.

Goldstein, B. 1971. 'Levi ben Gerson on Instrumental Errors and the Transversal Scale', *Journal for the History of Astronomy* 8, 101–112.

Gouk, P. 1988. *The Ivory Sundials of Nuremberg 1500–1700*, Cambridge: Whipple Museum.

1997. 'Natural Philosophy and Natural Magic', in Fucikova 1997b, 231–237.

Goulding, R. 1995. 'Henry Savile and the Tychonic World-System', *Journal of the Warburg and Courtauld Institutes* 58, 152–179.

Grafton, A. 1979. 'Rhetoric, Philosophy and Egyptomania in the 1570s: J. J. Scaliger's Invective Against M. Guilandinus's *Papyrus*', *Journal of the Warburg and Courtauld Institutes* 41, 167–194.

1980. 'The Importance of Being Printed', *Journal of Interdisciplinary History* 11, 265–286.

1983–1993. *Joseph Scaliger: A Study in the History of Classical Scholarship*, Oxford: Clarendon Press. 2 vols.

1989. 'Editing Neo-Latin Texts: Two Cases and their Implications', in J. Grant, ed., *Editing Greek and Latin Texts*, New York: AMS Press, 163–186.

1991. *Defenders of the Text: The Traditions of Scholarship in an Age of Science, 1450–1800*, Cambridge MA: Harvard University Press.

1992. 'Kepler as a Reader', *Journal for the History of Ideas* 53, 561–572.

1997a. *Commerce with the Classics: Ancient Books and Renaissance Readers*, Ann Arbor: The University of Michigan Press.

1997b. 'From Apotheosis to Analysis: Some Late Renaissance Histories of Classical Astronomy', in Kelley 1997, 261–276.

1999. *Cardano's Cosmos. The World and Works of a Renaissance Astrologer*, Cambridge MA: Harvard University Press.

Grafton, A., ed., 1993. *Rome Reborn: The Vatican Library and Renaissance Culture*, Washington, DC and New Haven: Library of Congress and Yale University Press.

Grafton, A. and L. Jardine, 1986. *From Humanism to the Humanities: Education and the Liberal Arts in Fifteenth and Sixteenth Century Europe*, London: Duckworth, 1986.

Graham, L. and W. Lepenies and P. Weingart, eds., 1983. *Functions and Uses of Disciplinary Histories*, Dordrecht: D. Reidel.

Granada, M. 1996. *El debate cosmológico en 1588: Bruno, Brahe, Rothmann, Ursus, Röslin*, Lezione della scuola di studi superiori in Napoli 18, Istituto Italiano per gli studi filosofici, Naples: Bibliopolis.

1999. 'Christoph Rothmann und die Auflösung der himmlischen Sphären. Die Briefe an den Landgrafen von Hessen-Kassel 1585', in Dick and Hamel 1999, 34–57.

2002. *Sfere solide e cielo fluido. Momenti del dibattito cosmologico nella seconda metà del Cinquecento*, Milan: Guerini e Associati.

2004. 'Astronomy and Cosmology in Kassel: The Contributions of Christoph Rothmann and His Relationship to Tycho Brahe and Jean Pena', in J. Zamrzlova, ed., *Science in Contact at the Beginning of Scientific Revolution*, Prague: National Technical Museum, 237–248.

2006. 'Did Tycho Eliminate the Celestial Spheres Before 1586?', *Journal for the History of Astronomy* 37, 125–145.

Granada, M. and J. Hamel and L. von Mackensen, eds., 2003. *Christoph Rothmanns Handbuch der Astronomie von 1589*, Thun and Frankfurt am Main: Harri Deutsch.

Grant, Edward 1978. 'Cosmology', in Lindberg 1978, 265–302.

1987. 'Celestial Orbs in the Latin Middle Ages', *Isis* 78, 153–173.

1994. *Planets, Stars and Orbs: The Medieval Cosmos 1200–1687*, Cambridge: Cambridge University Press.

Graves, R. 1955. *The Greek Myths*, London: Penguin. 2 vols.

Green, W. *et al.*, trans., 1960–1972. *Augustine of Hippo. De civitate Dei*, Loeb Classical Library, London: W. Heinemann Ltd. 6 vols.

Grendler, M. 1980. 'A Greek Collection in Padua: The Library of Gian Vincenzo Pinelli (1535–1601)', *Renaissance Quarterly* 33, 386–416.

Grendler, P. 1975. 'The Roman Inquisition and the Venetian Press, 1540–1605', *The Journal of Modern History* 47, 48–65.

1988. 'Printing and Censorship', in Schmitt and Skinner 1988, 25–53.

Groiss, E. 1980. 'The Augsburg Clockmakers' Craft', in Maurice and Mayr 1980, 57–86.

Gummere, R., trans., 1917–1925. *Seneca. Ad Lucilium Epistulae Morales*, Loeb Classical Library, London and New York: William Heinemann and G. P. Putnam's Sons. 3 vols.

Gundlach, F. 1927. *Catalogus Professorum Academiae Marburgensis. Die Akademischen Lehrer der Philipps-Universität in Marburg von 1527 bis 1910*, Veröffentlichtungen der Historischen Kommission für Hessen und Waldeck XV, Marburg: N. G. Elwert and G. Braun.

Günther, S. 1889. 'Der bayerische Staatskanzler Herwart von Hohenburg als Freund und Befördere der exacten Wissenschaften', *Jahrbuch für Münchener Geschichte* 3, 183–219.

Gutfleisch, B. and J. Menzhausen, 1989. '"How a *Kunstkammer* Should be Formed': Gabriel Kaltemarckt's Advice to Christian I of Saxony on the Formation of an Art Collection, 1587', *Journal of the History of Collections* 1, 3–32.

Guthrie, W. K. C., trans., 1939. *Aristotle. On the Heavens*, Cambridge MA: Harvard University Library.

Haber, F. 1981. 'The Clock as Intellectual Artifact', in Maurice and Mayr 1981, 9–18.

Hacking, I. 1975. *The Emergence of Probability: A Philosophical Study of Early Ideas about Probability, Induction, and Statistical Inference*, Cambridge: Cambridge University Press.

Hackmann, W. and A. Turner, eds., 1994. *Learning, Language and Invention: Essays Presented to Francis Maddison*, Aldershot and Paris: Variorum Press and Société Internationale de l'Astrolabe.

Hadravova, A., P. Hadrava and J. Shackelford, trans., 1996. *Tycho Brahe. Instruments of the Renewed Astronomy*, Prague: KLP.

Hagecius, T. 1574. *Dialexis de novae et prius incognitae stellae*, Frankfurt am Main: n. p.

1578. *Descriptio cometae, qui apparuit Anno M. D. LXXVII*, Prague: G. Melantrich.

Hagen, H. 1874. *Jacobus Bongarsius: ein Beitrag zur Geschichte der gelehrten Studien des 16.-17. Jahrhunderts*, Berne: A. Fischer.

Hale, J. 1994. *The Civilisation of Europe in the Renaissance*, London: Fontana Press.

Halleux, R. and A. C. Bernès, 1995. 'La Cour savante d'Ernest de Bavière', *Archives internationales d'histoire des sciences* 45, 3–29.

Halsberghie, N. 1994. 'The Resemblances and Differences of the Construction of Ferdinand Verbiest's Astronomical Instruments, as Compared to those of Tycho Brahe. A Study Based on Their Writings', in Witek 1994, 85–92.

Hamel, J. 1998. *Die astronomischen Forschungen in Kassel unter Wilhelm VI. Mit einer Teiledition der deutschen Übersetzung des Hauptwerkes von Copernicus um 1586*, Frankfurt am Main: Harri Deutsch.

Hammerstein, H. 1986. 'The Battle of the Booklets: Prognostic Tradition and the Proclamation of the Word in Early Sixteenth-century Germany', in Zambelli 1986, 129–151.

Hannaway, O. 1986. 'Laboratory Design and the Aim of Science: Andreas Libavius versus Tycho Brahe', *Isis* 77, 585–610.

Harlow, A. 1928. *Old Post Bags. The Story of Sending a Letter in Ancient and Modern Times*, New York: Appleton and Co.

Harrison, P. 1998. *The Bible, Protestantism and the Rise of Natural Science*, Cambridge: Cambridge University Press.

Heath, T. 1913. *Aristarchus of Samos*, Oxford: Clarendon Press.

    1921. *A History of Greek Mathematics*, Oxford: Clarendon Press. 2 vols.

Hegendorff, C. 1545. *Methodus Conscribendi Epistolas*, Paris: J. L. Tiletanus.

Helfricht, J. 1999. 'Fünf Briefe Tycho Brahes an den Görlitzer Astronomen Bartholomäus Scultetus (1540–1614)', in Dick and Hamel 1999, 11–33.

Hellinga, L. 1983. 'Manuscripts in the Hands of Printers', in Trapp 1983, 3–11.

Hellman, C. D. 1944. *The Comet of 1577: Its Place in the History of Astronomy*, Columbia Studies in the Social Sciences 510, New York: Columbia University Press.

    1963. 'Was Tycho as Influential as He Thought?', *British Journal for the History of Science* 1, 295–324.

Henderson, J. R. 1983a. 'Defining the Genre of the Letter: Juan Luis Vives' *De Conscribendis Epistolis*', *Renaissance and Reformation* 7, 89–105.

    1983b. 'Erasmus on the Art of Letter-Writing', in Murphy 1983, 331–355.

    1993. 'On Reading the Rhetoric of the Renaissance Letter', in H. Plett, ed., *Renaissance Rhetorik*, Berlin and New York: De Gruyter, 143–162.

    2002. 'Humanist Letter Writing: Private Conversation or Public Forum?', in Papy 2002, 17–38.

Henkel, A. and A. Schöne, 1996. *Emblemata: Handbuch zur Sinnbildkunst des XVI. und XVII. Jahrhunderts*, Stuttgart: J. B. Metzler.

Hesse, M. 1966. *Models and Analogies in Science*, Notre Dame: University of Notre Dame Press.

Heydon, C. 1603. *A Defence of Iudiciall Astrologie, In Answer to a Treatise lately published by M. Iohn Chamber*, Cambridge: J. Legat.

Hillard, D. and Poulle, E. 1971. 'Oronce Fine et l'horloge planètaire de la bibliothèque Sainte-Geneviève', *Bibliothèque d'Humanisme et Renaissance* 33, 311–351.

Hind, A. 1963. *A History of Engraving and Etching from the 15th Century to the Year 1914*, New York: Dover Publications. 3rd edition.

Hofmann, N. 1982. *Die Artistenfakultät an der Universität Tübingen 1534–1601*, Beiträge zur Geschichte der Eberhard-Karls-Universität Tübingen 28, Tübingen: J. C. B. Mohr, Paul Siebeck.

Holbrook, M. 1983. 'Beschreibung des Himmelsglobus von Henricus, Arnoldus und Jacobus van Langren und eines Planetariums von H. van Laun im Historischen Museum zu Frankfurt am Main', *Der Globusfreund* 31/32, 69–77.

Hooykaas, R. 1977. *Religion and the Rise of Modern Science*, Edinburgh: Scottish Academic Press. Corrected reprint of 1972 edition.

    1984. *G. J. Rheticus' Treatise on Holy Scripture and the Motion of the Earth*, Amsterdam: North-Holland Publishing Company.

Housden, J. A. J. 1903. 'Some Early English Posts', *English Historical Review* 18, 713–718.

Houzeau, J. and A. Lancaster, 1964. *Bibliographie Générale de l'Astronomie jusqu'en 1880*, London: The Holland Press. With additions by D. Dewhirst.

Howell, K. 2002. *God's Two Books. Copernican Cosmology and Biblical Interpretation in Early Modern Science*, Notre Dame: University of Indiana Press.

Hues, R. 1594. *Tractatus de Globis et eorum usu*, London: T. Dawson.

    1639. *A Learned Treatise of Globes, Both Coelestiall and Terrestriall: with their several uses. . . . Illustrated with Notes, by Io. Isa. Pontanus. And . . . made English, for the benefit of the Unlearned, by John Chilmead*, London: P. Stephens and C. Meredith.

Iannaccone, I. 1994. 'Syncretism between European and Chinese Culture in the Astronomical Instruments of Ferdinand Verbiest in the Old Beijing Observatory', in Witek 1994, 93–121.

Iliffe, R. 1992. ''In the Warehouse': Privacy, Property and Priority in the Early Royal Society', *History of Science* 30, 29–68.

Impey, O. and A. MacGregor, eds., 1985. *The Origins of Museums: The Cabinet of Curiosities in Sixteenth- and Seventeenth-Century Europe*, Oxford: Clarendon Press.

Israel, J. 1995. *The Dutch Republic: Its Rise, Greatness and Fall 1477–1806*, Oxford: Clarendon Press.

Jaki, S., trans., 1975. *Giordano Bruno. The Ash Wednesday Supper*, The Hague: Mouton.

Jankovics, J. and I. Monok, 1993. *András Dudith's Library: A Partial Reconstruction*, Szeged: Scriptum.

Janson, T. 1964. *Latin Prose Prefaces: Studies in Literary Conventions*, Studia Latina Holmiensa 13, Stockholm: Stockholm University.

Jardine, L. 1993. *Erasmus, Man of Letters: The Construction of Charisma in Print*, Princeton: Princeton University Press.

    1996. *Worldly Goods: A New History of the Renaissance*, London: Macmillan.

Jardine, N. 1979. 'The Forging of Modern Realism: Clavius and Kepler against the Sceptics', *Studies in History and Philosophy of Science* 10, 141–173.

    1982. 'The Significance of the Copernican Orbs', *Journal for the History of Astronomy* 13, 167–194.

    1987. 'Scepticism in Renaissance Astronomy: A Preliminary Study', in C. Schmitt and R. Popkin, eds., *Scepticism from the Renaissance to the Enlightenment*, Wiesbaden: O. Harrassowitz, 83–102.

    1988a. *The Birth of History and Philosophy of Science: Kepler's A Defence of Tycho against Ursus with Essays on its Provenance and Significance*, Cambridge: Cambridge University Press. Corrected edition.

    1988b. 'Epistemology of the Sciences', in Schmitt and Skinner 1988, 685–711.

    1991. 'Writing Off the Scientific Revolution', *Journal for the History of Astronomy* 22, 311–318.

    1998. 'The Places of Astronomy in Early-Modern Culture', *Journal for the History of Astronomy* 29, 49–62.

2006. 'Kepler as Castigator and Historian: His Preparatory Notes for *Contra Ursum*', *Journal for the History of Astronomy* 36, 257–297.

Jardine, N. and D. Launert, *et al.*, 2005. 'Tycho *v.* Ursus: The Build-up to a Trial', *Journal for the History of Astronomy* 36, 81–106, 125–165.

Jardine, N. and A. Segonds, 2001. 'A Challenge to the Reader: Ramus on Astrologia without Hypotheses', in M. Feingold, J. Freedman and W. Rothgang, eds., *The Influence of Petrus Ramus*, Stuttgart: Schwabe and Co., 248–266.

Jarrell, R. 1989. 'The Contemporaries of Tycho Brahe', in Taton and Wilson 1989, 22–32.

Jervis, J. 1980. 'Vögelin on the Comet of 1532: Error Analysis in the 16th Century', *Centaurus* 23, 216–229.

1985. 'Cometary Theory in Fifteenth-Century Europe', *Studia Copernicana* 26, 1–206.

Jöcher, C. 1750–1751. *Allgemeines Gelehrten-Lexikon*, Leipzig: Gleditschens Buchhandlung. 4 vols.

Johns, A. 1998a. 'Science and the Book in Modern Cultural Historiography', *Studies in History and Philosophy of Science* 29, 167–194.

1998b. *The Nature of the Book. Print and Knowledge in the Making*, Chicago: University of Chicago Press.

2003. 'The Ambivalence of Authorship in Early Modern Natural Philosophy', in M. Biagioli and P. Galison, eds., *Scientific Authorship: Credit and Intellectual Property in Science*, London and New York: Routledge, 66–90.

Johnson, F. 1953. 'Astronomical Text-books in the Sixteenth Century', in A. Underwood, ed., *Science, Medicine and History: Essays on the Evolution of Scientific Thought and Medical Practice Written in Honour of Charles Singer*, Oxford: Oxford University Press, 2 vols., vol. I, 285–302.

Jones, H. 1981. *Pierre Gassendi 1592–1655: An Intellectual Biography*, Bibliotheca Humanistica and Reformatorica 35, Nieuwkoop: B. de Graaf.

Joy, L. S. 1989. *Gassendi the Atomist: Advocate of History in an Age of Science*, Cambridge: Cambridge University Press.

Kamen, H. 2000. *Early Modern European Society*, London: Routledge.

Keil, I. 1992. 'Tycho Brahes Aufenthalt in Augsburg oder: Wer war der Augsburger Gastfreund Tychos?', *Zeitschrift des Historisches Vereins für Schwaben* 85, 357–358.

Kejlbo, I. R. 1969–1971. 'Tycho Brahe und seine Globen', *Der Globusfreund* 18/20, 57–66.

1995. *Rare Globes: A Cultural Historical Exposition of Selected Terrestrial and Celestial Globes made before 1850 – Especially connected with Denmark*, Copenhagen: Munskgaard/Rosinante.

Kellenbenz, H. 1954. 'German Aristocratic Entrepreneurship. Economic Activities of the Holstein Nobility in the Sixteenth and Seventeenth Centuries', *Explorations in Entrepreneurial History* 6, 103–114.

Keller, A. 1985. 'Mathematics, Mechanics, and the Origins of the Culture of Mechanical Invention', *Minerva* 23, 348–361.

Kelley, D. 1970. *Foundations of Modern Historical Scholarship: Language, Law and History in the French Renaissance*, New York and London: Columbia University Press.

Kelley, D., ed., 1997. *History and the Disciplines: The Reclassification of Knowledge in Early Modern Europe*, New York: The University of Rochester Press.

Kenney, E. J. 1974, *The Classical Text: Aspects of Editing in the Age of the Printed Book*, Berkeley and Los Angeles: University of California Press.

Kettering, S. 1988. 'Gift-giving and Patronage in Early Modern France', *French History* 2, 131–151.

Keuning, J. 1956. 'The van Langren Family', *Imago Mundi* 13, 101–109.

Kiessling, N. 1988. *The Library of Robert Burton*, Oxford: Oxford Bibliographical Society.

King, D. 1993. 'Some Medieval Astronomical Instruments and Their Secrets', in R. Mazzolini, ed., *Non-Verbal Communication in Science Prior to 1900*, Florence: Leo S. Olschki, 29–52.

1996. 'Islamic Astronomy', in Walker, ed., *Astronomy Before the Telescope*, 143–174.

King, D. and G. L'E. Turner, 1994. 'The Astrolabe Presented by Regiomontanus to Cardinal Bessarion in 1462', *Nuncius* 9, 165–206.

King, H. and J. Millburn, 1978. *Geared to the Stars: The Evolution of Planetariums, Orreries, and Astronomical Clocks*, Bristol: Adam Hilger.

Kirchvogel, P. 1967. 'William IV, Tycho Brahe, and Eberhard Baldewein – The Missing Instruments of the Kassel Observatory', *Vistas in Astronomy* 9, 109–121.

Koeman, I. 1970. *Joan Blaeu and His Grand Atlas*, Amsterdam: Theatrum Orbis Terrarum Ltd.

Kolb, R. 1977. 'Dynamics of Party Conflict in the Saxon Late Reformation: Gnesio-Lutherans vs. Philippists', *Journal of Modern History* 49, D1289–D1305.

Kovacs, D., trans., 2002. *Euripides. Helen. Phoenician women. Orestes*, Loeb Classical Library, Cambridge MA: Harvard University Press.

Koyré, A. 1973. *The Astronomical Revolution: Copernicus–Kepler–Borelli*, London: Methuen.

Krafft, F. and R. Schmitz, eds., 1980. *Humanismus und Naturwissenschaften*, Beiträge zur Humanismusforschung 6, Boppard am Rein: Harald Boldt.

Kren, C. 1989. 'Astronomical Teaching at the Late Medieval University of Vienna', *History of Universities* 3, 15–30.

Kristeller, P. 1990. *Renaissance Thought and the Arts*, Princeton: Princeton University Press. Expanded edition.

Kuhn, T. 1957. *The Copernican Revolution. Planetary Astronomy in the Development of Western Thought*, Cambridge MA: Harvard University Press.

Kühne, H. 1983, 'Kaspar Peucer. Leben und Werk eines großen Gelehrten an der Wittenberger Universität im 16. Jahrhundert', *Letopis* 30, 151–161.

Kusukawa, S. 1993. 'Aspectio divinorum operam: Melanchthon and Astrology for Lutheran Medics', in A. Cunningham and O. Grell, eds., *Medicine and the Reformation*, London: Routledge, 33–56.

1995. *The Transformation of Natural Philosophy: The Case of Philip Melanchthon*, Cambridge: Cambridge University Press.

Lamprey, J. 1997. 'An Examination of Two Groups of Georg Hartmann 16th-century Astrolabes and the Tables Used in their Manufacture', *Annals of Science* 54, 111–142.

Langebek, J., ed., 1747. 'Samlung verschiedener Briefe und Nachrichten welche des berühmten Mathematic Tychonis Brahe Leben Schriften und Schichsale betreffen und theils von ihm selbst; theils aber von andern verfasset sind', *Dänische Bibliothec oder Sammlung Von Alten und Neuen Gelehrten Sachen aus Dannemarck* 9, 229–280.

Lapidge, M. 1978. 'Stoic Cosmology', in J. Rist, ed., *The Stoics*, Berkeley: University of California Press, 161–185.

1988. 'The Stoic Inheritance', in P. Dronke, ed., *A History of Twelfth-Century Western Philosophy*, Cambridge: Cambridge University Press, 81–112.

Lattis, J. 1994. *Between Copernicus and Galileo: Christoph Clavius and the Collapse of Ptolemaic Astronomy*, Chicago: University of Chicago Press.

Launert, D. 1999. *Nicolaus Reimers (Raimarus Ursus): Günstling Rantzaus, Brahes Feind, Leben und Werk*, Munich: Institut für Geschichte der Naturwissenschaften.

Lee, H. D. P., trans., 1952. *Aristotle. Meteorologica*. Loeb Classical Library, Cambridge MA: Harvard University Press.

Lehman-Haupt, H., ed., 1967. *Homage to a Bookman: Essays in Manuscripts, Books and Printing Written for Hans P. Kraus on his 60th Birthday*, Berlin: Gebr. Mann Verlag.

Lemay, H. R. 1977. 'Science and Theology at Chartres: The Case of the Supracelestial Waters', *British Journal for the History of Science* 10, 226–236.

Lemay, R. 1976. 'The Teaching of Astronomy in Medieval Universities, Principally at Paris in the Fourteenth Century', *Manuscripta* 20, 197–217.

Leopold, J. 1986. *Astronomen Sterne Geräte: Landgraf Wilhelm IV. und seine sich selbst bewegenden Globen*, Lucerne: Joseph Fremersdorf.

1995. 'Collecting Instruments in Protestant Europe before 1800', *Journal of the History of Collections* 7, 151–157.

1997. 'Mechanical Globes *Circa* 1500–1650', *Bulletin of the Scientific Instrument Society* 53, 5–8.

Lerner, M.-P. 1989. 'Le problème de la matière céleste après 1550: aspects de la bataille des cieux fluides', *Revue d'Histoire des Sciences* 42, 255–280.

1996–1997. *Le Monde des Sphères*, Paris: Les Belles Lettres. 2 vols.

Levere, T. H. and W. Shea, eds., 1990. *Nature, Experiment, and the Sciences: Essays on Galileo and the History of Science in Honour of Stillman Drake*, Boston Studies in the History of Science 120, Dordrecht: Kluwer Academic Publishers.

Libbrecht, U. 1994. 'General Evaluation of the Scientific Work of Ferdinand Verbiest', in Witek 1994, 55–64.

Lindberg, D, ed., 1978. *Science in the Middle Ages*, Chicago: University of Chicago Press.

Lindberg, D. and R. Westman, eds., 1990. *Reappraisals of the Scientific Revolution*, Cambridge: Cambridge University Press.

Lindberg, S. 1979. 'Mobiles in Books. Volvelles, Inserts, Pyramids, Divinations and Children's Games', *The Private Library* 2, ser. 3, 49–82. Trans. W. Mitchell.

Lindner, K. 1987. 'German Globe-Makers, Especially in Nuremberg and Berlin', *Der Globusfreund* 35/37, 169–183.

Lippincott, K. 1999. 'Globes in Art: Problems of Interpretation and Representation', in Dekker 1999, 75–86.

Lohmeier, D. 2000. *Heinrich Rantzau. Humanismus und Renaissance in Schleswig-Holstein*, Heide: Boyens and Co.

Lohr, C. 1975. 'Renaissance Latin Aristotle Commentaries: Authors C', *Renaissance Quarterly* 28, 689–741.

1980. 'Renaissance Latin Aristotle Commentaries: Authors Pi-Sm', *Renaissance Quarterly* 33, 623–734.

Long, P. 1991a. 'The Openness of Knowledge: An Ideal and Its Context in Sixteenth-Century Writings on Mining and Metallurgy', *Technology and Culture* 32, 318–355.

1991b. 'Invention, Authorship, "Intellectual Property" and the Origin of Patents: Notes Toward a Conceptual History', *Technology and Culture* 32, 846–884.

1997. 'Power, Patronage and the Authorship of Ars: From Mechanical Know-how to Mechanical Knowledge in the Last Scribal Age', *Isis* 88, 1–41.

2001. *Openness, Secrecy, Authorship: Technical Arts and the Culture of Knowledge from Antiquity to the Renaissance*, Baltimore: Johns Hopkins University Press.

Longomontanus, C. 1622. *Astronomica Danica*, Amsterdam: G. I. Caesius. 2 parts in 1 vol., separately paginated.

Louthan, H. 1997. *The Quest for Compromise. Peacemakers in Counter-Reformation Vienna*, Cambridge: Cambridge University Press.

Love, H. 1987. 'Scribal Publication in Seventeenth-Century England', *Transactions of the Cambridge Bibliographical Society* 9, 130–154.

1998. *The Culture and Commerce of Texts: Scribal Publication in Seventeenth-Century England*, Amherst: University of Massachussetts Press.

Lux, D. and H. Cook, 1998. 'Closed Circles or Open Networks? Communicating at a Distance during the Scientific Revolution', *History of Science* 36, 179–211.

Lyby, T. and O. P. Grell, 1995. 'The Consolidation of Lutheranism in Denmark and Norway', in O. P. Grell, *The Scandinavian Reformation: From Evangelical Movement to Institutionalisation of Reform*, Cambridge: Cambridge Univeristy Press, 114–143.

Macdonald, A. and A. D. Morrison-Low, 1994. *A Heavenly Library: Treasures from the Royal Observatory's Crawford Collection*, Edinburgh: National Museums of Scotland.

Maestlin, M. 1578. *Observatio & demonstratio cometae aetherei, qui anno 1577 et 1578 constitutus in sphaera veneris*, Tübingen: G. Gruppenbach.

Manitius, C., ed., 1909. *Procli Diadochi Hypotyposis Astronomicarum Positionum (una cum scholiis antiquis e libris manu scriptis edidit germanica interpretatione et commentariis instruxit)*, Leipzig: Teubner.

Margolin, J.-C. 1976. 'L'enseignement des mathématiques en France (1540–70): Charles de Bovelle, Finé, Peletier, Ramus', in P. Sharratt, ed., *French Renaissance Studies, 1540–1570. Humanism and the Encyclopedia*, Edinburgh: Edinburgh University Press, 109–155.

Markham, C., ed., 1889. *Tractatus de Globis et eorum usu. A Treatise Descriptive of the Globes Constructed by Emery Molyneux, and Published in 1592. By Robert Hues*, London: The Hakluyt Society.

Martens, R. 2000. *Kepler's Philosophy and the New Astronomy*, Princeton: Princeton University Press.

Maurice, K. 1980. 'Jost Bürgi, or On Innovation', in Maurice and Mayr 1980, 87–102.

1985. *Sovereigns as Turners. Materials on a Machine Art by Princes*, Zurich: Verlag Ineichen. Trans. D. Schade.

Maurice, K. and O. Mayr, eds., 1980. *The Clockwork Universe: German Clocks and Automata 1550–1650*, New York: Neale Watson Academic Publications.

Mauss, M. 1990. *The Gift: The Form and Reason for Exchange in Archaic Societies*, London: Routledge.

Mayer, J. 1903. 'Der Astronom Cyprianus Leowitz', *Bibliotheca Mathematica* 4, 134–159.

McDonald, W. 1976. 'Maximilian of Habsburg and the Veneration of Hercules: On the Revival of Myth and the German Renaissance', *Journal of Medieval and Renaissance Studies* 6, 139–154.

McFarlane, I. D. 1981. *Buchanan*, London: Duckworth.

McKitterick, D. 2003. *Print, Manuscript and the Search for Order, 1450–1830*, Cambridge: Cambridge University Press.

McMullin, E. 1985. 'Openness and Secrecy in Science: Some Notes on Early History', *Science, Technology and Human Values* 10, 14–23.

Melanchthon, P. 1563. *Initia doctrinae physicae dictata in Academia Witebergensi*, Leipzig: Johannes Rhamba.

Menzhausen, J. 1985. 'Elector Augustus's *Kunstkammer*: An Analysis of the Inventory of 1587', in Impey and MacGregor 1985, 69–89.

Mercier, R. 1998. 'The Astronomical Tables of George Gemistus Plethon', *Journal for the History of Astronomy* 29, 117–127.

Methuen, C. 1996a. 'Maestlin's Teaching of Copernicus: The Evidence of his University Textbook and Disputations', *Isis* 87, 230–247.

1996b. 'The Role of the Heavens in the Thought of Philip Melanchthon', *Journal for the History of Ideas* 57, 385–403.

1997. '"This Comet or New Star": Theology and the Interpretation of the Nova of 1572', *Perspectives on Science* 5, 499–515.

1998. *Kepler's Tübingen*, Aldershot: Ashgate.

Michel, P. H. 1973. *The Cosmology of Giordano Bruno*, Paris: Hermann. Trans. R. E. W. Maddison.

Midelfort, H. C. E. 1992. 'Curious Georgics: The German Nobility and Their Crisis of Legitimacy in the Late Sixteenth Century', in A. Fix and S. Karant-Nunn, eds., *Germania Illustrata. Essays on Early Modern Germany Presented to Gerald Strauss*, Sixteenth Century Essays and Studies 18, Kirksville MO: Sixteenth Century Journal Publishers, 217–242.

Miller, F. J., trans., 1977–1985. *Ovid. Metamorphoses*, Loeb Classical Library, Cambridge MA: Harvard University Press. Revised by G. P. Goold. 2 vols.

Miniati, M. and M. Rudan, 1981. 'Il Quadrante Universale di Tobias Volckmer di Brunswick', *Annali dell'Instituto e Museo di Storia della Scienza di Firenze* 6.1, 241–246.

Moesgaard, K. 1972a. 'The Copernican Influence on Tycho Brahe', *Studia Copernicana* 5, 31–55.

1972b. 'How Copernicanism Took Root in Denmark and Norway', *Studia Copernicana* 5, 117–151.

1975. 'Tychonian Observations, Perfect Numbers, and the Date of Creation: Longomontanus' Solar and Precessional Theories', *Journal for the History of Astronomy* 6, 84–99.

1988. 'Refraction in Tycho Brahe's Small Universe', in S. Débarbat *et al.*, eds., *Mapping the Sky*, Dordrecht: Kluwer Academic Press, 87–93.

Molland, G. 1995. 'Scottish–Continental Relations as Mirrored in the Career of Duncan Liddel (1561–1613)', in P. Dukes, ed., *The Universities of Aberdeen and Europe: The First Three Centuries*, Aberdeen: Aberdeen University Press, 79–101.

Monfasani, J., ed., 1984. *Collectanea Trapezuntiana. Texts, Documents, and Bibliographies of George of Trebizond*, Medieval and Renaissance Texts and Studies 25, Binghamton: The Renaissance Society of America.

Montgomery, S. 2000. *Science in Translation. Movements of Knowledge through Cultures and Time*, Chicago and London: University of Chicago Press.

Moran, B. 1978. 'Science at the Court of Hesse-Kassel: Informal Communication, Collaboration, and the Role of the Prince-Practitioner in the Sixteenth Century', Ph.D. Thesis, Los Angeles: University of California.

1980. 'William IV of Hesse-Kassel: Informal Communication and the Aristocratic Context of Discovery', in T. Nickles, ed., *Scientific Discovery: Case Studies*, Dordrecht: Reidel, 67–96.

1981. 'German Prince-Practitioners: Aspects in the Development of Courtly Science, Technology, and Procedures in the Renaissance', *Technology and Culture* 22, 253–274.

1982. 'Christoph Rothmann, The Copernican Theory, and Institutional and Technical Influences on the Criticism of Aristotelian Cosmology', *Sixteenth Century Journal* 13, 85–108.

1985. 'Privilege, Communication, and Chemiatry: The Hermetic-Alchemical Circle of Moritz of Hessen-Kassel', *Ambix* 32, 110–126.

Moran, B., ed., 1991. *Patronage and Institutions: Science, Technology and Medicine at the European Court, 1500–1750*, Woodbridge: The Boydell Press.

Morsing, E. O. 1586. *Diarium Astrologicum et Metheorologicum Anni a Nato Christi 1586*, Uraniborg: n. p.

Mortensen, L. 1994. 'The Printed Dedication: Its Functions and some Danish Statistics from the Nordic Neo-Latin Database', in Moss 1994, 699–709.

Mosley, A. 2002a. 'Tycho Brahe's Epistolae astronomicae: A Reappraisal', in Papy 2002, 449–468.

2002b. 'Tycho Brahe and John Craig: The Dynamic of a Dispute', in Christianson and Hadravova *et al.*, 2002, 70–83.

2004. 'Tycho Brahe, Erasmus Oswald Schreckenfuchs, and the Mysterious Planetary Motion Machine', in P. Boner and C. Eagleton, eds., *Instruments of Mystery*, Cambridge: Whipple Museum, 14–26.

2006. 'Objects of Knowledge: Mathematics and Models in Sixteenth-Century Cosmology and Astronomy', in S. Kusukawa and I. Maclean, eds., *Transmitting Knowledge: Words, Images, and Instruments in Early Modern Europe*, Oxford: Oxford University Press, 193–216.

Mosley, A. and N. Jardine and K. Tybjerg, 2003. 'Epistolary Culture, Editorial Practices, and the Propriety of Tycho's *Astronomical Letters*', *Journal for the History of Astronomy* 29, 421–451.

Moss, A. *et al.*, eds., 1994. *Acta Conventus Neo-Latini Hafniensis*, Medieval and Renaissance Texts and Studies 120, New York: Binghamton.

Mozley, J. H., trans., 1943. *Valerius Flaccus. Argonautica*, Cambridge MA: Harvard University Press.

Müller, U. 1995. 'Zum Stand der Rekonstruktion der Praetorius-Saxonicus-Bibliothek', *Sudhoffs Archiv* 79, 120–123.

Müller, U., ed., 1993. *450 Jahre Copernicus 'De revolutionibus'. Astronomische und mathematische Bücher aus Schweinfurter Bibliotheken*, Schweinfurt: Stadtarchiv Schweinfurt.

Multhauf, R. 1954. 'John of Rupescissa and the Origin of Medical Chemistry', *Isis* 45, 359–367.

Murphy, J., ed., 1983. *Renaissance Eloquence*, Berkeley: University of California Press.

Murray, A. T., trans., 1995. *Homer. The Odyssey*, Loeb Classical Library, Cambridge MA: Harvard University Press. 2 vols. 2nd edition, rev. G. Dimock.

Naiden, J. 1952. *The Sphera of George Buchanan (1506–1582): A Literary Opponent of Copernicus and Tycho Brahe*, n. p: n. p.

Neal, K. 1999. 'The Rhetoric of Utility: Avoiding Occult Associations for Mathematics through Profitability and Pleasure', *History of Science* 37, 151–178.

Nelles, P. 1997. 'The Library as an Instrument of Discovery: Gabriel Naudé and the Uses of History', in Kelley 1997, 41–57.

Neugebauer, O. 1949. 'The Early History of the Astrolabe', *Isis* 11, 240–256.

Neugebauer, O. and N. Swerdlow, 1984. *Mathematical Astronomy in Copernicus's De Revolutionibus*, New York and Berlin: Springer-Verlag.

Neuschel, K. 1989. *Word of Honor: Interpreting Noble Culture in Sixteenth-Century France*, Ithaca: Cornell University Press.

Newton, R. 1977. *The Crime of Claudius Ptolemy*, Baltimore: Johns Hopkins University Press.

Niccoli, O. 1990. *Prophecy and People in Renaissance Italy*, Princeton: Princeton University Press.

Nørlind, W. 1954. 'A Hitherto Unpublished Letter from Tycho Brahe to Christoph Clavius', *The Observatory* 74, 20–23.

    1970. *Tycho Brahe: En levnadsteckning med nya bidrag belysande hans liv och verk*, Lund: C. W. K. Gleerup.

Nørlind, W., ed. and trans., 1926. *Ur Tycho Brahes Brevväksling*, Lund: C. W. K. Gleerup.

North, J. 1969. 'A Post-Copernican Equatorium', *Physis* 11, 418–457.

Oestmann, G. 2002. 'Cyprianus Leovitius, der Astronom und Astrologe Ottheinrichs', in *Pfalzgraf Ottheinrich: Politik, Kunst und Wissenschaft im 16. Jahrhundert*, Regensburg: Friedrich Pustet, 348–359.

    2004. *Heinrich Rantzau und die Astrologie*, Braunschweig: Braunschweigisches Landesmuseum.

Oldfather, C. *et al.*, trans., 1933–1967. *Diodorus of Sicily. The Library of History*, Loeb Classical Library, Cambridge MA: Harvard University Press. 12 vols.

Olmi, G. 1985. 'Science-Honour-Metaphor: Italian Cabinets of the Sixteenth and Seventeenth Centuries', in Impey and MacGregor 1985, 5–16.

Ornstein, M. 1913. *The Role of Scientific Societies in the Seventeenth Century*, New York: n. p.

Panofsky, E. 1955. *The Life and Art of Albrecht Dürer*, Princeton: Princeton University Press.

Pantin, I. 1988. 'Les problèmes de l'édition des livres scientifiques: l'exemple de Guillaume Cavellat', in P. Aquilon and H.-J. Martin, eds., *Le Livre dans l'Europe de la Renaissance*, Actes du XXVIIe Colloque international d'Etudes humanistes de Tours, Tours: Promodis, 240–251.

    1995. *La poèsie du ciel en France dans la seconde moitié du seizième siècle*, Geneva: Librairie Droz.

    1999. 'New Philosophy and Old Prejudices: Aspects of the Development of Copernicanism in a Divided Europe', *Studies in History and Philosophy of Science* 30, 237–262.

Papy, J. *et al.*, eds., 2002. *Self-Presentation and Social Identification. The Rhetoric and Pragmatics of Letter Writing in Early-Modern Times*, Leuven: Leuven University Press.

Patrides, C. 1982. *Premises and Motifs in Renaissance Thought and Literature*, Princeton: Princeton University Press.

Pedersen, O. 1978a. 'Astronomy', in Lindberg 1978, 303–337.

    1978b. 'The Decline and Fall of the *Theorica Planetarum*: Renaissance Astronomy and the Art of Printing', *Studia Copernicana* 16, 157–185.

    1996. 'European Astronomy in the Middle Ages', in Walker 1996, 175–186.

Pena, J. 1557. *Euclidis Optica et Catoptrica*, Paris: A. Wechel.

Petersen, E. L. 1968. 'La crise de la noblesse danoise entre 1580 et 1660', *Annales* 23, 1237–1261.

Pettegree, A. 2005. *Reformation and the Culture of Persuasion*, Cambridge: Cambridge University Press.

Pettegree, A. and M. Hall, 2004. 'The Reformation and the Book: A Reconsideration', *Historical Journal* 47, 785–808.

Pomian, K. 1986. 'Astrology as a Naturalistic Theology of History', in Zambelli 1986, 29–43.

Pontanus, J. J. 1512. *Commentationes Super Centum Sententiis Ptolemaei*, Naples: S. Mayr.

Popkin, R. 1988. 'Theories of Knowledge', in Schmitt and Skinner 1988, 668–684.

Pörtner, R. 2001. *The Counter-Reformation in Central Europe: Styria 1580–1630*, Oxford: Oxford University Press.

Pottinger, D. 1958. *The French Book Trade in the Ancien Régime, 1500–1791*, Cambridge MA: Harvard University Press.

Poulle, E. 1980. *Equatoires et Horlogerie Planétaire du XIIIe au XVIe Siècle*, Hautes Etudes Médiévales et Modernes 42, Centre de Recherches d'Histoire et de Philologie, Geneva: Librairie Droz.

1988. 'The Alphonsine Tables and Alfonso X of Castile', *Journal for the History of Astronomy* 19, 97–113.

Prandtl, W. 1932. 'Die Bibliothek des Tycho Brahes', *Philobiblon* 5, 291–299, 321–329.

Principe, L. 1992. 'Robert Boyle's Alchemical Secrecy: Codes, Ciphers, and Concealments', *Ambix* 39, 64–74.

Procter, E. S. 1945. 'The Scientific Works of the Court of Alfonso X of Castile: The King and his Collaborators', *The Modern Language Review* 40, 12–29.

Pumfrey, S. 1991. 'The History of Science and the Renaissance Science of History', in S. Pumfrey, P. Rossi and M. Slawinski, eds., *Science, Culture and Popular Belief in Renaissance Europe*, Manchester: Manchester University Press, 48–70.

Rackham, H., trans., 1933. *Cicero. De Natura Deorum. Academica*, Loeb Classical Library, Cambridge MA: Harvard University Press. Revised and reprinted 1951.

Rackham, H. *et al.*, trans., 1938–1963. *Pliny. Natural History*, Loeb Classical Library, Cambridge MA: Harvard University Press. 10 vols.

Raeder, H. and B. Strömgren and E. Strömgren, eds. and trans., 1946. *Tycho Brahe's Description of his Instruments and Scientific Work*, Copenhagen: Munksgaard.

Raggio, O. and A. Wilmering, 1996. *The Liberal Arts Studiolo from the Ducal Palace at Gubbio*, New York: The Metropolitan Museum of Art.

Ramus, P. 1567. *Prooemium mathematicum*, Paris: A. Wechel.

Randles, W. G. L. 1999. *The Unmaking of the Medieval Christian Cosmos, 1500–1760: From Solid Heavens to Boundless Aether*, Aldershot: Ashgate.

Ransom, H. 1951. 'The Personal Letter as Literary Property', *Studies in English* 30, 116–131.

Rantzau, H. 1580. *Catalogus imperatorum, regum ac principum qui astrologicam artem amarunt, ornarunt and exercuerunt*, Antwerp: C. Plantin.

1584. *Catalogus imperatorum, regum ac principum qui astrologicam artem amarunt, ornarunt and exercuerunt*, Leipzig: G. Defner.

Recorde, R. 1556. *The Castle of Knowledge*, London: R. Wolfe.

Reeve, M. 1983. 'Manuscripts Copied from Printed Books', in Trapp 1983, 12–20.

Reeves, E. 1991. 'Augustine and Galileo on Reading the Heavens', *Journal of the History of Ideas* 52, 563–579.

Regiomontanus, J. 1537. *Oratio de introductoria in omnes scientias Mathematicas*, Nuremberg: J. Petreius.

Reisch, G. 1503. *Margarita philosophica*, Freiburg: J. Schott.

Riccioli, G. B. 1651. *Almagestum novum astronomiam veterem novamque complectens observationibus aliorum et propriis novisque theorematibus, problematibus ac talibus promotam; in tres tomos distributam quorum argumentum sequens pagina explicabit*, Bologna: Ex typographia haeredis Victorii Benatii.

Richardson, B. 1999. *Printing, Writers and Readers in Renaissance Italy*, Cambridge: Cambridge University Press.

Risk, R. 1982. *Erhard Ratdolt, Master Printer*, Francestown: Typographeum.

Risner, F. 1572. *Opticae Thesaurus*, Basle: Officina Episcopiana.

Robbins, F. E., trans., 1940. *Ptolemy. Tetrabiblos*, Loeb Classical Library, Cambridge MA: Harvard University Press. Reprinted 1994.

Rolfe, J. 1927–1928. *The Attic Nights of Aulus Gellius*, Loeb Classical Library, London and New York: William Heinemann and G. P. Putnam's Sons. 3 vols.

Ronca, I. and M. Curr, trans., 1997. *William of Conches. A Dialogue on Natural Philosophy (Dragmaticon Philosophiae)*, Notre Dame: University of Indiana Press.

Roob, A. 1997. *The Hermetic Museum: Alchemy and Mysticism*, Cologne: Taschen.

Root, R. 1913. 'Publication before Printing', *Publications of the Modern Language Association* 28, 417–431.

Rose, M. 1993. *Authors and Owners: The Invention of Copyright*, Cambridge MA: Harvard University Press.

Rose, P. L. 1969. 'Certitudo mathematicarum from Leonardo to Galileo', in C. Maccagni, ed., *Atti del Simposio Internazionale Leonardo da Vinci nella Scienza e nella Tecnica*, Florence: Giunti Barbèra, 43–49.

1975. *The Italian Renaissance of Mathematics: Studies on Humanists and Mathematicians from Petrarch to Galileo*, Geneva: Droz.

Rosen, E. 1961. 'Calvin's Attitude Towards Copernicus', *Journal of the History of Ideas* 21, 431–441.

1976. 'Reply to N. Swerdlow', *Archives internationales d'histoire des sciences* 26, 301–304.

1984. 'Francesco Patrizi and the Celestial Spheres', *Physis* 26, 305–324.

1985a. 'The Dissolution of the Solid Celestial Spheres', *Journal of the History of Ideas* 46, 13–31.

1986. *Three Imperial Mathematicians: Kepler Trapped Between Tycho Brahe and Ursus*, New York: Abaris Books.

Rosen, E., ed. and trans., 1959. *Three Copernican Treatises*, New York: Dover Publications. Revised edition; first published 1939.

1967. *Kepler's Somnium. The Dream, or Posthumous Work on Lunar Astronomy*, Madison: The University of Wisconsin Press.

1985b. *Nicholas Copernicus: Complete Works*, Vol. III, *Minor Works*, Warsaw and Cracow: Polish Scientific Publishers.

Ross, R. 1974. 'Oronce Finé's Printed Works. Additions to Hillard and Poulle's Bibliography', *Bibliothèque d'Humanisme et Renaissance* 36, 83–85.

1975. 'Oronce Finé's *De sinibus libri II*: The First Printed Trigonometric Treatise of the French Renaissance', *Isis* 66, 379–386.

Sabra, A. I. 1967. 'The Authorship of the *Liber de crepusculis*', *Isis* 58, 77–85.

Saunders, J. W. 1951. 'From Manuscript to Print: A Note on the Circulation of Poetic Manuscripts in the Sixteenth Century', *Proceedings of the Leeds Philosophical and Literary Society* 6, 507–528.

Scaliger, J. C. 1557. *Exotericum Exercitationum Liber Quintus Decimus, De Subtilitate ad Hieronymus Cardanum*, Paris: M. Vascosanus.

Schechner Genuth, S. 1998. 'Astrolabes: A Cross-Cultural and Social Perspective', in M. Webster and R. Webster, eds., *Western Astrolabes*, Historic Scientific Instruments of the Adler Planetarium and Astronomy Museum 1, Chicago: Adler Planetarium and Astronomy Museum, 2–25.

Scheicher, E. 1985. 'The Collection of Archduke Ferdinand II at Schloss Ambras: Its Purpose, Composition and Evolution', in Impey and MacGregor 1985, 29–38.

Schillinger, K. 1990. 'The Development of Saxon Scientific Instrument-Making Skills from the Sixteenth Century to the Thirty Years War', *Annals of Science* 47, 277–289.

Schmitt, C. and Q. Skinner, eds., 1988. *The Cambridge History of Renaissance Philosophy*, Cambridge: Cambridge University Press.

Schöbel, J. 1975. *Princely Arms and Armour. A Selection from the Dresden Collection*, London: Barrie and Jenkins. Trans. M. O. A. Stanton.

Schofield, C. 1981. *Tychonic and Semi-Tychonic World Systems*, New York: Arno Press.

Schöner, A. 1562. *Gnomonice*, Nuremberg: I. Montanus and U. Neuberus.

Schöner, J. 1533. *Opusculum geographicum*, Nuremberg: J. Petreius.

1551. *Opera Mathematica*, Nuremberg: J. Montanus and U. Neuberus.

1561. *Opera Mathematica*, Nuremberg: J. Montanus and U. Neuberus.

Schottenloher, K. 1907. 'Johann Schöner und seine Hausdruckerei', *Zentralblatt für Bibliothekswesen* 24, 145–155.

1933. 'Die Druckpriviligien des 16. Jahrhunderts', *Gutenberg-Jahrbuch* 8, 89–110.

1953. *Die Widmungsvorrede im Buch des 16. Jahrhunderts*, Reformationsgeschichtliche Studien und Texte 76 and 77, Münster: Aschendorffsche Verlagsbuchhandlung.

Schröter, K. 1998. *Die Terminologie der italienischen Buchdrucker im. 15. und 16. Jahrhundert*, Beihefte zur Zeitschrift für Romanische Philologie 290, Tübingen: Max Niemeyer Verlag.

Seelig, L. 1985. 'The Munich Kunstkammer, 1565–1807', in Impey and MacGregor 1985, 76–89.

1995. *Silver and Gold: Courtly Splendour from Augsburg*, Bayerisches National-museum, Munich: Prestel-Verlag.

Segonds, A. 1993. 'Tycho Brahe et l'Alchimie', in J.-C. Margolin and S. Matton, eds., *Alchimie et Philosophie à la Renaissance*, Paris: J. Vrin, Paris, 365–378.

1994. 'A Propos d'un emblème de Tycho Brahe dans les *Mechanica*', in Hack-mann and Turner 1994, 261–272.

Seznec, J. 1953. *The Survival of the Pagan Gods: The Mythological Tradition and its Place in Renaissance Humanism and Art*, New York: Pantheon Books. Trans. B. Sessions.

Shackelford, J. 1989. 'Paracelsianism in Denmark and Norway in the Sixteenth and Seventeenth Centuries', Ph.D. Thesis, University of Wisconsin.

1991. 'Paracelsianism and Patronage in Early Modern Denmark', in Moran 1991, 88–109.

1993. 'Tycho Brahe, Laboratory Design, and the Aim of Science: Reading Plans in Context', *Isis* 84, 211–230.

2002. 'Providence, Power and Cosmic Causality in Early Modern Astronomy: The Case of Tycho Brahe and Petrus Severinus', in Christianson and Hadravova 2002, 46–69.

Shank, M. 1994. 'Galileo's Day in Court', *Journal of the History of Astronomy* 25, 236–243.

1996. 'How Shall We Practice History? The Case of Mario Biagioli's *Galileo, Courtier*', *Early Science and Medicine* 1, 106–150.

Shapin, S. and S. Schaffer, 1989. *Leviathan and the Air-Pump: Hobbes, Boyle and the Experimental Life*, Princeton: Princeton University Press. Corrected reprint of 1985 edition.

Sharratt, M. 1996. *Galileo, Decisive Innovator*, Cambridge: Cambridge University Press.

Sharratt, P. 1976. 'Peter Ramus and the Reform of the University: The Divorce of Philosophy and Eloquence', in P. Sharratt, ed., *French Renaissance Studies, 1540–1570. Humanism and the Encyclopedia*, Edinburgh: Edinburgh University Press, 4–20.

Shea, W. 1990. 'Galileo Galilei: An Astronomer at Work', in Levere and Shea 1990, 51–76.

Sherman, W. 1995. *John Dee: The Politics of Reading and Writing in the English Renaissance*, Amherst: University of Massachusetts Press.

Shipman, J. 1967. 'Johannes Petreius, Nuremberg Publisher of Scientific Works 1524–1580', in Lehman-Haupt 1967, 147–162.

Siegert, B. 1999. *Relays: Literature as an Epoch of the Postal System*, Stanford: Stanford University Press.

Sima, Z. 1993. 'Prague Sextants of Tycho Brahe', *Annals of Science* 50, 445–453.

Simms, D. 1995. 'Archimedes the Engineer', *History of Technology* 17, 45–111.

Siraisi, N. 1991. 'Girolamo Cardano and the Art of Medical Narrative', *Journal of the History of Ideas* 52, 581–602.

Smith, A. M. 1990. 'Alhazen's Debt to Ptolemy's *Optics*', in Levere and Shea 1990, 147–164.

2003. 'Ptolemy, Alhacen, and Ibn Mu'adh and the Problem of Atmospheric Refraction', *Centaurus* 45, 100–115.

Smith, J. 1983. *Nuremberg, A Renaissance City, 1500–1618*, Austin: University of Texas Press.

Smoller, L. 1994. *History, Prophecy and the Stars. The Christian Astrology of Pierre d'Ailly 1350–1420*, Princeton: Princeton University Press.

Snellius, W. 1619. *Descriptio Cometæ Qui Anno 1618 Mense Novembri Primum Effulsit. Huc accessit Christophori Rhotmanni Ill. Princ. Wilhelmi Hassiae Lantgravii Mathematici descriptio accurata cometae anni 1585*, Leiden: Officina Elziviriana.

Sonnino, L. 1968. *A Handbook to Sixteenth-century Rhetoric*, London: Routledge and Kegan Paul.

Staikos, K. 2000. *The Great Libraries From Antiquity to the Renaissance (3000 BC to AD 1600)*, New Castle DE and London: Oak Knoll Press and The British Library.

Steinmetz, W. 1991. *Heinrich Rantzau (1526–1598): ein Vertreter des Humanismus in Nordeuropa und seine Wirkungen als Förderer der Künste*, Frankfurt am Main: Peter Lang. 2 vols.

Stöffler, J. 1524. *Elucidatio fabricae ususque astrolabii*, Oppenheim: Jakob Köbel.

Strauss, G. 1976. *Nuremberg in the Sixteenth Century: City Politics and Life Between Middle Ages and Modern Times*, London: Indiana University Press. Revised edition.

Strong, R. 1984. *Art and Power: Renaissance Festivals 1450–1650*, Woodbridge: The Boydell Press.

Sturlese, M. R. P. 1985. 'Su Bruno e Tycho Brahe', *Rinascimento*, 2nd ser., 25, 309–333.

Swerdlow, N. 1974. 'The Holograph of *De Revolutionibus* and the Chronology of its Composition', *Journal for the History of Astronomy* 5, 186–198.

1976. 'Pseudodoxia Copernicana', *Archives internationales d'histoire des sciences* 26, 108–158.

1990. 'Regiomontanus on the Critical Problems of Astronomy', in Levere and Shea 1990, 165–195.

1993a. 'Science and Humanism in the Renaissance: Regiomontanus' Oration on the Dignity and Utility of the Mathematical Sciences', in P. Horwich, ed., *World-Changes: Thomas Kuhn and the Nature of Science*, Cambridge MA and London: MIT Press, 131–168.

1993b. 'The Recovery of the Exact Sciences of Antiquity: Mathematics, Astronomy, Geography', in Grafton 1993, 125–165.

1996. 'Astronomy in the Renaissance', in Walker 1996, 187–230.

1999. 'Regiomontanus's Concentric-sphere Models for the Sun and Moon', *Journal for the History of Astronomy* 30, 1–23.

Szczucki, L. and T. Szepessy, eds., 1992– . *Epistulae Andreas Dudithius*, Budapest: Akademiai Kiado. In progress.

Tanner, M. 1993. *The Last Descendant of Aeneas: The Hapsburgs and the Mythic Image of the Emperor*, New Haven: Yale University Press.

Tanselle, G. T. 1995. 'Printing History and Other History', *Studies in Bibliography* 48, 269–289.

Taton, R. and C. Wilson, eds. 1989. *Planetary Astronomy from the Renaissance to the Rise of Astrophysics. Part A: Tycho Brahe to Newton*, The General History of Astronomy 2, Cambridge: Cambridge University Press.

Taub, L. 1993. *Ptolemy's Universe: The Natural Philosophical and Ethical Foundations of Ptolemy's Astronomy*, Chicago: Open Court.

Taylor, A., ed., 1950. *Advice on Establishing a Library by Gabriel Naudé*, Berkeley and Los Angeles: University of California Press.

Taylor, E. G. R. 1954. *The Mathematical Practitioners of Tudor and Stuart England*, Cambridge: Cambridge University Press.

Tennant, E. 1996. 'The Protection of Invention: Printing Privileges in Early Modern Germany', in S. Schindler and G. Williams, eds., *Knowledge, Science, and Literature in Early Modern Germany*, Chapel Hill: University of North Carolina Press, 7–48.

Tester, S. J. 1987. *A History of Western Astrology*, Woodbridge: Boydell.

Thill, O. 2002. *The Life of Copernicus (1473–1543)*, Fairfax VA: Xulon Press.

Thompson, J. W., ed. and trans., 1911. *The Frankfort Book Fair. The Francofordiense Emporium of Henri Estienne*, Chicago: The Caxton Club.

Thoren, V. 1973. 'New Light on Tycho's Instruments', *Journal for the History of Astronomy* 4, 25–45.

1979. 'Tycho Brahe's System of the World and the Comet of 1577', *Archives internationales d'histoire des sciences* 29, 53–67.

1985. 'Tycho Brahe as the Dean of a Renaissance Research Institute', in P. L. Farber and M. Osler, eds., *Religion, Science and Worldview: Essays in Honour of Richard S. Westfall*, Cambridge: Cambridge University Press, 275–296.

1988. 'Prosthaphaeresis Revisited', *Historia Mathematica* 15, 32–39.

1989. 'Tycho Brahe', in Taton and Wilson, eds., *Planetary Astronomy From the Renaissance to the Rise of Astrophysics*, 3–21.

1990. *The Lord of Uraniborg*, Cambridge: Cambridge University Press.

Thorndike, L. 1923–1958. *A History of Magic and Experimental Science*, New York: Macmillan. 8 vols.

Thornton, D. 1997. *The Scholar in His Study: Ownership and Experience in Renaissance Italy*, New Haven: Yale University Press.

Tihon, A. 1998. 'The Astronomy of George Gemistus Plethon', *Journal for the History of Astronomy* 29, 109–116.

Toomer, G., trans., 1998. *Ptolemy's Almagest*, Princeton: Princeton University Press. 2nd edition.

Topsell, E. 1607. *The History of Four-footed beastes Describing the true and liuely figure of euery beast, with a discourse of their seuerall names, conditions, kindes, vertues (both naturall and medicinall) countries of their breed, their love and hate to mankinde, and the wonderfull worke of God in their creation, preseruation, and destruction*, London: W. Iaggard.

Trapp, J., ed., *Manuscripts in the Fifty Years after the Invention of Printing*, London: The Warburg Institute.

Tredennick, H., trans., 1933–1935. *Aristotle. The Metaphysics*, Loeb Classical Library, Cambridge MA: Harvard University Press. 2 vols.

Treutlerus, H. 1592. *Oratio Historica de Vita et Morte Illustrissimi et Potentissimi Cattorum Principis, ac D. D. Wilhelmi Hassiae Landtgravii, Comitis Cattimelibocorum, Deciorum, Zigenhaini Niddae, &c.*, Marburg: P. Egenolphus.

Trevor-Roper, H. 1991. *Princes and Artists: Patronage and Ideology at Four Habsburg Courts 1517–1633*, New York: Thames and Hudson.

1998. 'Paracelsianism made Political 1600–1650', in O. P. Grell, ed., *Paracelsus: The Man and his Reputation, his Ideas and their Transformation*, Leiden: Brill, 119–133.

Trinkaus, C. 1985. 'The Astrological Cosmos and Rhetorical Culture of Giovanni Gioviano Pontano', *Renaissance Quarterly* 38, 446–472.

Tupman, G. 1900. 'A Comparison of Tycho Brahe's Meridian Observations of the Sun with Leverrier's Solar Tables', *The Observatory* 23, 132–135, 165–171.

Turner, A. 1986. *Astrolabes and Related Devices*, The Time Museum Catalogue of the Collection, vol. I: Time-Measuring Instruments, part 1, Rockford IL: The Time Museum.

1987. *Early Scientific Instruments: Europe 1400–1800*, London: Sotheby's Publications.

1995a. 'From Mathematical Practice to the History of Science. The pattern of collecting scientific instruments', *Journal of the History of Collections* 7, 135–150.

Turner, G. L'E. 1979. 'Johann Daniel von Berthold: A Clerical Craftsman and his Universal Ring-Dial', *Annali dell'Instituto e Museo di Storia della Scienza di Firenze* 4, 2, 15–20.

1988. 'Charles Whitwell's Addition c. 1595, to a Fourteenth-Century Astrolabe', *The Antiquaries Journal* 65, 454–455, 476.

1994. 'The Three Astrolabes of Gerard Mercator', *Annals of Science* 51, 329–353.

1995b. 'The Florentine Workshop of Giovan Battista Giusti, 1556–c.1575', *Nuncius* 10, 131–171.

1996. 'Later Medieval and Renaissance Instruments', in Walker 1996, 231–244.

2000. *Elizabethan Instrument Makers: The Origins of the London Trade in Precision Instrument Making*, Oxford: Clarendon Press.

Ursus, N. 1588. *Fundamentum astronomicum*, Strasbourg: B. Iobin.

1597. *De astronomicis hypothesibus seu systemate mundano, tractatus*, Prague: n. p.

Valerio, V. 1987. 'Historiographic and Numeral Notes on the Atlante Farnese and its Celestial Sphere', *Der Globusfreund* 35/37, 97–124.

van Cleempoel, K. 2002. *A Catalogue Raisonné of Scientific Instruments from the Louvain School*, Turnhout: Brepols.

Vanden Broecke, S. 2006. 'Teratology and the Publication of Tycho Brahe's New World System (1588)', *Journal for the History of Astronomy* 37, 1–17.

van der Krogt, P. 1993. *Globi Neerlandici: The Production of Globes in the Low Countries*, Utrecht: HES Publishers.

van Helden, A. 1989a. 'The Telescope and Cosmic Dimensions', in Taton and Wilson 1989, 106–118.

van Helden, A. ed. and trans., 1989b. *Sidereus Nuncius or The Sidereal Messenger*, Chicago: University of Chicago Press.

van Helden, A. and M. Winkler, 1993. 'Johannes Hevelius and the Visual Language of Astronomy', in J. Field and F. James, eds., *Renaissance and Revolution: Humanists, Scholars, Craftsmen and Natural Philosophers in Early Modern Europe*, Cambridge: Cambridge University Press, 97–116.

van Nouhuys, T. 1998. *The Age of Two-Faced Janus: The Comets of 1577 and 1618 and the Decline of the Aristotelian World View in the Netherlands*, Leiden: Brill.

Verdet, J.-P. 1989. 'La diffusion de l'héliocentrisme', *Revue d'Histoire des Sciences* 42, 241–253.

Vickery, B. C. 2000. *Scientific Communication in History*, London: Scarecrow.

Viellard, J. 1973. 'Instruments d'astronomie conservés à la bibliothèque du college de Sorbonne au XVe et XVe siècles', *Bibliothèque de l'Ecole des Chartres* 131, 587–593.

Vivanti, C. 1967. 'Henri IV, the Gallic Hercules', *Journal of the Warburg and Courtauld Institutes* 30, 176–197.

Voelkel, J. 1999. 'Publish or Perish: Legal Contingencies and the Publication of Kepler's "Astronomia nova"', *Science in Context* 12, 33–59.

von Bertele, H. 1988. 'Bürgi als Uhrmacher und Ingenieur', in von Mackensen 1988, 42–60.

von Braunmühl, A. 1900. *Vorlesungen über Geschichte der Trigonometrie. Erster Teil. Von den Ältesten Zeiten bis Zur Erfindung Der Logarithmen*, Lepizig: Teubner.

von Gschliesser, O. 1942. *Der Reichshofrat*, Vienna: Adolf Holzhausens.

von Mackensen, L. 1991. *Die Naturwissenschaftlich-Technische Sammlung. Geschichte, Bedeutung und Ausstellung in der Kasseler Orangerie*, Kassel: Georg Wenderoth Verlag.

von Mackensen, L., ed., 1988. *Die erste Sternwarte Europas mit ihren Instrumenten und Uhren: 400 Jahre Jost Bürgi in Kassel*, Munich: Callway Verlag. 3rd edition.

von Philippovich, E. 1969–1971. 'Karl V. als Atlas', *Der Globusfreund*, 18/20, 108–111.

von Stromer, W. 1997. 'Nuremberg as Epicentre of Invention and Innovation towards the End of the Middle Ages', *History of Technology* 19, 19–45.

Walker, C., ed., 1996. *Astronomy Before the Telescope*, London: British Museum Press.

Wallis, C. G., trans., 1995. *Epitome of Copernican Astronomy and Harmonies of the World*, Amherst: Prometheus Books.

Warner, D. 1979. *The Sky Explored: Celestial Cartography 1500–1800*, New York and Amsterdam: Alan R. Liss Inc. and Theatrum Orbis Terrarum Ltd.

Watanabe-O'Kelly, H. 2002. *Court Culture in Early Modern Dresden*, Basingstoke: Palgrave.

Way, A., trans., 1912. *Euripides*, Loeb Classical Library, Cambridge MA: Harvard University Press. 4 vols.

Weisheipl, J. A. 1978. 'The Nature, Scope, and Classification of the Sciences', in Lindberg 1978, 461–482.

Werner, J. 1522. *In hoc opere continentur. Libellus Ioannis Verneri Nurembergen super vigintiduobus elementis conicis . . . de motu octavae Sphaerae*, Nuremberg: Fridericus Peypus.

Wesley, W. 1978. 'The Accuracy of Tycho Brahe's Instruments', *Journal for the History of Astronomy* 9, 42–53.

Westman, R. 1972a. 'Kepler's Theory of Hypothesis and the "Realist Dilemma"', *Studies in History and Philosophy of Science* 3, 233–264.

1972b. 'The Comet and the Cosmos: Kepler, Mästlin, and the Copernican System', *Studia Copernicana* 5, 7–30.

1975a. 'The Melanchthon Circle, Rheticus, and the Wittenberg Interpretation of the Copernican Theory', *Isis* 66, 165–193.

1975b. 'Three Responses to the Copernican Theory: Johannes Praetorius, Tycho Brahe, and Michael Maestlin', in R. Westman ed., *The Copernican Achievement*, 285–345.

1980a. 'Humanism and Scientific Roles in the Sixteenth Century', in Krafft and Schmitz 1980, 83–99.

1980b. 'On Communication and Cultural Change', *Isis* 71, 474–477.

1980c. 'The Astronomer's Role in the Sixteenth Century: A Preliminary Study', *History of Science* 18, 105–147.

1990. 'Proof, Poetics and Patronage: Copernicus's Preface to *De revolutionibus*', in Lindberg and Westman 1990, 167–205.

Westman, R., ed., 1975c. *The Copernican Achievement*, Los Angeles: University of California Press.

Whitfield, P. 1995. *The Mapping of the Heavens*, London: The British Library.

Widmalm, S. 1992. 'A Commerce of Letters: Astronomical Communication in the 18th Century', *Science Studies* 5, 43–58.

Wightman, W. P. D. 1962. *Science and the Renaissance. An Annotated Bibliography of the Sixteenth-Century Books Relating to the Sciences in the Library of the University of Aberdeen*, Edinburgh: Oliver and Boyd. 2 vols.

Williams, A. 1948. *The Common Expositor: An Account of the Commentaries on Genesis, 1527–1633*, Chapel Hill: University of North Carolina Press.

Willis, J., ed., 1983. *Martianus Capella*, Bibliotheca Scriptorum Graecorum et Romanorum Teubneriana, Leipzig: Teubner.

Wingen-Trennhaus, A. 1991. 'Regiomontanus als Frühdrucker in Nürnberg', in *Mitteilungen des Vereins für die Geschichte der Stadt Nürnberg* 78, 17–87.

Witek, J., ed., 1994. *Ferdinand Verbiest (1623–1688). Jesuit Missionary, Scientist, Engineer and Diplomat*, Monumenta Serica Monograph Series 30, Nettetal: Steyler Verlag.

Witt, R. 1982. 'Medieval "Ars Dictaminis" and the Beginnings of Humanism: A New Construction of the Problem', *Renaissance Quarterly* 35, 1–35.

Witt, R. *et al.*, 1999. *Heinrich Rantzau (1526–1598). Königlicher Statthalter in Schleswig und Holstein. Ein Humanist beschreibt sein Land*, Husum: Landesarchiv Schleswig-Holstein.

Yates, F. 1964. *Giordano Bruno and the Hermetic Tradition*, London: Routledge and Kegan Paul.

    1967. 'The Hermetic Tradition in Renaissance Science', in C. Singleton, ed., *Art, Science, and History in the Renaissance*, Baltimore: Johns Hopkins Press, 255–274.

Zambelli, P., ed., 1986. *Astrologi hallucinati. Stars and the End of the World in Luther's Time*, Berlin: Walter de Gruyter.

Zeeberg, P. 1994. 'Alchemy, Astrology and Ovid: A Love Poem by Tycho Brahe', in Moss 1994, 997–1007.

    2004. *Heinrich Rantzau. A Bibliography*, Copenhagen: C. A. Reitzel.

Zinner, E. 1964. *Geschichte und Bibliographie der Astronomischen Literatur in Deutschland zur Zeit der Renaissance*, Stuttgart: Anton Hiersemann. 2nd edition.

    1968. *Leben und Werken des Joh. Müller von Königsberg, genannt Regiomontanus*, Osnabrück: O. Zeller. 2nd edition.

    1979. *Deutsche und Niederländische Astronomische Instrumente des 11.–18. Jahrhunderts*, Munich: C. H. Beck. 2nd edition, reprinted.

    1988. *Entstehung und Ausbreitung der Coppernicanischen Lehre*, Munich: Verlag C. H. Beck. 2nd, enlarged edition.

General index

Index of correspondence